VOLUME THREE HUNDRED AND EIGHTY SIX

# INTERNATIONAL REVIEW OF CELL AND MOLECULAR BIOLOGY

Targeting Signaling Pathways in Solid Tumors - Part B

# INTERNATIONAL REVIEW OF CELL AND MOLECULAR BIOLOGY

Series Editors

| | |
|---|---|
| GEOFFREY H. BOURNE | *1949–1988* |
| JAMES F. DANIELLI | *1949–1984* |
| KWANG W. JEON | *1967–2016* |
| MARTIN FRIEDLANDER | *1984–1992* |
| JONATHAN JARVIK | *1993–1995* |
| LORENZO GALLUZZI | *2016–* |

Editorial Advisory Board

AARON CIECHANOVER
SANDRA DEMARIA
SILVIA FINNEMANN
KWANG JEON
CARLOS LOPEZ-OTIN

WALLACE MARSHALL
SHIGEKAZU NAGATA
MOSHE OREN
ANNE SIMONSEN

VOLUME THREE HUNDRED AND EIGHTY SIX

# INTERNATIONAL REVIEW OF CELL AND MOLECULAR BIOLOGY

Targeting Signaling Pathways in Solid Tumors - Part B

Edited by

**SUMIT MUKHERJEE**
*Albert Einstein College of Medicine,
Bronx, NY, United States*

**KAUSHIKI CHATTERJEE**
*Weill Cornell Medicine,
New York, NY, United States*

Academic Press is an imprint of Elsevier
125 London Wall, London, EC2Y 5AS, United Kingdom
50 Hampshire Street, 5th Floor, Cambridge, MA 02139, United States
525 B Street, Suite 1650, San Diego, CA 92101, United States

First edition 2024

Copyright © 2024 Elsevier Inc. All rights are reserved, including those for text and data mining, AI training, and similar technologies.

Publisher's note: Elsevier takes a neutral position with respect to territorial disputes or jurisdictional claims in its published content, including in maps and institutional affiliations.

No part of this publication may be reproduced or transmitted in any form or by any means, electronic or mechanical, including photocopying, recording, or any information storage and retrieval system, without permission in writing from the publisher. Details on how to seek permission, further information about the Publisher's permissions policies and our arrangements with organizations such as the Copyright Clearance Center and the Copyright Licensing Agency, can be found at our website: www.elsevier.com/permissions.

This book and the individual contributions contained in it are protected under copyright by the Publisher (other than as may be noted herein).

**Notices**
Knowledge and best practice in this field are constantly changing. As new research and experience broaden our understanding, changes in research methods, professional practices, or medical treatment may become necessary.

Practitioners and researchers must always rely on their own experience and knowledge in evaluating and using any information, methods, compounds, or experiments described herein. In using such information or methods they should be mindful of their own safety and the safety of others, including parties for whom they have a professional responsibility.

To the fullest extent of the law, neither the Publisher nor the authors, contributors, or editors, assume any liability for any injury and/or damage to persons or property as a matter of products liability, negligence or otherwise, or from any use or operation of any methods, products, instructions, or ideas contained in the material herein.

ISBN: 978-0-443-23548-1
ISSN: 1937-6448

For information on all Academic Press publications
visit our website at https://www.elsevier.com/books-and-journals

Publisher: Zoe Kruze
Acquisitions Editor: Leticia M. Lima
Editorial Project Manager: Dewwart Chauhan
Production Project Manager: James Selvam
Cover Designer: Vicky Pearson

Typeset by MPS Limited, India

# Contents

*Contributors*  ix

## 1. Cellular signaling in glioblastoma: A molecular and clinical perspective  1
Debarati Ghosh, Brett Pryor, and Nancy Jiang

1. Introduction  1
2. Molecular subclasses of *Glioblastoma Multiforme*  2
3. Clinical prognosis and signaling pathways in *Glioblastoma multiforme*  12
4. Therapeutic intervention in signaling pathways in *Glioblastoma multiforme*  20
5. Future direction  35
References  37

## 2. Therapeutic potential of hedgehog signaling in advanced cancer types  49
Richa Singh and Anindita Ray

1. Overview of hedgehog signaling  51
   1.1 Hedgehog protein  51
   1.2 Components of hedgehog signaling  51
   1.3 Canonical hedgehog signaling  52
   1.4 Non-canonical hedgehog signaling  53
2. Hedgehog signaling in cancer  55
   2.1 Mechanisms of hedgehog signaling in cancer  55
   2.2 Hedgehog signaling in metastatic tumors  59
   2.3 Relevance of hedgehog signaling in neuroendocrine tumors  62
   2.4 Clinical prevalence of hedgehog signaling in different cancers  63
3. Therapeutic relevance of hedgehog signaling in cancer  64
   3.1 Targeting hedgehog signaling components  64
   3.2 Combination with epigenetic inhibitors  69
4. Perspective  70
References  71

## 3. An Overview of the Unfolded Protein Response (UPR) and Autophagy Pathways in Human Viral Oncogenesis — 81

Shovan Dutta, Anirban Ganguly, and Sounak Ghosh Roy

1. Cross talk between autophagy and UPR — 82
2. Autophagy and UPR in oncogenic viruses — 84
3. Epstein-Barr virus (EBV) and its relationship to cancer — 84
   - 3.1 ERS/UPR and autophagy in EBV-induced solid tumors — 85
4. Human papillomavirus (HPV) and its relationship to cancer — 87
   - 4.1 ERS/UPR and autophagy in HPV-induced solid tumors — 89
5. Human immunodeficiency virus (HIV) and its relationship to cancer — 90
   - 5.1 HIV-mediated non-Hodgkin lymphoma (NHL) — 91
   - 5.2 ERS/UPR and autophagy in non-Hodgkin lymphoma (NHL) — 91
6. Human herpes virus-8 (HHV-8) and its relationship to cancer — 92
   - 6.1 Important oncoproteins in HHV-8 — 93
   - 6.2 Autophagy in KSHV-induced solid tumors — 94
   - 6.3 ERS/UPR in KSHV-induced solid tumors — 95
7. HTLV-1 (human T-cell lymphotropic virus type 1) and its relationship to cancer — 97
   - 7.1 Important oncoproteins in HTLV-1 — 98
   - 7.2 Autophagy in HTLV-1 induced solid tumors — 100
   - 7.3 ERS/UPR in HTLV-1 induced solid tumors — 101
8. Hepatitis B (HBV) and its relationship to cancer — 102
   - 8.1 ERS/UPR in HBV-induced solid tumors — 103
   - 8.2 Autophagy in HBV-induced solid tumors — 107
9. Discussion — 117

References — 118

## 4. The crosstalk between miRNAs and signaling pathways in human cancers: Potential therapeutic implications — 133

Ritu Shekhar, Sujata Kumari, Satyam Vergish, and Prajna Tripathi

1. Introduction — 134
   - 1.1 Biogenesis of miRNAs — 135
   - 1.2 Regulation of gene expression by miRNAs — 136
2. Overall action of miRNAs in the regulation of physiological processes — 137
3. MicroRNAs as potential oncogenes and tumor suppressors — 138
4. MicroRNAs as modulators of various signaling pathways in human cancers — 140
   - 4.1 TGF-β signaling pathway — 140
   - 4.2 MAPK signaling pathway — 143
   - 4.3 PI3K/Akt signaling — 149

|   |   |   |
|---|---|---|
| 5. | Discussion | 151 |
|   | 5.1 Diagnostic applications of miRNAs in cancer | 153 |
|   | 5.2 Therapeutic applications of miRNAs in cancer | 153 |
|   | 5.3 Limitations | 154 |
| 6. | Conclusion | 155 |
|   | References | 155 |

## 5. Targeting KRAS and SHP2 signaling pathways for immunomodulation and improving treatment outcomes in solid tumors     167
Priyanka Sahu, Ankita Mitra, and Anirban Ganguly

|   |   |   |
|---|---|---|
| 1. | RAS signaling pathway and its brief biological activity and importance in solid tumors | 168 |
| 2. | KRAS structure and function | 169 |
| 3. | KRAS signaling in health and disease | 170 |
|   | 3.1 RAF-MEK-ERK pathway | 171 |
|   | 3.2 PI3K pathway | 173 |
|   | 3.3 Post translational modification | 173 |
| 4. | KRAS mutant solid tumors | 174 |
|   | 4.1 Non-small cell lung cancer | 174 |
|   | 4.2 Colorectal carcinoma | 175 |
|   | 4.3 Pancreatic ductal adenocarcinoma | 175 |
| 5. | KRAS inhibitors | 176 |
| 6. | KRAS targeted resistance | 178 |
|   | 6.1 Intrinsic resistance | 178 |
|   | 6.2 Acquired resistance | 178 |
| 7. | KRAS and SHP2 pathways linked with immunomodulation and TME | 179 |
|   | 7.1 Regulation of cytokines and chemokine by Ras signaling | 180 |
|   | 7.2 Modulation of immune cells in the TME by Ras signaling | 188 |
|   | 7.3 Alternate targets for Ras mutant cancers in the immune-community | 193 |
| 8. | SHP-2 tyrosine phosphatase | 196 |
|   | 8.1 Structure and biological function of SHP2 | 196 |
|   | 8.2 SHP-2 at the interface of KRAS and immune signaling | 197 |
|   | 8.3 SHP2 inhibitors emerge as potential therapy for KRAS-mutant cancers | 199 |
|   | 8.4 SHP2 inhibitors used as monotherapy | 201 |
| 9. | Combination of SHP2 inhibitors with other drugs in modulating KRAS driven solid tumours | 202 |
|   | 9.1 SHP2 in combination with MEK (mitogen activated ERK kinase) inhibitors | 202 |

| | |
|---|---|
| 9.2  SHP2 in combination with KRAS inhibitors | 203 |
| 9.3  SHP2 in combination with anti-CXCR1/2 | 205 |
| 9.4  SHP2 in combination with other drugs | 206 |
| 10. Future directions | 207 |
| References | 207 |

## 6. Mitochondria driven innate immune signaling and inflammation in cancer growth, immune evasion, and therapeutic resistance     223

Sanjay Pandey, Vandana Anang, and Michelle M. Schumacher

| | |
|---|---|
| 1. Introduction | 224 |
| 2. Mitochondrial outer membrane permeabilization (MOMP) driven inflammation and therapeutic resistance | 226 |
| 3. Mitochondrial ROS and the mitochondrial DAMPS in cancer | 227 |
| 4. mtDNA as immune activator and cytosolic DAMP | 229 |
| 5. Immune sensing of mtDNA by cGAS sting pathway | 230 |
| 6. ATP and mtDNA activate inflammaomes | 231 |
| 7. mtDNA drives TLR9 signaling | 232 |
| 8. mtRNAsensing by RIG-1 signaling | 233 |
| 9. Cardiolipin (CL) in innate inflammation | 233 |
| 10. Formyl peptides driven inflammation | 234 |
| 11. Innate immune signaling and cancer growth and acquired resistance | 235 |
| 12. Modulation of mitochondria-associated inflammation and future directions | 237 |
| 13. Conclusion | 241 |
| References | 242 |

# Contributors

**Vandana Anang**
International Center for Genetic Engineering and Biotechnology (ICGEB), New Delhi, India

**Shovan Dutta**
Center for Immunotherapy & Precision Immuno-Oncology (CITI), Lerner Research Institute, Cleveland Clinic, Cleveland, OH, United States

**Anirban Ganguly**
Department of Biochemistry, All India Institute of Medical Sciences, Deoghar, Jharkhand, India

**Sounak Ghosh Roy**
Henry M Jackson for the Advancement of Military Medicine, Naval Medical Research Command, Silver Spring, MD, United States

**Debarati Ghosh**
McGovern Institute for Brain Research, Massachusetts Institute of Technology, Cambridge, MA, United States

**Nancy Jiang**
Wellesley College, Wellesley, MA, United States

**Sujata Kumari**
Department of Zoology, Magadh Mahila College, Patna University, Patna, India

**Ankita Mitra**
Laura and Isaac Perlmutter Cancer Center, New York University Langone Medical Center, New York, NY, United States

**Sanjay Pandey**
Department of Radiation Oncology, Montefiore Medical Center, Bronx, NY, United States

**Brett Pryor**
McGovern Institute for Brain Research, Massachusetts Institute of Technology, Cambridge, MA, United States

**Anindita Ray**
Neurodegenerative Diseases Research Unit, National Institute of Neurological Disorders and Stroke, Bethesda, MD, United States

**Priyanka Sahu**
Laura and Isaac Perlmutter Cancer Center, New York University Langone Medical Center, New York, NY, United States

**Michelle M. Schumacher**
Department of Radiation Oncology, Montefiore Medical Center; Department of Pathology, Albert Einstein College of Medicine, Bronx, NY, United States

**Ritu Shekhar**
Department of Molecular Genetics and Microbiology, University of Florida, Gainesville, FL, USA

**Richa Singh**
Department of Pathology and Laboratory Medicine, Weill Cornell Medicine, New York, NY, United States

**Prajna Tripathi**
Department of Microbiology and Immunology, Weill Cornell Medical College, New York, USA

**Satyam Vergish**
Department of Plant Pathology, University of Florida, Gainesville, FL, USA

CHAPTER ONE

# Cellular signaling in glioblastoma: A molecular and clinical perspective

**Debarati Ghosh[a,*,1], Brett Pryor[a,1], and Nancy Jiang[b]**
[a]McGovern Institute for Brain Research, Massachusetts Institute of Technology, Cambridge, MA, United States
[b]Wellesley College, Wellesley, MA, United States
*Corresponding author. e-mail address: ghosh.debarati.genetics@gmail.com

## Contents

| | |
|---|---|
| 1. Introduction | 1 |
| 2. Molecular subclasses of *Glioblastoma Multiforme* | 2 |
| 3. Clinical prognosis and signaling pathways in *Glioblastoma multiforme* | 12 |
| 4. Therapeutic intervention in signaling pathways in *Glioblastoma multiforme* | 20 |
| 5. Future direction | 35 |
| References | 37 |

## Abstract

Glioblastoma multiforme (GBM) is the most aggressive brain tumor with an average life expectancy of less than 15 months. Such high patient mortality in GBM is pertaining to the presence of clinical and molecular heterogeneity attributed to various genetic and epigenetic alterations. Such alterations in critically important signaling pathways are attributed to aberrant gene signaling. Different subclasses of GBM show predominance of different genetic alterations and therefore, understanding the complex signaling pathways and their key molecular components in different subclasses of GBM is extremely important with respect to clinical management. In this book chapter, we summarize the common and important signaling pathways that play a significant role in different subclasses and discuss their therapeutic targeting approaches in terms of preclinical studies and clinical trials.

## 1. Introduction

Glioblastoma multiforme (GBM) is the common term for most gliomas given a CNS WHO Grade IV rating (Brennan et al., 2009; Ghosh, Nandi, & Bhattacharjee, 2018; Louis et al., 2021; Paolillo, Boselli, & Schinelli, 2018; Shergalis et al., 2018; Szopa et al., 2017). It is the most

[1] Equal authorship.

frequently occurring malignant central nervous system tumor. The global incidence of GBM is 5.26 per 100,000 people (Omuro & DeAngelis, 2013), and it accounts for 14% of all primary brain tumors (Faris, 2022). It has a poor prognosis, with a 5-year survival rate of <5% after diagnosis (Tan et al., 2020; Verhaak et al., 2010). The poor prognosis and high mortality of GBM may be attributed to its complex pathogenesis, which in turn is governed by various cellular signaling pathways. This chapter will summarize different cell signaling pathways of GBM pathogenesis based on the disease prognosis and describe its clinical management through several preclinical models and clinical trials.

## 2. Molecular subclasses of *Glioblastoma Multiforme*

GBM is broadly classified into two groups: sporadically occurring primary (de novo) glioma, or secondary glioma developed from a lower grade astrocytoma. Primary GBM makes up ~90% of cases; however, primary and secondary GBM are almost histologically indistinguishable (Furnari et al., 2007). Histologically similar glioblastomas can differ in their clinical outcome, with several genetic, epigenetic, and molecular factors beyond the primary/secondary GBM distinction contributing to such events. Thus, stratifying them in additional subclasses proves helpful for determining clinical outcome.

Epigenetic changes, especially methylation of certain gene promoters, are among the most important contributors to the heterogeneity of GBM pathology and markers for refined GBM diagnosis (Capper et al., 2018). Many of these will be discussed later with relevance to specific signaling pathways, but the clinical implications of one particular epigenetic change makes it especially important for GBM subclassification. O6-methylguanine-DNA methyl-transferase (MGMT) is a protein which repairs DNA damage caused by alkylating agents, such as the chemotherapy drug temozolomide (TMZ). Methylation of the *MGMT* promoter silences the gene, causing increased tumor susceptibility to TMZ. *MGMT* promoter methylation thus improves median patient survival time by about 9–12 months compared to unmethylated *MGMT* promoter cases where both are treated with concomitant radiotherapy and TMZ followed by adjuvant TMZ, commonly known as the "Stupp regimen". When comparing the Stupp regimen to radiotherapy alone in cases with *MGMT* promoter methylation, the Stupp regimen increases median survival by 5–10 months

(Alnahhas et al., 2020; Hegi et al., 2005; Stupp et al., 2005). More recent analysis indicates that the Stupp regimen may still provide some benefits over radiotherapy alone to those with an unmethylated *MGMT* promoter, but more data is needed to determine the true significance of any such benefit (Alnahhas et al., 2020). Furthermore, some distinct subtypes of GBM do not see improved prognosis with *MGMT* promoter methylation to the same degree as others, which is covered in the following sections (Verhaak et al., 2010). The methylation status of the *MGMT* promoter is therefore a generally useful prognostic factor and important for treatment planning, but should be considered a supplemental subclassification for GBM and not used as a binary determinant of treatment efficacy.

In 2010, Verhaak et al. proposed four GBM subtypes based on genetic profiles that correlate with prognosis and response to treatment (Verhaak et al., 2010): (i) Classical GBM; (ii) Mesenchymal GBM; (iii) Proneural GBM; (iv) Neuronal GBM.

i. **Classical GBM:** Classical GBM is characterized by *EGFR* mutation and gene amplification. This subtype also distinctly lacks any *TP53* mutations. Homozygous deletion of *CDKN2A* inactivates the RB pathway and is exclusive with aberrations in other RB-pathway genes found in other subtypes (Table 1). Morphologically, Classical GBM is astrocytic-like. MGMT-methylation can occur in Classical GBM and appears to have a positive effect on outcome with Stupp regimen treatment, but no unique association between *MGMT*-methylation status and Classical GBM exists (Nørøxe, Poulsen, & Lassen, 2016; Verhaak et al., 2010).

ii. **Mesenchymal GBM:** Mesenchymal GBM is largely defined by alterations to the *NF1* and *PTEN* genes, especially when they are co-mutant (Table 1). *NF1* typically has a heterozygous deletion, low expression, and/or various mutations, while *PTEN* alterations are most commonly mutations (Cancer Genome Atlas Research Network, 2008; Verhaak et al., 2010). *PTEN* mutations can activate the RAS pathway, while disruption of both genes impacts the Akt pathway (Table 1) (Nørøxe et al., 2016; Verhaak et al., 2010). Morphologically, mesenchymal GBM is astrocytic-like and additionally characterized by a higher degree of necrosis and inflammation than other GBM subtypes. Mesenchymal GBM may present with MGMT methylation, and Mesenchymal GBM does generally have improved prognosis with the Stupp regimen of treatment (Verhaak et al., 2010).

Table 1 Cell signaling pathways in different subtypes of GBM.

| Pathway | Relevant classifications | Defining gene alterations | Additional notes |
|---|---|---|---|
| p53-MDM2-p14ARF signaling pathway | Primary GBM | *TP53*: Mu 20% *MDM2*: 10% amplified if TP53Mu⁻ | |
| | Secondary GBM | *TP53*: Mu 80% *TP14*: Promoter methylation 30% | |
| | Classical GBM | ***TP53: Mu 0%*** | |
| | Mesenchymal GBM | *TP53*: Mu 32% | |
| | Proneural GBM | ***TP53: Mu 53%, LOH*** | |
| | IDH-mutant GBM (Louis et al., 2016) | *TP53*: Mu 82% | |
| | Astrocytoma, IDH-mutant grade 4 (Louis et al., 2021) | ***TP53: Mu 82%*** | |
| | Glioblastoma, IDH-wildtype (Louis et al., 2021) | *TP53*: Mu 27% | |
| | General | *p14*: Altered 76% | |

| | | |
|---|---|---|
| pRB-p16(INK4a) signaling pathway | Primary GBM | *RB1*: promoter methylation (14%)<br>*CDKN2A/B*: HomDel prevalent |
| | Secondary GBM | *p16(INK4a)*: HomDel<br>*RB1*: promoter methylation (43%) |
| | Classical GBM | **CDKN2A: HomDel 94% coincidence with EGFR amplification** |
| | Astrocytoma, IDH-mutant grade 4 (Louis et al., 2021) | **CDKN2A/B: HomDel** *CDKN2A/B* deletion higher in IDH WT and lower in IDH mutant |
| | Glioblastoma, IDH-WT (Louis et al., 2016, 2021) | *CDKN2A/B*: HomDel — Uncommon, but one criterion for diagnosis |
| | | More common in IDH-WT than IDH-mutant |
| | General | 78% of GBM cases show some pathway abnormality |

*(continued)*

**Table 1** Cell signaling pathways in different subtypes of GBM. (cont'd)

| Pathway | Relevant classifications | Defining gene alterations | Additional notes |
|---|---|---|---|
| EGFR–PI3K–PTEN–AKT–mTOR signaling pathway | Primary GBM | *EGFR*: Mu, increased expression (40%–60%) *PTEN*: Mu | |
| | Classical GBM | ***EGFR*: Mu 50% (*EGFRvIII* 25%), increased gene expression, copy number amplification** *PTEN*: Mu 23% | |
| | Mesenchymal GBM | ***PTEN*: Mu 32%, LOH** | |
| | Proneural GBM | ***EGFR*: Mu 0%** ***PTEN*: Mu 0%** | |
| | Glioblastoma, IDH-Mutant (Louis et al., 2016, 2021) | *EGFR*: Mu 0% | IDH mutant has distinctly no *EGFR* alternations or *PTEN* mutations |
| | Glioblastoma, IDH-wildtype (Louis et al., 2021) | ***EGFR*: Mu, increased gene expression Chromosomal alteration:+7/−10** | Chromosome 10 contains PTEN |

| | | | |
|---|---|---|---|
| RTKs and EGFR signaling | Primary GBM | Alterations in 66% of cases (EGFR 40%–60%) | |
| | Glioblastoma, IDH-wildtype (Louis et al., 2016) | *EGFR*: Mu, increased gene expression, gene amplification | |
| | Glioblastoma, IDH-wildtype (Louis et al., 2021) | **EGFR: Mu, increased gene expression, gene amplification** | Pathway is often altered by alterations in other pathways |
| RTKs and VEGF, PDGF signaling | Mesenchymal GBM | *PDGFRA*: decreased expression | Caused by NF-κB pathway activation |
| | Proneural GBM | *PGDFRA*: Mu or amplified 11% | |
| | Glioblastoma, IDH-Mutant (Louis et al., 2016) | *PGDFRA*: Higher rate of amplification, lower rate of mutation | |
| RAS/MAPK signaling pathway | Primary GBM | *RAS*: Mu 2% | |

*(continued)*

Table 1 Cell signaling pathways in different subtypes of GBM. (cont'd)

| Pathway | Relevant classifications | Defining gene alterations | Additional notes |
|---|---|---|---|
| STAT3 and ZIP4 signaling pathway | Classical GBM | High pathway activation | |
| | Mesenchymal GBM | High pathway activation STAT3: increased expression/activation | STAT3 effects caused by NF-κB pathway activation |
| | Glioblastoma, IDH-wildtype (Louis et al., 2016, 2021) | High pathway activation | |
| TGF-β signaling pathway | Glioblastoma, IDH-wildtype (Louis et al., 2016, 2021) | TGFb1 & 2: gene upregulated | |
| | Glioblastoma, IDH-mutant (Louis et al., 2016) | TGFb1 & 2: gene upregulated | |
| | Astrocytoma, IDH-mutant grade 4 (Louis et al., 2021) | TGFb1 & 2: gene upregulated | |
| WNT signaling pathway | Primary GBM | sFRP1, sFRP2, NKD2: promoter hypermethylation 40% | |
| | General | Beta-catenin: 50% upregulation TCF1: increased expression 51.6% LEF1: increased expression 71% GSK3β: promoter methylation 10.34% DKK1: promoter methylation 37.93% DKK3: promoter methylation 44.83% pGSK3β-Y216: increased expression | |

| | | | |
|---|---|---|---|
| Nrf2 and cell signaling | Glioblastoma, IDH-wildtype (Louis et al., 2016, 2021) | *Nrf2*: increased expression | |
| | Mesenchymal GBM | *Nrf2*: increased expression | |
| NF-κB and cell signaling | Mesenchymal GBM | *CD44*: increased expression<br>*YKL40*: increased activation<br>*C/EBPβ*: increased expression/activation<br>*TAZ*: increased expression/activation<br>*OLIG2*: decreased expression | Factors associated with Proneural-to-Mesenchymal transition by induced activation of NF-κB pathway. This transition causes tumor radioresistance in a NF-κB-dependent manner |
| | General | *RELA (p65)*: increased expression<br>*eva1*: increased expression<br>*MGMT*: increased expression | *MGMT* expression increased independent of *MGMTp* methylation |

Summary of cell signaling pathways implicated in various GBM subtypes and associated pathway alterations. **Bolded text**: hallmark or diagnostic-criteria pathway alterations. *Mu*, Mutation. *Amp*, Gene Amplification. *HomDel*, Homozygous Deletion. *LOH*, Loss of Heterozygosity. Percentages represent the percentage of cases in a subtype which contain the associated alteration.

iii. **Proneural GBM:** Proneural GBM's major hallmarks are alterations of *PDGFRA* and *IDH1* point mutations (Table 1). *PDGFRA* exhibits both a focal amplification, common to most GBMs, and uniquely elevated gene expression. This can cause activation of PI3K and RAS pathways (Cancer Genome Atlas Research Network, 2008; Nørøxe et al., 2016; Verhaak et al., 2010). *IDH1* point mutations are very common in Proneural cases, yet mostly exclusive with *PDGFRA* abnormalities. *PIK3CA/PIK3R1* mutations also follow this trend (Verhaak et al., 2010). *TP53* mutations and loss of heterozygosity are most common in Proneural GBM, though not exclusive (Table 1) (Cancer Genome Atlas Research Network, 2008; Verhaak et al., 2010). Morphologically, Proneural GBM is oligodendrocyte-like. It can present with MGMT methylation, but this does not correlate to any improved treatment response as Proneural GBM does not appear to respond to the Stupp regimen (Nørøxe et al., 2016; Verhaak et al., 2010).

iv. **Neural GBM:** Lastly, the Neural subtype is defined by expression of neuron markers and genes associated with neuron characteristics such as neuron projections, axons, and synaptic transmission. Verhaak et al. points out NEFL, GABRA1, SYT1, and SLC12A5 as examples of such genes. Neural GBM appears to have a positive response to Stupp regimen treatment, but this data is not significant and the contribution of *MGMT* promoter methylation to this response is not known (Verhaak et al., 2010).

In 2016, the WHO officially re-categorized GBM types based on the mutation status of the isocitrate dehydrogenase (IDH) genes. Three new classifications were created: Glioblastoma *IDH*-wildtype, Glioblastoma *IDH*-mutant, and Glioblastoma NOS (not otherwise specified) (Louis et al., 2016). These largely follow the primary/secondary GBM distinction, with about 94% of Primary GBMs being *IDH*-wildtype and 85% of Secondary GBMs being *IDH*-mutant (Yan et al., 2009). Glioblastomas NOS are cases where genetic testing was not performed or was inconclusive. *IDH*-mutant GBMs contain mutually-exclusive point mutations in either the *IDH1* or *IDH2* genes, with the most common being mutation of *IDH1* residue 132 (Parsons et al., 2008; Yan et al., 2009). *IDH*-mutant GBM is genetically distinct, displaying no *EGFR*, *PTEN*, *NF1*, or *RB1* mutations which are common in *IDH*-wildtype GBM (Parsons et al., 2008). *IDH*-mutants also have a significantly higher incidence of *TP53* mutation and a much lower incidence of *CDKN2A/B* deletion than *IDH*-wildtype GBM (Table 1) (Parsons et al., 2008; Yan et al., 2009). The prognosis of *IDH*-mutant cases is generally improved over *IDH*-wildtype (Yan et al., 2009; Zhang et al., 2020).

In 2021, the WHO again re-classified many gliomas based on clinicopathological grading (Louis et al., 2021). This led to new, more specific criteria for diagnosis of Grade 4 gliomas which constitute GBM. These are described below:

i. **Glioblastoma, *IDH*-wildtype:** This classification has replaced the 2016 *IDH*-wildtype designation, with the only changes being more specific diagnostic criteria. Diagnosis requires *IDH*-wildtype status and one of the following traits: microvascular proliferation, necrosis, *TERTp* mutation, *EGFR* gene amplification, or co-incidental gain of chromosome 7 and loss of chromosome 10 (Table 1). It is generally accepted that prognosis of glioblastoma, *IDH*-wildtype is improved in cases with *MGMT* promoter methylation when treated with the Stupp regimen (Brawanski et al., 2023). *EGFRvIII* expression may further enhance prognosis when coincident with *MGMTp* methylation (Struve et al., 2020). In cases of *IDH*-wildtype glioblastomas harboring *TERTp* mutations, the methylation status of the *MGMT* promoter has a more nuanced impact on prognosis. A comprehensive study of 456 *IDH*-wildtype GBM cases indicated that *TERTp* mutations can compound the sensitivity to chemotherapy conferred by *MGMT* promoter methylation; however, cases with *TERTp* mutations and an unmethylated *MGMT* promoter had the lowest median survival time of all conditions by a significant degree (Arita et al., 2016). This implies that *TERTp* mutation may be a positive prognostic factor when coinciding with *MGMTp* methylation, but a negative prognostic factor without *MGMTp* methylation.

ii. **Astrocytoma, *IDH*-mutant grade 4:** Many grade 2 and 3 astrocytoma have a high rate (~80%) of *IDH* mutation, consistent with that in *IDH*-mutant GBM as defined in the 2016 WHO designation (Yan et al., 2009). This led to the 2021 WHO reclassification of some such astrocytoma into the Astrocytoma, *IDH*-mutant class, with internal grades 2-4 each having distinct diagnostic criteria. Astrocytoma, *IDH*-mutant grade 4 are those with either *ATRX* expression loss or *TP53* mutations, and which display microvascular proliferation and/or necrosis, or a homozygous *CDKN2A/B* deletion (Table 1) (Louis et al., 2021; Torp, Solheim, & Skjulsvik, 2022). Studies show that *MGMT* promoter methylation also significantly improves prognosis with Stupp regimen treatment in grade 4 *IDH*-mutant astrocytoma, but not in lower grade *IDH*-mutant astrocytoma (Chai et al., 2021; Lam et al., 2022).

The classification and reclassification of the GBM into various subclasses is intended to predict disease aggressiveness and their response to therapy. Therefore, understanding cellular signaling in different prognostic subgroups may clarify our understanding of the course of disease development and thus manage GBM more efficiently.

## 3. Clinical prognosis and signaling pathways in *Glioblastoma multiforme*

Genetic, epigenetic, and transcriptional heterogeneity leads to complex cell signaling systems with clinical manifestations in GBM. Understanding those pathways in the light of their clinical prognosis and molecularly different GBM subgroups (Table 1) are extremely important to understand the disease pathology.

**A. Signaling pathways in GBM:**
  **i. p53-MDM2-p14ARF signaling pathway in GBM:** Alterations of the p53 pathway are very common in all cancers, but none so much as GBM (Ohgaki et al., 2004). p53 is a transcription factor that responds to cellular stress and DNA damage. It acts as a tumor suppressor, halting cell cycle progression and promoting apoptosis, by triggering p21's inhibition of Cdk4/Cyclin D and CDK2/Cyclin E complexes (He et al., 2005). Binding with p14$^{ARF}$ stabilizes p53, whereas binding with protein mouse double minute 2 homolog (MDM2) degrades p53 (Stott et al., 1998). De-regulation of p53 is found in 65% of secondary GBM cases, but only 28% of de-novo primary GBM cases (Watanabe et al., 1996). Mutations of the *TP53* gene itself also arise in more secondary GBM cases (>90%) than primary GBM cases (<35%) (Watanabe et al., 1996). Co-mutation of *TP53* and *PTEN* in a mouse model study showed elevated c-Myc levels and increased tumorigenesis (Zheng et al., 2008). *MDM2* gene amplification occurs in 10% of primary GBM cases that lack any p53 mutation (Reifenberger et al., 1996). A total of 76% of GBM cases have altered p14$^{ARF}$, with little difference in frequency between primary and secondary GBM; however, the p14$^{ARF}$ promoter methylation frequency is much higher in secondary GBM (30% cases) than primary (Nakamura et al., 2001). Furthermore, a recent *in vivo* and *in vitro* study suggests a sex-biased role of different p53 mutations: the p53R172H mutation increased transformation ability

in females' astrocytes, the p53Y202C mutation transformed both males' and females' astrocytes with a small male bias, and the p53Y217C mutation shows gain-of-function transformation in only males' astrocytes (Rockwell et al., 2021).

ii. **pRB-p16$^{INK4a}$ signaling pathway in GBM:** Under physiological condition, the cyclin dependent kinase (CDK4)/cyclin D1 complex phosphorylates the RB protein (pRB) and triggers its release from the transcription factor E2F, in turn promoting the G1-S phase transition. On the other side, p16$^{INK4}$ binds to CDK4 and thus inhibits pRB phosphorylation and release from E2F, preventing G1-S transition. The alteration of such function leads to aberrant E2F1 function and uncontrolled cell growth (Gomez-Manzano et al., 2001; Sherr, 2001). Overall, the pRB pathway is altered in 78-79% of total GBM cases studied (Cancer Genome Atlas Research Network, 2008; Pearson & Regad, 2017). Homozygous deletion of p16$^{INK4}$ is more frequently found in secondary GBM (Nakamura et al., 2001), and promoter methylation of the *RB1* gene is more frequently found in secondary GBM (43%) than in primary GBM (14%) (Nakamura et al., 2001). However, it is not frequently found in low grade astrocytoma, indicating its late onset during gliomagenesis.

iii. **EGFR-PI3K-PTEN-AKt-mTOR signaling pathway in GBM:** In normal conditions, the phosphatidylinositol 3-kinase (PI3K) is activated by epidermal growth factor (EGF) and its receptor (EGFR), and upon activation it generates phosphatidylinositol-3,4,5-triphosphate (PIP3). This in turn activates Akt, which then activates mammalian target of rapamycin (mTOR) (Laplante & Sabatini, 2012) to trigger multiple downstream pathways controlling cell growth and division (Colardo, Segatto, & Di Bartolomeo, 2021). The PI3K/AKT/mTOR axis is altered in various human cancers by promoting tumor growth and survival (Courtney, Corcoran, & Engelman, 2010; Dienstmann et al., 2014). The PI3K pathway is altered in 70% of GBM cases and attributed to overexpression of EGFR, a tyrosine kinase growth factor receptor (Watanabe et al., 1996). A ligand-independent activated EGFR receptor (EGFRvIII mutation) is the most common mutation in GBM and it leads to poor prognosis (Shinojima et al., 2003). The phosphatase and tensin homolog (PTEN), a tumor suppressor, negatively regulates the PI3K/AKT/

PKB pathway (Stambolic et al., 1998) by blocking Akt signaling via reduction of PIP3 levels (Maehama & Dixon, 1998). A mutation in the *PTEN* gene is found in 40% of GBM, (Lucifero & Luzzi, 2022) with moderate frequency in primary GBM and low frequency in secondary GBM (Ohgaki et al., 2004). Loss of heterozygosity of chromosome 10, which harbors *PTEN*, is very commonly found in GBM.

iv. **RTK and EGFR signaling pathway in GBM:** The receptor tyrosine kinases (RTKs) are a family of cell-surface receptors which are involved in cellular proliferation, differentiation, migration, metabolism, and many other functions (Blume-Jensen & Hunter, 2001). Altered RTK expression, due to mutation or amplifications, occurs in 66% of primary GBM samples tested in the pan-cancer project TCGA. Among all of these, EGFR alteration is most common, seen in 40%–60% of primary GBMs but rarely in secondary GBM (Watanabe et al., 1996; Wong et al., 1992). Besides EGFR, platelet derived growth factor receptor alpha (PDGFRA) alterations appear in 13% of GBM cases (Cameron, Roel, Aaron, Benito, & Houtan, 2013; Cancer Genome Atlas Research Network, 2008), whereas MET amplification and fibroblast growth factor receptor (FGFR) alterations are co-expressed in some GBM cases (Velpula et al., 2012).

v. **RTK and VEGF, PDGF pathway in GBM:** Angiogenesis remains part of tumorigenesis and without exception plays a pivotal role in GBM pathogenesis. Among several factors, two of the most critical are vascular endothelial growth factor (VEGF) and platelet derived growth factor (PDGF), which work via the RAS/RAF/MEK/MAPK signaling pathways and PI3K/AKT/mTOR signaling pathways (Ahir, Engelhard, & Lakka, 2020). Under hypoxic conditions, hypoxia inducible factors (HIF1α and HIF1β) translocate to the nucleus and activate VEGF genes (Pearson & Regad, 2017). In GBM, VEGF is the most important pro-angiogenic cytokine that binds to VEGF receptors on endothelial cells. These cells secrete MMP into surrounding tissues to break down ECM and thus promote cell proliferation and migration (Lamalice, Le Boeuf, & Huot, 2007). VEGF-A works synergistically with FGF-2 and PDGF-BB to induce angiogenesis *in vitro* and *in vivo* (Richardson et al., 2001). In GBM the PDGF autocrine loop is regularly exhibited, which is absent in normal human brain tissue

(Pearson & Regad, 2017). Furthermore, all PDGF ligands (PDGF-A, PDGF-B, PDGF-C and PDGF-D) and two surface receptors: PDGFR-α and PDGFR-β are expressed in GBM tissue (Nazarenko et al., 2012) with PDGFR-α expression accounting for 10-13% of GBM cases as per the TCGA database (Cancer Genome Atlas Research Network, 2008).

vi. **RAS/MAPK signaling pathway in GBM:** Human Ras genes (Rat Sarcoma) belong to the G-protein family and are transforming oncogenes that include H-Ras, N-Ras and K-Ras. Upon activation, Ras activates RAF kinase through direct binding, which in turn regulates downstream signaling pathways like the mitogen-activated protein kinase (MAPK) pathway (Moodie et al., 1993; Thomas et al., 1992). Ras also regulates other signaling pathways like PI3K for regulating cell growth and tumorigenesis. Furthermore, growth factor receptors (EGFRs, PDGFRs and RTKs) also activate Ras, which in turn influences downstream regulator RAF and then MAPK (Warner et al., 1993).

vii. **STAT3 and ZIP4 signaling pathway in GBM:** Signal transducers and activators of transcription (STAT) protein complexes are SH2 domain-containing cytoplasmic proteins that act as transcription factors to help regulate the cellular response to cytokines and growth factors via activities like cell proliferation, invasion, and apoptosis (Abal et al., 2006). This pathway is activated by EGF, and upregulation leads to many cancers including GBM (Rahaman, Vogelbaum, & Haque, 2005). A more recent study suggests the role of STAT3 in the autophagy pathway and tumor microenvironment (TME) in the pathogenesis of GBM (Laribee et al., 2023; Piperi, Papavassiliou, & Papavassiliou, 2019). Zinc is an essential trace element in cellular function and alterations in zinc transporter ZIP4 are also linked with malignancies and cellular invasion in GBM via STAT3 regulation (Mao et al., 2012).

viii. **TGF-β signaling pathway in GBM:** Overexpression or altered signaling of growth factors and their pathways are implicated in GBM. TGF-β acts as a tumor suppressor in GBM that inhibits or downregulates the expression of CDKs by inducing expression of CDK inhibitors p15, p27, and Cip/WAF1/p21 (Johnson et al., 1993; Matsuura et al., 2004). TGF-β plays a critical role in cell adhesion, migration, and invasion, and thus is involved in the epithelial to mesenchymal transformation (Bryukhovetskiy & Shevchenko, 2016;

Joseph et al., 2021; Merzak et al., 1994; Paulus et al., 1995; Yang et al., 2022). TGF-β1 and TGF-βRII are expressed in GBM but not in low grade glioma (Yamada et al., 1995). TGF-β also induces expression of PDGF-A, MAPK (Ras-ErK) and SAPK (Rho-JNKL, TAK-1-p38 kinase) and works upstream of Smad pathways. Chemoresistant GBM expresses high levels of microtubules in the cytoplasm, as compared to the chemosensitive IDH mutant, 1p19q deleted subtype (now considered a lower grade astrocytoma). *In vitro* and *in vivo* study reveals the TGF-β signaling pathway is involved in such processes (Joseph et al., 2021).

ix. **Wnt signaling pathway in GBM:** Wnt signaling plays a crucial role in CNS development and differentiation. Aberrant Wnt signaling is implicated in GBM (Gong & Huang, 2012; Yu et al., 2013; Zhang et al., 2012). Under normal cellular function, the Wnt proteins bind to frizzled (FZD) receptors and low-density lipoprotein receptor-related protein/alpha 2-macroglobulin receptor (LRP) families. Upon binding, the signal is transduced to *β*-catenin, and activates transcription of Wnt target genes. Homozygous deletion of FAT Atypical Cadherin 1(FAT1) in GBM, which is a negative regulator of Wnt signaling, and FAT1 gene mutation is also found in 1% of GBM cases, as per the TCGA data set (Lee et al., 2016). Promoter hypermethylation of the sFRP1, sFRP2 and NKD2 genes involved in Wnt signaling pathways are present in >40% of primary GBM specimens (Roth et al., 2000). Also, there is evidence of Wnt signaling components in GBM stem cell maintenance (Guan, Zhang, & Guo, 2020; Zheng et al., 2010), invasiveness (Jin et al., 2011), and therapeutic resistance (Auger et al., 2006). Among the gene set involved in the Wnt/b-catenin pathway, high expression of the Dickkopf-3 (DKK3) gene is negatively associated with increased anti-tumor immunity, especially in CD8+ and CD4+ T cells. The gene set enrichment analysis (GSEA) from the Cancer Gene Database revealed an association between high DKK3 expression, poor survival, and disease progression (Han et al., 2022). Another study based on *in vitro* and *in vivo* analysis demonstrates reduced *REIC/ Dkk-3* expression and its involvement in cell growth through caspase-dependent apoptosis and reduction in b-catenin (Mizobuchi et al., 2008). A more recent study found involvement of DKK1 and DKK1 methylation specifically in high grade GBM (Kafka et al., 2021).

x. **Nrf2 and cell signaling in GBM:** Nuclear Factor Erythroid 2-related Factor 2 (Nrf2) in normal cellular function regulates protective responses to oxidative stress. Upon detection of reactive oxygen species (ROS), Nrf2 migrates to the nucleus where it promotes antioxidant enzyme expression and maintains homeostasis (Fan et al., 2017). Recent evidences show elevated expression of Nrf2 in IDH1-wildtype GBM cases (Fan et al., 2017; Haapasalo et al., 2018), where its overexpression leads to aggressive development of mesenchymal GBM (Pölönen et al., 2019). Another study shows an increased presence of Nrf2 in CD133+ GBM stem cells in comparison to CD133-stem cells, suggesting its role in malignant growth and GBM stem cell differentiation (Zhu et al., 2014). A study by Ahmad et al. (2016), indicates involvement of Nrf2-TERT in maintaining defense mechanisms against oxidative stress in GBM (Ahmad et al., 2016). Furthermore, there is a suggested role of the Nrf2-ARE mediated pathway in apoptotic regulation in GBM (Pan et al., 2013). There is evidence of Nrf2 involvement in multiple additional cell-signaling pathways, including HIF-1 alpha, TGF-β, ERK and PI3K (Awuah et al., 2022).

xi. **NF-κB and cell signaling in GBM:** Nuclear factor-κB (NF-κB) is a family of dimeric transcription factors which binds to DNA and regulates several important cellular functions, including cellular proliferation, DNA repair, apoptosis, immune response, and inflammatory response (Soubannier & Stifani, 2017). The NF-κB-family proteins p65 (RelA), RelB, c-Rel, p105/p50, and p100/p52 can homo- and heterodimerize into functional complexes to regulate transcription of target genes. NF-κB is activated in both canonical and non-canonical pathways. In the canonical pathway, p65-p50 dimers are bound to IκBα in the cytoplasm, and upon TNF-α or IL-1 stimulation the IKK complex gets activated and phosphorylates and degrades the IκBα complex. This leads to release of NF-κB and its accumulation in the nucleus. In the non-canonical pathway, RelB is bound in the cytoplasm by p100, which acts as an IκB molecule upon stimulation by BAFF or CD40. Upon activation, the RelB-p52 dimerize and regulate transcription of downstream target genes (Bradford & Baldwin, 2014; Hayden & Ghosh, 2008). A study by Rinkenbaugh et al. (2016), suggests a role for NF-κB in stem cell maintenance, as they found RelA (p65) upregulation in CD133/GBM⁺ cells as compared to CD133/GBM⁻ cells (Rinkenbaugh et al., 2016). The same study also showed the TGF-β pathway mediates NF-κB activation via transforming growth factor-β-activated kinase 1 (TAK1), which again plays a role in tumor growth and survival (Rinkenbaugh et al., 2016). Another study showed involvement of epithelial V-like antigen 1 (Eva1) in non-canonical activation of NF-κB (Ohtsu et al., 2016). Furthemore, NF-κB activation is found more in the mesenchymal subtype of GBM where a RelB-mediated NF-κB activation leads to cell elongation, migration, and invasion in GBM via SMAC (second mitochondrial activator of caspases) (Bhat et al., 2013; Ohtsu et al., 2016; Tchoghandjian et al., 2013). Also, a canonical TNFα/NF-κB signaling pathway is associated with proneural-to-mesenchymal transition in a subset of GSCs (Bredel et al., 2011; Garner et al., 2013; Kim et al., 2016; Xu et al., 2015). The NF-κB binds to ZEB1 and the *FN14* promoter, which are involved in glioma cell invasiveness (Edwards et al., 2011; Tran et al., 2006), and activates IL-8, monocyte chemoattractant protein 1, and cxc chemokine receptor 4, which are involved in cell migration (Tchoghandjian et al., 2013). Thus, NF-κB is implicated in the epithelial-to-mesenchymal transition (EMT). Apart from its role in EMT, NF-κB plays a critical role in resistance to chemotherapy by binding to the *MGMT* promoter and activating gene expression (Lavon et al., 2007). A pharmacological inhibition of NF-κB in the presence of TMZ leads to reduced cell proliferation, increased apoptosis, and reduced cell motility via cytoskeleton pathway changes (Avci et al., 2020). On the other hand, NF-κB-mediated radioresistance is mediated partly by maintaining stemness in the subpopulation, as well as enhancing damage repair and unperturbed cell cycle progression (Soubannier & Stifani, 2017).

**B. Subtype-specific signaling pathways in GBM:**
  **i. Signaling pathway in classical GBM:** In classical GBM, EGFR amplification or EGFRvIII mutation is observed in 97% of cases. Focal 9p21.3 homozygous deletion, harboring CDKN2A (encoding both $p16^{INK4A}$ and $p14^{ARF}$) is frequent in classical GBM, co-occurring with EGFR amplification in 94% of classical GBM. The sonic hedgehog (SMO, GAS1 and GLI2) and Notch (NOTCH3, JAG1 and LFNG) signaling pathways are highly expressed in classical

subtype GBM (Verhaak et al., 2010). Using a GSEA and genome-wide Cox regression from a TCGA dataset containing 608 GBM patients' data revealed the involvement of cell cycle (TGFB1, MNAT1, SMC4, TFDP1), DNA repair (ATRX, BRCA1, MGMT and SMUG1), and Janus kinase/signal transducers and activators of transcription (JAK-STAT) pathways (Park et al., 2018).

ii. **Signaling pathways in mesenchymal GBM:** Focal deletion of 17q11.2 harboring *NF1* gene and mutations of *NF1* that reduce expression are predominantly present in the mesenchymal subtype. *NF1* and *PTEN* are often co-mutant, and are both part of the Akt pathway. Furthermore, TNF superfamily genes and NF-κB pathway genes are highly expressed in this subtype, where they control overall necrosis and inflammatory function (Verhaak et al., 2010). The "Notch" signaling pathways are prognostic indicators of this subclass. Furthermore, PIK3R1/PCL0 mutations and the PI3K/MAPK/Wnt pathway are associated within this subclass (Park et al., 2018). Non-canonical Wnt signaling pathway molecule WNT5a is highly expressed in the mesenchymal subtype, compared to the classical, proneural, and neural subtypes (Tompa et al., 2019), and are involved in cell motility and thus the aggressiveness of GBM tumors (Kalluri & Weinberg, 2009).

iii. **Signaling pathways in G-CIMP GBM:** The GSEA of the G-CIMP, or global CpG island hypermethylation subclass, demonstrate 15 overexpressed prognostic gene sets that belong to cellular metabolism or transport (Park et al., 2018). Among the 15 gene sets, 6 gene sets show negative correlation between gene expression and CpG island methylation. The more recurrently mutated genes for this GBM subtype are *TP53, IDH1,* and *ATRX*, which are associated with worse prognosis, whereas loss of 10q, 14q, and/or gain of 12p leads to favorable prognosis (Park et al., 2018).

iv. **Signaling pathways in proneural and neural GBM:** The G-CIMP subsets with favorable outcomes are grouped in the proneural subtype (Noushmehr et al., 2010). According to TCGA pan-glioma analysis, the non-G-CIMP (without IDH1 mutation) are clustered into two transcriptomic groups: LGr1 and LGr4. Although the proneural-subtype genes are more involved in epithelial to mesenchymal transition, due to a lack of validated datasets and heterogeneity in these subtypes they are not considered for further subgroup-wide pathway analysis (Park et al., 2018).

## 4. Therapeutic intervention in signaling pathways in Glioblastoma multiforme

### A. p53-MDM2-p14ARF signaling pathway in GBM:

p53 mutation and dysfunction of p53-mediated cell signaling occur in the majority of GBM (Kandoth et al., 2013) patients. Therapeutic management of this pathway plays a critical role in GBM disease progression. This is generally accomplished either by reactivating p53 or inhibiting its negative regulators.

**i. Preclinical model:** MDM2 negatively regulates p53, and the compound AMG 232 inhibits such p53-MDM2 interactions by binding to MDM2 with picomolar affinity, thus increasing p53 activity that leads to cell cycle arrest and cell proliferation. Preclinical xenograft models, when treated with AMG 232 in combination with chemotherapy, showed DNA damage and p53 activity that resulted in tumor regression (Canon et al., 2015). Various preclinical studies involved medicinal chemistry analysis, including optimization strategies for pharmacodynamics and kinetics using structure-activity relationship (SAR) for MDM2/MDMX inhibitors (RG7112, RG7388, MI77301, CGM097, MK8242, AMG232) and MDM2 proteolysis targeting chimera (PROTAC) degraders, as summarized in Fang Y et al. (Fang, Liao, & Yu, 2020). Out of these compounds, AMG232 and RG7388 were tested for GBM (Fig. 1). Another study with patient-derived GBM cell lines and xenografts showed RG7112 mediates restoration of p53 function, reduction in tumor growth, and longer animal survival (Verreault et al., 2016) (Fig. 1). Restoration of WT-p53 function is another avenue to rescue the pathway from loss-of-function *TP53* mutations. Cancer cell-based screening from the NCI drug library uncovered the role of PRIMA-1 (2,2-bis(hydroxymethyl)-1-azabicyclo[2.2.2]octan-3-one) in reactivating WT-p53 properties from select *TP53* missense mutants (Bykov et al., 2002). Subsequently, PRIMA-1 and PRIMA-1$^{MET}$/APR-246 have been shown to reduce tumor growth in GBM mouse models (Bykov et al., 2002, 2005; Perdrix et al., 2017) (Fig. 1). Delivering wildtype p53 to cells using the SGT-53 nanocomplex can increase treatment sensitivity in TMZ-resistant tumors (Kim et al., 2015) (Fig. 1). Combinatorial tumor-suppressor peptide treatments (p14$^{ARF}$ + p16$^{INK4a}$ or p16$^{INK4a}$ + p21$^{CIP1}$) targeted to *TP53*-mutant GBM cell lines led to a drastic reduction of tumor growth and

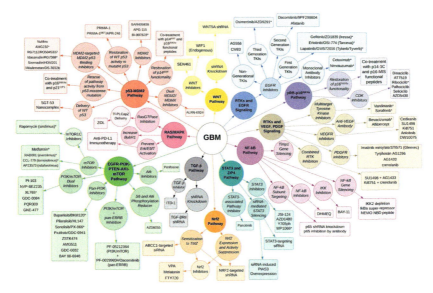

**Fig. 1** Summary of pre-clinical targets and therapies for the rescue of signaling pathways altered in *Glioblastoma multiforme*. *Progressed to clinical trials (NIH approved only).

increased survival significantly ($P > 0.0001$) in a murine brain tumor model of GBM (Kondo et al., 2008) (Fig. 1). A more recent review summarizes a list of MDM2-p53 binding inhibitors (Nutilins, RG7112/RO5045337, Idasanutlin/RG7388, Siremadlin/HDM201, Milademetan/DS-3032b), MDM2 inhibitors (SAR405838, APG-115, BI-907828), and the MDM2/MDMX dual inhibitor ALRN-6924 (Pellot Ortiz et al., 2023) (Fig. 1).

ii. **Clinical trial:** AMG-232 is under clinical trials both as a monotherapy and in combination with standard-of-care cytotoxicity (Canon et al., 2015; Fang et al., 2020) (NCT03107780) (Table 2). Another MDM2 inhibitor, RG7388, is currently under clinical trial (Fang et al., 2020) (NCT03158389) (Table 2). A Phase I trial using adenovirus (Ad-p53) mediated intratumoral delivery caused apoptosis of transfected cells (Lang et al., 2003) (Table 2). A Phase I trial with MDM2 inhibitor BI 907828 is underway in combination with radiotherapy (NCT05376800) (Table 2).

**B. pRB-p16$^{INK4a}$ signaling pathway in GBM:**

Genetic alteration in CDKN2A/B, amplification of CDK4/6, and mutations in *RB1* and *TP16* are found in 79% GBM cases, which make them major contributors to disease pathophysiology and therapeutic targets (Cameron et al., 2013).

Table 2 Clinical Trials with different signaling pathways inhibitors in *Glioblastoma multiforme*.

| Clinical trial ID | Pathway affected | Target | Phase | Monotherapy/combination therapy |
|---|---|---|---|---|
| NCT03107780 | p53-MDM2-p14ARF | P53-MDM2 | Phase I | AMG 232+ TMZ + radiotherapy |
| NCT03158389 | p53-MDM2-p14ARF | MDM2 | Phase I/II | N2M2 |
| NCT05376800 | p53-MDM2-p14ARF | MDM2 | Phase I | BI-907828+ radiotherapy |
| NCT03355794 | pRB-p16$^{INK4a}$ | CDKN2 | Phase I | Radiotherapy + Ribociclib + Everolimus |
| NCT01473901 | EGFR-PI3K-PTEN-AKt-mTOR | PI3K | Phase I | BKM120+TMZ+/− radiotherapy |
| NCT01259869 | EGFR-PI3K-PTEN-AKt-mTOR | PI3K | Phase II | PX-866 |
| NCT00112736 | EGFR-PI3K-PTEN-AKt-mTOR | EGFR | Phase I/II | Erlotinib + sirolimus/temsirolimus |
| NCT00805961 | EGFR-PI3K-PTEN-AKt-mTOR | mTOR | Phase II | Everolimus + bevacizumab |
| NCT01430351 | EGFR-PI3K-PTEN-AKt-mTOR | mTOR | Phase I | Memantine hydrochloride + TMZ + metformin hydrochloride |
| NCT00387400 | EGFR-PI3K-PTEN-AKt-mTOR | mTOR | Phase I | Everolimus + TMZ |

| | | | | |
|---|---|---|---|---|
| NCT01062399 | EGFR-PI3K-PTEN-AKt-mTOR | mTOR | Phase I/II | Everolimus + TMZ + radiation |
| NCT00553150 | EGFR-PI3K-PTEN-AKt-mTOR | mTOR | Phase I/II | Everolimus + TMZ + radiation |
| NCT00805961 | EGFR-PI3K-PTEN-AKt-mTOR | mTOR | Phase II | Bevacizumab + TMZ + radiation followed by everolimus + bevacizumab |
| NCT00047073 | EGFR-PI3K-PTEN-AKt-mTOR | mTORC1 | Phase II | Rapamycin |
| NCT00316849 | EGFR-PI3K-PTEN-AKt-mTOR | mTORC1 | Phase II | Temsirolimus + TMZ + radiotherapy |
| NCT01019434 | EGFR-PI3K-PTEN-AKt-mTOR | mTORC1 | Phase II | Radiotherapy + Temsirolimus/TMZ |
| NCT00704080 | EGFR-PI3K-PTEN-AKt-mTOR | PI3K/mTOR | Phase I | XL765 |
| NCT01062425 | RTK and EGFR signaling | EGFR | Phase II | Cediranib Maleate + TMZ + radiotherapy |
| NCT02861898 | RTK and EGFR signaling | EGFR | Phase I/II | Cetuximab + TMZ + radiotherapy + mannitol |
| NCT03388372 | RTK and EGFR signaling | EGFR | Phase II | Nimotuzumab + TMZ + radiotherapy |

(*continued*)

**Table 2** Clinical Trials with different signaling pathways inhibitors in *Glioblastoma multiforme*. (cont'd)

| Clinical trial ID | Pathway affected | Target | Phase | Monotherapy/combination therapy |
|---|---|---|---|---|
| NCT02928575 | RTK and EGFR signaling | EGFR | Phase I/II | Sunitinib + TMZ + radiotherapy |
| NCT00441142 | RTK and VEGF, PDGF signaling | VEGF, PDGF | Phase II | Vandetanib + TMZ + radiotherapy |
| NCT00704288 | RTK and VEGF, PDGF signaling | VEGFR2 | Phase II | Cabozantinib |
| NCT00884741 | RTK and VEGF, PDGF signaling | VEGFR | Phase III | Bevacizumab + TMZ + radiotherapy |
| NCT01122888 | RTK and VEGF, PDGF signaling | VEGFR2- + EGFR | Phase I | cilengitide + sunitinib malate |
| NCT01124240 | RTK and VEGF, PDGF signaling | VEGFR2 | Phase II | cilengitide |
| NCT00050986 | RAS/MAPK | Ras | Phase I/II | Zarnestra + TMZ |
| NCT03535350 | STAT3 and ZIP4 | STAT3 | Phase I | Ibrutinib + TMZ |
| NCT05879250 | STAT3 and ZIP4 | STAT3 | Phase II | WP1066+ radiotherapy |
| NCT00431561 | TGF-signaling | TGF-β | Phase IIb | AP 12009 |
| NCT00761280 | TGF-signaling | TGF-β | Phase III | AP 12009 |
| NCT01220271 | TGF-signaling | TGF-βR1 | Phase I/II | LY2157299+ TMZ + radiotherapy |

i. **Preclinical model:** A preclinical study using CDK inhibitors (ribociclib, palbociclib, seliciclib, AZD5438, and dinaciclib) on glioma cell line U87 showed that nanomolar concentrations of dinaciclib caused high cytotoxicity against Bcl-xL-silenced cells in a time and concentration dependent manner (Fig. 1). Dinaciclib down-regulated cyclin D1, D3 and total Rb, causing cell cycle arrest (Premkumar et al., 2018) (Fig. 1). A high throughput screening of the Drug Library and Clinical Compound Library (2718 compounds) showed a small-molecular multi-CDK inhibitor, named AT7519, inhibits cell growth and viability in multiple GBM cell lines (U87MG, U251, patient derived primary GBM cells), occurring via induction of apoptosis and pyroptosis through caspase-3-mediated cleavage of gasdermin E (GSDME). Such an effect was replicated in GBM xenograft assays, where AT7519 appears to reduce the tumor volume significantly (W. Zhao et al., 2023) (Fig. 1).

ii. **Clinical trial:** CDKN2 inhibitors Ribociclib (NCT03355794), and Abemaciclib (NCT04238819) are being evaluated as both monotherapies and in combination with TMZ (Table 2).

C. **EGFR-PI3K-PTEN-AKt-mTOR signaling pathway in GBM:**

Alterations in this pathway lead to abnormal cell signaling, and thus studies of its management through animal and cell culture models or human studies play a crucial role in identifying new treatments.

i. **Preclinical model:** Preclinical studies investigating the effect of pan-PI3K inhibitors (Buparlisib/BKM120, Pilaralisib/XL147, Sonolisib/PX-866, Pictilisib/GDC-0941, ZSTK474, AMG511, GDC-0032 and BAY 80-6946) on GBM cell lines and xenograft mouse model showed antitumorigenic effects on cell cycle arrest, reduction in tumor growth, and prolonged survival (Zhao et al., 2017) (Fig. 1). Preclinical studies have also been conducted on both mTOR inhibitors [RAD001 (everolimus), CCL-779 (temsirolimus), AP23573 (ridaforolimus)] and the mTORC1 inhibitor rapamycin (sirolimus) (Rivera et al., 2011) (Fig. 1). Another mTOR inhibitor, Metformin, reduced tumor growth (Yu et al., 2015) in GBM cell lines and is currently undergoing clinical trials (Fig. 1). The Akt inhibitor Perifosine works by preventing Akt translocation to the cell membrane (Momota, Nerio, & Holland, 2005) (Fig. 1). AZD8055 is a small molecular inhibitor that reduces S6 and Akt phosphorylation, effectively reducing tumor growth (Chresta et al., 2010) (Fig. 1). PI3K/mTOR dual inhibitors (PI-103, NVP-BEZ235, XL765,

GDC-0084, PQR309) have been pre-clinically tested in U87MG, U251MG, and U373 MG cells (Fig. 1). Treated cells displayed increased radiosensitivity, reduction in elevated G1 arrest, and in a xenograft study demonstrated reduced tumor growth and enhanced survival (Liu et al., 2009; Zhao et al., 2017). A new PI3K/mTOR inhibitor, GNE-477, causes significant inhibition of cell proliferation, migration, invasion, and induction of apoptosis in U87 and U251 GBM cell lines (Wang et al., 2021) (Fig. 1). A combination of dual PI3K/mTOR inhibitor (PF-05212384) and pan-ERBB inhibitor (dacomitinib/PF-00299804) led to more apoptosis in GBM cells with EGFR amplification and PI3K activation than as single-agent treatments (Zhu & Shah, 2014) (Fig. 1).

ii. **Clinical trial:** The pan-PI3K inhibitor BKM120 along with TMZ underwent clinical trials with and without combinatorial radiotherapy until 2017 on newly diagnosed GBM cases (NCT01473901) (Table 2). Another clinical trial with the pan-PI3K inhibitor PX-866 (NCT01259869) (Table 2) underwent a clinical Phase II trial for GBM cases at first relapse or progression. The drug combinations of EGFR-inhibitor erlotinib with sirolimus or temsirolimus (NCT00112736) (Table 2) both underwent clinical trial testing. The combination of mTOR inhibitor everolimus with angiogenic inhibitor bevacizumab was also tested, progressing to a Phase II clinical trial (NCT00805961) (Table 2). Another combination approach using TMZ with memantine hydrochloride (mTOR inhibitor), and metformin hydrochloride (NCT01430351) (Table 2), underwent a Phase I clinical trial. Again, the combination of everolimus (mTOR blocker) with TMZ was used for a Phase I clinical trial (NCT00387400) (Table 2) in GBM and everolimus (mTOR blocker), TMZ, and radiation was tested in combination as another council Phase I/II clinical trial (NCT01062399, NCT00553150) (Table 2). The combination of RT, TMZ, and bevacizumab, followed by bevacizumab/everolimus underwent a Phase II clinical trial (NCT00805961) (Table 2) in newly diagnosed GBM cases. The mTORC1 inhibitor rapamycin was used in GBM patients before surgical tumor resection as a Phase I/II trial (NCT00047073) (Table 2). Another mTORC1 inhibitor, temsirolimus, was applied along with TMZ and radiation in a Phase I trial (NCT00316849) (Table 2). A separate randomized Phase II trial was designed to again study temsirolimus together with radiation therapy, but compared this treatment to combined TMZ and radiation in patients with newly diagnosed glioblastoma (NCT01019434) (Table 2). A PI3K/mTOR dual inhibitor (XL765) was tested with TMZ in a Phase I trial for safety and tolerability, along with growth restriction and apoptosis (NCT00704080) (Table 2).

## D. RTKs and EGFR signaling pathway in GBM:

Receptor tyrosine kinase (RTK) misregulation is commonly observed in GBM cases, and clinical management of RTKs in by a variety of treatments is important for improving prognosis.

**i. Preclinical model:** *In vitro* and *in vivo* studies of EGFR small molecular inhibitors identified gefitinib/ZD1839 (Iressa), erlotinib/OSI-774 (Tarceva), and lapatinib/GW572016 (Tykerb/Tyverb) as potential clinical treatments (Fig. 1). Each binds to EGFR in a reversible fashion, and showed promising results (Blackledge & Averbuch, 2004; Karpel-Massler & Halatsch, 2014; Wood et al., 2004); however, they showed only moderate success in clinical trials. A second generation of EGFR inhibitors were then developed that irreversibly bind the receptor target. In that group, dacomitinib/9PF299804 showed an enhanced effect on GBM cell lines and tumor growth reduction (Zahonero et al., 2015) (Fig. 1). It then proceeded to clinical trials for recurrent GBM, which is beyond the scope of this review. Third generation EGFR inhibitors are mutant-targeting, wildtype-sparing, irreversible receptor antagonists which trigger less resistance than 1st and 2nd generation inhibitors. Most target the EGFR T790M mutation. Among them, osimertinib/AZD9291 displays better blood brain barrier penetration and >10 times improved reduction in GBM cell proliferation over first generation inhibitors (Fig. 1). It further shows significant reduction in tumor volume and enhanced survival in orthotopic GBM models (Kwatra et al., 2017; Liu et al., 2019). Another study using high-grade glioma cell lines and treating them with EGFR small molecule inhibitor (AG556), in combination with radiotherapy, yielded different results, despite the similar origin of osimertinib and AG556 (Alexandru et al., 2018) (Fig. 1). Cetuximab and nimotuzumab are monoclonal antibodies against EGFR that display *in vitro* and *in vivo* receptor inhibition in GBM cells (Combs et al., 2008; Eller et al., 2002, 2005) (Fig. 1). Afatinib (EGFR inhibitor, 2nd Generation) in combination with TMZ caused effective cell growth inhibition in U87 and U251 cells and delayed tumor growth in a preclinical mouse study (Vengoji et al., 2019) (Fig. 1). A novel covalent-binding EGFR inhibitor, CM93, inhibited EGFR phosphorylation in GBM tumors *in vitro* and *in vivo*, suggesting it as a potential GBM therapy (Rodriguez et al., 2023) (Fig. 1).

ii. **Clinical trial:** Very few RTK-based therapies have shown success in clinical trials for GBM. The EGFR inhibitor AZD2171 (Cediranib Maleate) was used in combination with TMZ and radiation therapy in a Phase II clinical trial (NCT01062425) (Table 2) for adult GBM. Another Phase I/II trial with antibody-based EGFR inhibitor Cetuximab (CTX) underwent a clinical trial in combination with radiotherapy, chemotherapy, and mannitol (NCT02861898) (Table 2). Another antibody-based EGFR inhibitor, Nimotuzumab (NCT03388372) (Table 2), underwent a Phase II clinical trial, in combination with radio and chemotherapy, for GBM to evaluate efficacy and safety. The EGFR inhibitor sunitinib was used in combination with TMZ and radiotherapy to determine if the combination yields a better outcome than monotherapy (NCT02928575) (Table 2).

E. **RTKs and VEGF, PDGF pathways signaling in GBM:**
There are several ways to block the VEGF and PDGF pathways and here we summarize the preclinical and clinical aspects:

i. **Preclinical models:** A preclinical study using *in vitro* assays on GBM cell lines revealed Vandetanib, a multitarget tyrosine kinase inhibitor that blocks VEGF and PDGF receptors, works synergistically in combination with histone deacetylase inhibitors (HDACIs) to induce apoptosis via inhibition of MAPK, Akt and other downstream effectors (Jane et al., 2009) (Fig. 1). Another small molecule kinase inhibitor, STI571, inhibits GBM cells via disrupting PDGF ligand:receptor autocrine loop (Kilic et al., 2000) (Fig. 1). Tryphostin AG 1296 is a PDGFR inhibitor and using *in vitro* studies, it was shown that AG1296 induces apoptosis by reducing cell migration and promoting cell apoptosis (Li et al., 2015) (Fig. 1). Bevacizumab and Aflibercept are humanized murine monoclonal antibodies that 'trap' ligands that activate VEGF receptors (Pearson & Regad, 2017) (Fig. 1). Another study using GBM models showed dual inhibition of PDGFR and VEGFR with Ki8751 and crenolanib showed enhanced mouse survival and reduced tumor growth than the single agent (Liu et al., 2018) (Fig. 1). An *in vitro* study using AG1433 (a PDGFR inhibitor) and SU1498 (a VEGFR inhibitor) on low passage GB9B cell line showed activation of caspase 8, 9, and 3 mediated killing of GBM cells but such effect is milder as compared to using dual inhibitor (BEZ235) against PI3K/Akt/mTOR pathways (Popescu et al., 2015) (Fig. 1). Monotherapy using Sorafenib, a multitarget kinase inhibitor against both VEGF (VEGF-2 and -3) and PDGF (PDGF$\beta$ and Kit) showed better survival in GBM cell line than the vehicle treated controls (Siegelin et al., 2010) (Fig. 1). Pharmacological characterization of a small molecule inhibitor of VEGFR2, Anlotinib, showed broader and stronger antitumor efficacy in a well tolerated manner as compared to RTK inhibitor Sunitinib (Xie et al., 2018) (Fig. 1). Another small molecular inhibitor of VEGFR, DW10075, showed high selectivity and reduced cell migration *in vitro* and suppression of angiogenesis *in vivo* (Li et al., 2016) (Fig. 1).

ii. **Clinical trials:** Vendetanib was studied in a randomized Phase II clinical trial in combination with chemotherapy and radiation in newly diagnosed GBM and was early terminated due to no significant effect on elongating overall survival (NCT00441142) (Lee et al., 2015) (Table 2). Cabozantinib, a VEGFR2 inhibitor, showed modest effect on recurrent or progressive GBM (NCT00704288) (Cloughesy et al., 2018). Bevacizumab or placebo treatment along with radiotherapy and maintenance chemotherapy did not reveal a significant effect on progression-free survival (NCT00884741) (Gilbert et al., 2014) (Table 2). Sorafenib and TMZ have been used in combination in a Phase II clinical trial, and showed limited effect (Reardon et al., 2011). There is no study, however, on newly diagnosed GBM patients. Clinical trial of VEGFR2 inhibitor, cilengitide, in combination with EGFR inhibitor sunitinib malate, underwent clinical trial in GBM (NCT01122888) (Table 2). The same VEGFR2 inhibitor was used in a Phase II clinical trial in a combination therapy approach with TMZ and procarbazine for newly diagnosed non-methylated GBM (NCT01124240) (Table 2).

F. **RAS/MAPK signaling pathway in GBM:**

Constitutive Ras activity is one of the major targets in GBM pathogenesis, and thus understanding its preclinical and clinical aspects are important for disease management.

i. **Preclinical model:** A recent finding shows upregulation of RAS/MAPK activity leads to chromosome missegregation, and the requirement for kinetochore complex protein BubR1 is enhanced. It

has been found that aneuploid tumors show greater upregulation of RAS/MAPK genes than diploid tumors. Different GBM isolates show enhanced RAS/MAPK expression relative to the neural stem cell lines and enhanced BubR1 requirement, suggesting BubR1 as a therapeutic target for GBM (Herman et al., 2022) (Fig. 1). Another preclinical study showed GBM cell malignancy via intracellular programmed death ligand-1 (PD-L1), which binds to Ras and activates Erk-EMT signaling, thus providing useful information about improving immunotherapy in GBM (Qiu et al., 2018) (Fig. 1). A nanoparticle-based targeting of treatment for GBM was shown by Salzano G et al., in which they used transferrin (Tf) targeted self-assembling nanoparticles, Tf-PLCapZ, to deliver zoledronic acid (ZOL; inhibitor of farnesyl pyrophosphate synthase) to GBM cells and inhibit RasGTPase activity (Fig. 1). Treated GBM cell lines and animal models showed significant anticancer activity (Porru et al., 2014; Salzano et al., 2016).
  ii. **Clinical trial:** A farnesylation inhibitor, Zarnestra/R115777, that is targeted against Ras underwent clinical trial Phase I/II in combination with TMZ (NCT00050986) (Table 2) in GBM patients.
G. **STAT3 and ZIP4 signaling pathway in GBM:**
  STAT3 and ZIP4 are implicated in GBM cell proliferation, invasion, and treatment resistance, and their clinical management plays a very important role for better prognosis and disease management.
  i. **Preclinical model:** In GBM-patient exosome analysis, onco-miRNA-21 (miR-21) is upregulated. The STAT3-associated pathway inhibitor Pancitinib reduces miR-21-enriched exosomes in GBM cell lines. Furthermore, in the same study, Pancitinib was used on LN18-bearing mouse models with and without TMZ combination treatment, and the resulting reduction in tumor volume suggests STAT3/miR-21/PDCD4 signaling as a therapeutic target (Chuang et al., 2019) (Fig. 1). Analysis of human GBM versus normal tissue revealed higher STAT-3 activation with reduced PIAS3 protein expression in GBM cases compared to controls. Following that, an *in vitro* siRNA-mediated silencing of PIAS3 or PIAS3 overexpression in GBM cells demonstrated enhanced cell proliferation for the former, and inhibition of STAT-3 transcription activity and cell proliferation for the latter, suggesting the role of PIAS3 as a STAT-3 inhibitor in GBM (Brantley et al., 2008) (Fig. 1). Another study with STAT3 siRNA-based inhibition showed G1

phase arrest of GBM cell cycle increase and an associated decrease in cyclin D1 (Li et al., 2009). Another study with STAT3 inhibitor JSI-124 on a GBM cell line showed reduced cell density and morphological deformation (McFarland et al., 2013; Su et al., 2008) (Fig. 1). A STAT3 gene-signature study in GBM patient samples categorized the gliomas into STAT3-high [mesenchymal, classical, Glioblastoma IDH-WT (Louis et al., 2016, 2021), IDH-mutant 1p/19q non-codeletion (WHO 2021 Astrocytoma, IDH-mutant grade 4)] and STAT3-low groups [IDH-Mutant low-grade gliomas, proneural GBM with 1p/19q co-deletion]. *In vitro* and xenograft animal model study showed that STAT3-high cell lines and mouse models respond better to the inhibitors (AZD1480), displaying reduced cell proliferation, invasion, self-renewal, and gliomasphere formation ability as compared to STAT3 low group (Fig. 1). Thus, stratification of GBM patient groups with STAT3 markers holds potential promise for improved clinical management (Tan et al., 2019). Different classes of STAT3 inhibitors are summarized in Fu et al., 2023 (Fu et al., 2023), as natural inhibitors, siRNA targeting STAT3, exosome-based targeting agents, or pharmacological targeting agents. Furthermore, preclinical studies with STAT3 inhibitors in combination with radiotherapy led to TMZ reprogramming the immune response (Wang et al., 2019) and prolong survival with reduced tumor growth (Ott et al., 2020). In TMZ resistant cell lines, STAT3 inhibition or knockdown led to downregulation of MGMT gene expression and thus increased TMZ sensitivity (Han et al., 2016). STAT3 inhibitors can disrupt the blood brain barrier and thus, when in combination with chemotherapy, leads to prolonged survival in mice (Zhou et al., 2017). Also, STAT3 inhibition-targeted therapy can be combined with anti-PD-L1 immunotherapy to treat GBM (Noman et al., 2014). Another combination approach with a STAT3 inhibitor (AZD1480) and VEGFR inhibitor (cediranib) reduced GBM volume and microvessel density in murine models (de Groot et al., 2012) (Fig. 1). A small molecule inhibitor of STAT3 that blocks Y705 phosphorylation showed higher potency than other STAT3 inhibitors for reductions in cell growth, invasion, and proliferation significantly (Wang et al., 2022) (Fig. 1). STAT3 inhibition can also lead to induction of an immune response and changes from a "cold" tumor response to a "hot" immunologic tumor response (Gangoso et al., 2021).

**ii. Clinical trial:** A combination therapy with TMZ and STAT3 inhibitor ibrutinib in MGMT-unmethylated GBM and combination of TMZ, radiotherapy, and ibrutinib in MGMT-methylated GBM is currently undergoing Phase I clinical trial (NCT03535350) (Table 2). Another combination approach with STAT3 inhibitor WP1066 and radiation therapy is approved for Phase II clinical trials (NCT05879250) (Table 2).

## H. TGF-β-signaling pathway in GBM:

TGF-β signaling pathway has been implicated in tumor inversion, growth, chemoresistance, and various aspect of GBM growth and therefore, management of the pathway through preclinical research and clinical trials will play an important role in future treatments.

**i. Preclinical model:** TGF-β blocking in patient-derived tumor cells expressing high CD44 and Id1 (markers of cancer initiating cells) causes a reduction in their expression and inhibits tumor initiation (Anido et al., 2010). A shRNA-based knockdown of TGF-βRII leads to impairment of TGF-β-induced cell inversion and migration in GBM cell lines (Wesolowska et al., 2008) (Fig. 1). *In vitro* and *in vivo* studies with TGF-β stimulation and inhibition in a cell line and xenograft tumor model showed involvement of thrombospondin 1 (TSP1) as a potential mediator of microtubule (MT) formation downstream of the TGF-β signaling pathway. This acts further through SMAD activation and causes tumor resistance via enhanced MT formation, and could be a therapeutic target (Joseph et al., 2021) (Fig. 1). Another study with a murine cell line and animal data revealed upregulation and nuclear transport of Claudin-4 (CLDN4), highly expressed in aggressive GBM cases, via the TGF-β signaling pathway. ITD-1 (TGF-β inhibitor) can downregulate CLDN4 and reduce tumor invasiveness (Yan et al., 2022) (Fig. 1). Analysis of 64 newly diagnosed GBM patients' data suggested the involvement of three TGF-β isoforms that could be used as potential biomarkers for anti-TGF-β therapies (Frei et al., 2015). Another preclinical GBM cell line study provides new insight into Sox2 as a downstream target of TGF-β signaling and as a probable clinical target (Chao et al., 2020).

**ii. Clinical trial:** TGF-β2 inhibitor drug AP12009, i.e. phosphorothioate antisense oligodeoxynucleotides specific for the mRNA of human transforming growth factor-beta2 (TGF-beta2) completed a Phase IIb clinical trial (NCT00431561) (Table 2) for glioblastoma

(Bogdahn et al., 2011), whereas a Phase III trial with the same inhibitor used for secondary GBM has been terminated early (NCT00761280) (Table 2). Another TGF-βR1 (ALK-5) kinase inhibitor, LY2152799, was used as a monotherapy or in combination with radio and chemotherapy (NCT01220271) (Table 2) and has completed Phase II trials (Wick et al., 2020).

## I. WNT signaling pathway in GBM

Wnt signaling pathways are implicated in aggressive GBM cases where Wnt signaling aberrations make this pathway an important therapeutic target.

**i. Preclinical model:** shRNA-mediated knockdown of WNT5A in GBM-05 and U76MG yielded reduced cell proliferation (Yu et al., 2006), motility, and invasiveness (Kamino et al., 2011) (Fig. 1). An endogenous Wnt signaling antagonist, Wnt inhibitor factor (WIF1), selectively attenuates the Wnt/$Ca^{2+}$ pathway (Vassallo et al., 2015) (Fig. 1). A study of human patient samples showed 75% of GBM tumors downregulated WIF1 (Lambiv et al., 2011). A small molecule-based Wnt/b-catenin inhibitor, SEN461, decreased cell viability in a xenograft mouse model (De Robertis et al., 2013) (Fig. 1). An *in vitro* study used pharmacological PORCN-inhibitor LGK974's synergistic effect with TMZ to decrease cell growth in glioma. Transcription data from a study using the same inhibitor shows reduced expression of stem cell markers CD133, Nestin, and Sox2, (Suwala et al., 2018) (Fig. 1) supporting its role in improving TMZ effectiveness and decreased glioma stem cell growth.

**ii. Clinical trial:** Many studies indicate robust effectiveness of Wnt pathway inhibition at reducing GBM progression and increasing sensitivity to TMZ treatment. However, no clinical trials have been successfully conducted so far for GBM cases managing Wnt signaling pathways.

## J. Nrf2 and cell signaling in GBM

Nrf2 is a redox-sensitive transcription factor, and constitutive activation of Nrf2 leads to enhanced tumor survival and resistance to anticancer therapy, especially TMZ resistance. Therefore, further research on Nrf2 inhibitors will be immensely helpful for improving GBM treatment.

**i. Preclinical model:** Using cell-line-based Nrf2 overexpression and inhibition models, a study showed inhibition of Nrf2 leads to decreased mRNA and proteins related to anti-oxidative enzymes via

Ras/Raf/MEK signaling pathways, finally leading to inhibited cell proliferation. This study provides evidence that Nrf2 inhibition may enhance the effect of TMZ on GBM cells (Sun et al., 2020). Another study showed that the compound FTY720 suppressed Nrf2 protein and mRNA levels in GBM cell lines U251MG and U87MG. This effect is enhanced in Nrf2 knock-down cells and reversed in Nrf2-activated cells. Furthermore, FTY720 sensitizes GBM cells to TMZ treatment (Zhang & Wang, 2017) (Fig. 1). Using GBM cell line T98G, a study showed *NRF2* silencing via *NRF2* shRNA leads to increased sensitivity to TMZ treatment, as does silencing of *ABCC1* (an Nrf2 target) via siRNA, suggesting a new therapeutic strategy in GBM cells displaying high NRF2 and/or ABCC1 expression (de Souza et al., 2022) (Fig. 1). *In vitro* and *in vivo* study with shRNA-based knockdown of *NRF2* showed the downregulation of BMI-1, Sox2, and Cyclin E, resulting in cell cycle arrest and reduced GBM stem cell growth. This study demonstrates the role of Nrf2 in GBM stem cell maintenance (Zhu et al., 2013) (Fig. 1). Another study on U251-TMZ cells treated with the Nrf2 inhibitor valproic acid (VPA) and melatonin (MEL) demonstrates an increase in U251-TMZ chemosensitivity, suggesting the involvement of Nrf2-antioxidant response element (ARE) in GBM resistance to TMZ (Pan et al., 2017) (Fig. 1). Additionally, the Nrf2-Keap1 pathway (Fan et al., 2017), AMPK-Nrf2 pathway (Awuah et al., 2022), ERK-Nrf2 pathways (Wang et al., 2018), and PI3K-Nrf2 pathways (Awuah et al., 2022) are found to be potential targets for the clinical management of GBM.

  ii. **Clinical trial:** A Phase Ib/II study has been performed only on recurrent GBM patients using PI3K inhibitor combinations (buparlisib plus carboplatin or lomustine), and did not show enough anti-tumor activity to proceed to further trials (NCT01934361) (Table 2). None of the other preclinical data has been extended to the clinical trial stage.

K. **NF-κB and cell signaling in GBM**

NF-κB is involved in various critical cellular events starting from maintaining cell stemness, cell growth, proliferation, invasion and resistance to radio and chemotherapy. Thus, therapeutic targeting will be immensely helpful.

  i. **Preclinical models:** A lentivirus induced mouse model of GBM showed enrichment of nuclear NF-κB and enhanced expression of its target genes. Inhibition of NF-κB gene expression by IKK2 depletion, expressing IκBa super repressor, or expressing NEMO (NF-κB)-binding domain (NBD) peptide in cell lines reduces cell proliferation and enhance survival in murine model (Fig. 1). *Timp1* is one of the NF-κB target genes that is involved in such cell proliferation, and either inhibition of NF-κB or silencing *Timp1* can attenuate the tumor growth (Friedmann-Morvinski et al., 2016) (Fig. 1). Another preclinical study using glioma cell lines and human tumor samples using forced expression of NF-κB increased *MGMT* gene expression and suggested the role of MGMT-mediated chemoresistance in GBM (Lavon et al., 2007). This evidence implicates NF-κB as a potential therapeutic target of GBM. A study on GBM cell lines suggests a role of BAY-11 (IKK inhibitor) as a NF-κB pathway inhibitor (Fig. 1). This inhibitor also exerts an *in vitro* and *in vivo* role in GBM cell senescence and reverses chemoresistance (Coupienne et al., 2011; Nogueira et al., 2011; Shukla et al., 2013). A small molecule NF-κB inhibitor, dehydroxymethylepoxyquinomicin (DHMEQ), inhibits NF-κB function and its translocation to the nucleus, thus reducing cell proliferation *in vitro* and tumor growth *in vivo* (Fukushima et al., 2012) (Fig. 1). Furthermore, a preclinical study showed that the same inhibitor plays a significant role in chemosensitivity when used in combination with TMZ and thus provides a synergistic effect (Brassesco et al., 2013). Another way to block NF-κB function is by directly targeting its subunits. As an example, knockdown of p65, a NF-κB subunit, leads to cytotoxicity in GBM cells and shRNA mediated inhibition of the same subunit leads to reduced tumor growth in mouse models (Bonavia et al., 2012; Zanotto-Filho et al., 2011) (Fig. 1). An antibody-based inhibition of p65 leads to reduced tumor growth in xenograft mouse models (Li et al., 2007) (Fig. 1).

ii. **Clinical trials:** Despite a lot of preclinical trials, there are no clinical trials performed on newly diagnosed GBM patients using inhibitors of the NF-κB pathway.

## 5. Future direction

GBM is a highly aggressive form of tumor that displays extensive heterogeneity, where different genetically altered signaling pathways play critical roles in tumorigenesis, tumor aggressiveness, cellular invasion, migration, and resistance to available therapies. A detailed understanding of such pathways with respect to their preponderance in different subclasses of GBM is extremely important. In this chapter, we have summarized different signaling pathways and stratified them based on different subclasses of GBM. We have also summarized the pre-clinical evidence of potential targets for clinical management. Although many pre-clinical studies using small molecular inhibitors, pharmacological inhibitors, antibodies, etc. successfully showed reduction in cell proliferation and *in vivo* tumor growth in animal models, many of them could not be successfully translated to clinical trials due to therapeutic resistance and lack of efficacy. Some have been successfully applied, but mostly in combination with chemotherapy, radiotherapy, both, or other targeted therapies. Our summary demonstrates that a limited number of preclinical studies were recapitulated in clinical trials, and there is a great need for combinatorial therapeutic approaches to make significant clinical impact on patient health. This requires understanding such signaling pathways with respect to their cross-talk. Analyzing these overlapping signal transduction pathways will better illustrate the source of past therapeutic failures and how to better manage GBM from a clinical perspective. Therefore, more detailed work on therapeutic resistance in terms of cell signaling pathways and their overlap in different GBM subclasses is a necessary future direction in this area of research.

Another major challenge in GBM therapy is created by inefficient drug delivery, intracranial drug distribution, and poor drug retention as a result of the blood-brain-barrier (BBB). There are recent studies that use a nanoparticle-based drug delivery system which can better penetrate the BBB, thus showing promise in this area of research. For example: there is a preclinical study targeting microRNA 21 (miR-21) in GBM using two different nanoparticle-based delivery systems, where the anti-miR RNA is delivered either via cationic poly(amino-co-ester) (PACE) or anti-miR is delivered via block copolymer of poly(lactic acid) and hyperbranched polyglycerol (PLA-HPG). Such delivery systems enable miRNA-based suppression of the PTEN pathway and reduced GBM cell proliferation. Further, *in vivo* administration of anti-miR by conventional enhanced delivery (CAD) in mice with intracranial GBM showed significantly better miRNA gene knockdown and enhanced chemosensitivity (Seo et al., 2019). Another preclinical study has used a short (8-mer) γ-modified peptide nucleic acid (sγPNAs), that targets the seed region of oncomiRs 10b and 21 and is delivered by PLA-HPG which showed improved preclinical outcome (Wang et al., 2023). A recent study on PARP inhibitor delivery to cerebrospinal fluid in a xenograft mouse model of medulloblastoma used PLA-HPG-based delivery and showed better tumor regression than the control (Khang et al., 2023). Taken together, these studies show the promising side of nanoparticle-based delivery for antitumorigenic drugs. This technology should be further explored for signaling pathway-targeted therapies enabling better GBM management.

## References

Abal, M., et al. (2006). Molecular pathology of endometrial carcinoma: Transcriptional signature in endometrioid tumors. *Histology and Histopathology, 21*(2), 197–204.

Ahir, B. K., Engelhard, H. H., & Lakka, S. S. (2020). Tumor development and angiogenesis in adult brain tumor: Glioblastoma. *Molecular Neurobiology, 57*(5), 2461–2478.

Ahmad, F., et al. (2016). Nrf2-driven TERT regulates pentose phosphate pathway in glioblastoma. *Cell Death & Disease, 7*(5), e2213 –e2213.

Alexandru, O., et al. (2018). The influence of EGFR inactivation on the radiation response in high grade glioma. *International Journal of Molecular Sciences, 19*(1), https://doi.org/10.3390/ijms19010229.

Alnahhas, I., et al. (2020). Characterizing benefit from temozolomide in MGMT promoter unmethylated and methylated glioblastoma: A systematic review and meta-analysis. *Neuro-Oncology Advances, 2*(1), vdaa082.

Anido, J., et al. (2010). TGF-β receptor inhibitors target the CD44(high)/Id1(high) glioma-initiating cell population in human glioblastoma. *Cancer Cell, 18*(6), https://doi.org/10.1016/j.ccr.2010.10.023.

Arita, H., et al. (2016). A combination of TERT promoter mutation and MGMT methylation status predicts clinically relevant subgroups of newly diagnosed glioblastomas. *Acta Neuropathologica Communications, 4*(1), 1–14.

Auger, N., et al. (2006). Genetic alterations associated with acquired temozolomide resistance in SNB-19, a human glioma cell line. *Molecular Cancer Therapeutics, 5*(9), 2182–2192.

Avci, N. G., et al. (2020). NF-κB inhibitor with Temozolomide results in significant apoptosis in glioblastoma via the NF-κB(p65) and actin cytoskeleton regulatory pathways. *Scientific Reports, 10*(1), 1–14.

Awuah, W. A., et al. (2022). Exploring the role of Nrf2 signaling in glioblastoma multiforme. *Discover. Oncology, 13*. https://doi.org/10.1007/s12672-022-00556-4.

Bhat, K. P. L., et al. (2013). Mesenchymal differentiation mediated by NF-κB promotes radiation resistance in glioblastoma. *Cancer Cell, 24*(3), https://doi.org/10.1016/j.ccr.2013.08.001.

Blackledge, G., & Averbuch, S. (2004). Gefitinib ("Iressa", ZD1839) and new epidermal growth factor receptor inhibitors. *British Journal of Cancer, 90*(3), 566.

Blume-Jensen, P., & Hunter, T. (2001). Oncogenic kinase signalling. *Nature, 411*(6835), 355–365.

Bogdahn, U., et al. (2011). Targeted therapy for high-grade glioma with the TGF-β2 inhibitor trabedersen: Results of a randomized and controlled phase IIb study. *Neuro-Oncology, 13*(1), https://doi.org/10.1093/neuonc/noq142.

Bonavia, R., et al. (2012). EGFRvIII promotes glioma angiogenesis and growth through the NF-κB, interleukin-8 pathway. *Oncogene, 31*(36), https://doi.org/10.1038/onc.2011.563.

Bradford, J. W., & Baldwin, A. S. (2014). IKK/nuclear factor-kappaB and oncogenesis: Roles in tumor-initiating cells and in the tumor microenvironment. *Advances in Cancer Research, 121.* https://doi.org/10.1016/B978-0-12-800249-0.00003-2.

Brantley, E. C., et al. (2008). Loss of protein inhibitors of activated STAT-3 expression in glioblastoma multiforme tumors: Implications for STAT-3 activation and gene expression. *Clinical Cancer Research: An Official Journal of the American Association for Cancer Research, 14*(15), 4694–4704.

Brassesco, M. S., et al. (2013). Inhibition of NF-κB by dehydroxymethylepoxyquinomicin suppresses invasion and synergistically potentiates temozolomide and γ-radiation cytotoxicity in glioblastoma cells. *Chemotherapy Research and Practice, 2013*. https://doi.org/10.1155/2013/593020.

Brawanski, K. R., et al. (2023). Influence of MMR, MGMT promotor methylation and protein expression on overall and progression-free survival in primary glioblastoma patients treated with temozolomide. *International Journal of Molecular Sciences, 24*(7), https://doi.org/10.3390/ijms24076184.

Bredel, M., et al. (2011). NFKBIA deletion in glioblastomas. *The New England Journal of Medicine, 364*(7), https://doi.org/10.1056/NEJMoa1006312.

Brennan, C., et al. (2009). Glioblastoma subclasses can be defined by activity among signal transduction pathways and associated genomic alterations. *PLoS One, 4*(11), e7752.

Bryukhovetskiy, I., & Shevchenko, V. (2016). Molecular mechanisms of the effect of TGF-β1 on U87 human glioblastoma cells. *Oncology Letters, 12*(2), 1581–1590.

Bykov, V. J. N., et al. (2002). Restoration of the tumor suppressor function to mutant p53 by a low-molecular-weight compound. *Nature Medicine, 8*(3), 282–288.

Bykov, V. J. N., et al. (2005). PRIMA-1(MET) synergizes with cisplatin to induce tumor cell apoptosis. *Oncogene, 24*(21), 3484–3491.

Cameron, W. B., Roel, G. W. V., Aaron, M., Benito, C., & Houtan, N. (2013). The somatic genomic landscape of glioblastoma. *Cell, 155*(2), 462–477.

Cancer Genome Atlas Research Network. (2008). Comprehensive genomic characterization defines human glioblastoma genes and core pathways, *Nature, 455*(7216), 1061–1068.

Canon, J., et al. (2015). The MDM2 inhibitor AMG 232 demonstrates robust antitumor efficacy and potentiates the activity of p53-inducing cytotoxic agents. *Molecular Cancer Therapeutics, 14*(3), 649–658.

Capper, D., et al. (2018). DNA methylation-based classification of central nervous system tumours. *Nature, 555*(7697), 469–474.

Chai, R., et al. (2021). Predictive value of MGMT promoter methylation on the survival of TMZ treated IDH-mutant glioblastoma. *Cancer Biology & Medicine, 18*(1), https://doi.org/10.20892/j.issn.2095-3941.2020.0179.

Chao, M., et al. (2020). TGF-β signaling promotes glioma progression through stabilizing Sox9. *Frontiers in Immunology, 11*. https://doi.org/10.3389/fimmu.2020.592080.

Chresta, C. M., et al. (2010). AZD8055 Is a potent, selective, and orally bioavailable ATP-competitive mammalian target of rapamycin kinase inhibitor with in vitro and in vivo antitumor activity. *Cancer Research, 70*(1), 288–298.

Chuang, H. Y., et al. (2019). Preclinical evidence of STAT3 inhibitor pacritinib overcoming temozolomide resistance via downregulating miR-21-enriched exosomes from M2 glioblastoma-associated macrophages. *Journal of Clinical Medicine Research, 8*(7), https://doi.org/10.3390/jcm8070959.

Cloughesy, T. F., et al. (2018). Phase II study of cabozantinib in patients with progressive glioblastoma: Subset analysis of patients with prior antiangiogenic therapy. *Neuro-Oncology, 20*(2), 259.

Colardo, M., Segatto, M., & Di Bartolomeo, S. (2021). Targeting RTK-PI3K-mTOR axis in gliomas: An update. *International Journal of Molecular Sciences, 22*(9), 4899.

Combs, S. E., et al. (2008). Phase I/II study of cetuximab plus temozolomide as radio-chemotherapy for primary glioblastoma (GERT)—Eudract number 2005–003911–63; NCT00311857. *Journal of Clinical Oncology: Official Journal of the American Society of Clinical Oncology.* https://doi.org/10.1200/jco.2008.26.15_suppl.2077.

Coupienne, I., et al. (2011). NF-kappaB inhibition improves the sensitivity of human glioblastoma cells to 5-aminolevulinic acid-based photodynamic therapy. *Biochemical Pharmacology, 81*(5), https://doi.org/10.1016/j.bcp.2010.12.015.

Courtney, K. D., Corcoran, R. B., & Engelman, J. A. (2010). The PI3K pathway as drug target in human cancer. *Journal of Clinical Oncology: Official Journal of the American Society of Clinical Oncology, 28*(6), 1075–1083.

de Groot, J., et al. (2012). Modulating antiangiogenic resistance by inhibiting the signal transducer and activator of transcription 3 pathway in glioblastoma. *Oncotarget, 3*(9), 1036–1048.

De Robertis, A., et al. (2013). Identification and characterization of a small molecule inhibitor of WNT signaling in glioblastoma cells. *Molecular Cancer Therapeutics, 12*(7), 1180.

de Souza, I., et al. (2022). High levels of NRF2 sensitize temozolomide-resistant glioblastoma cells to ferroptosis via ABCC1/MRP1 upregulation. *Cell Death & Disease, 13*(7), 1–13.

Dienstmann, R., et al. (2014). Picking the point of inhibition: A comparative review of PI3K/AKT/mTOR pathway inhibitors. *Molecular Cancer Therapeutics, 13*(5), 1021–1031.

Edwards, L. A., et al. (2011). Effect of brain- and tumor-derived connective tissue growth factor on glioma invasion. *JNCI Journal of the National Cancer Institute, 103*(15), 1162.

Eller, J. L., et al. (2002). Activity of anti-epidermal growth factor receptor monoclonal antibody C225 against glioblastoma multiforme. *Neurosurgery, 51*(4), https://doi.org/10.1097/00006123-200210000-00028.

Eller, J. L., et al. (2005). Anti-epidermal growth factor receptor monoclonal antibody cetuximab augments radiation effects in glioblastoma multiforme in vitro and in vivo. *Neurosurgery, 56*(1), https://doi.org/10.1227/01.neu.0000145865.25689.55.

Fan, Z., et al. (2017). Nrf2-Keap1 pathway promotes cell proliferation and diminishes ferroptosis. *Oncogenesis, 6*(8), e371.

Fang, Y., Liao, G., & Yu, B. (2020). Small-molecule MDM2/X inhibitors and PROTAC degraders for cancer therapy: Advances and perspectives. *Yao Xue Xue Bao = Acta Pharmaceutica Sinica, 10*(7), 1253–1278.

Faris, J. (2022). Glioblastoma (GBM), American Brain Tumor Association. ⟨https://www.abta.org/tumor_types/glioblastoma-gbm/⟩ Accessed 11.09.23.

Frei, K., et al. (2015). Transforming growth factor-β pathway activity in glioblastoma. *Oncotarget, 6*(8), 5963.

Friedmann-Morvinski, D., et al. (2016). Targeting NF-κB in glioblastoma: A therapeutic approach. *Science Advances, 2*(1), https://doi.org/10.1126/sciadv.1501292.

Fu, W., et al. (2023). Roles of STAT3 in the pathogenesis and treatment of glioblastoma. *Frontiers in Cell and Developmental Biology, 11*, 1098482.

Fukushima, T., et al. (2012). Antitumor effect of dehydroxymethylepoxyquinomicin, a small molecule inhibitor of nuclear factor-κB, on glioblastoma. *Neuro-Oncology, 14*(1), https://doi.org/10.1093/neuonc/nor168.

Furnari, F. B., et al. (2007). Malignant astrocytic glioma: Genetics, biology, and paths to treatment. *Genes & Development, 21*(21), 2683–2710.

Gangoso, E., et al. (2021). Glioblastomas acquire myeloid-affiliated transcriptional programs via epigenetic immunoediting to elicit immune evasion. *Cell, 184*(9), 2454–2470.e26.

Garner, J. M., et al. (2013). Constitutive activation of signal transducer and activator of transcription 3 (STAT3) and nuclear factor κB signaling in glioblastoma cancer stem cells regulates the Notch pathway. *The Journal of Biological Chemistry, 288*(36), https://doi.org/10.1074/jbc.M113.477950.

Ghosh, D., Nandi, S., & Bhattacharjee, S. (2018). Combination therapy to checkmate Glioblastoma: Clinical challenges and advances. *Clinical and Translational Medicine, 7*(1), e33.

Gilbert, M. R., et al. (2014). A randomized trial of bevacizumab for newly diagnosed glioblastoma. *The New England Journal of Medicine, 370*(8), https://doi.org/10.1056/NEJMoa1308573.

Gomez-Manzano, C., et al. (2001). Transfer of E2F-1 to human glioma cells results in transcriptional up-regulation of Bcl-21. *Cancer Research, 61*(18), 6693–6697.

Gong, A., & Huang, S. (2012). FoxM1 and Wnt/β-catenin signaling in glioma stem cells. *Cancer Research, 72*(22), 5658–5662.

Guan, R., Zhang, X., & Guo, M. (2020). Glioblastoma stem cells and Wnt signaling pathway: Molecular mechanisms and therapeutic targets. *Chinese Neurosurgical Journal, 6*, 25.

Haapasalo, J., et al. (2018). NRF2, DJ1 and SNRX1 and their prognostic impact in astrocytic gliomas. *Histology and Histopathology, 33*(8), 791–801.

Han, M.-H., et al. (2022). High DKK3 expression related to immunosuppression was associated with poor prognosis in glioblastoma: Machine learning approach. *Cancer Immunology, Immunotherapy: CII, 71*(12), 3013–3027.

Han, T. J., et al. (2016). Inhibition of STAT3 enhances the radiosensitizing effect of temozolomide in glioblastoma cells in vitro and in vivo. *Journal of Neuro-oncology, 130*(1), https://doi.org/10.1007/s11060-016-2231-9.

Hayden, M. S., & Ghosh, S. (2008). Shared principles in NF-kappaB signaling. *Cell, 132*(3), https://doi.org/10.1016/j.cell.2008.01.020.

He, G., et al. (2005). Induction of p21 by p53 following DNA damage inhibits both Cdk4 and Cdk2 activities. *Oncogene, 24*(18), 2929–2943.

Hegi, M. E., et al. (2005). *MGMT gene silencing and benefit from temozolomide in glioblastoma.* https://doi.org/10.1056/NEJMoa043331.

Herman, J. A., et al. (2022). Hyper-active RAS/MAPK introduces cancer-specific mitotic vulnerabilities. *Proceedings of the National Academy of Sciences of the United States of America, 119*(41), e2208255119.

Jane, E. P., et al. (2009). Abrogation of mitogen-activated protein kinase and Akt signaling by vandetanib synergistically potentiates histone deacetylase inhibitor-induced apoptosis in human glioma cells. *The Journal of Pharmacology and Experimental Therapeutics, 331*(1), 327.

Jin, X., et al. (2011). Frizzled 4 regulates stemness and invasiveness of migrating glioma cells established by serial intracranial transplantation. *Cancer Research, 71*(8), 3066–3075.

Johnson, M. D., et al. (1993). Transforming growth factor-beta in neural embryogenesis and neoplasia. *Human Pathology, 24*(5), 457–462.

Joseph, J. V., et al. (2021). TGF-β promotes microtube formation in glioblastoma through thrombospondin 1. *Neuro-Oncology, 24*(4), 541–553.

Kafka, A., et al. (2021). Methylation patterns of DKK1, DKK3 and GSK3β are accompanied with different expression levels in human astrocytoma. *Cancers, 13*(11), 2530.

Kalluri, R., & Weinberg, R. A. (2009). The basics of epithelial-mesenchymal transition. *The Journal of Clinical Investigation, 119*(6), 1420–1428.

Kamino, M., et al. (2011). Wnt-5a signaling is correlated with infiltrative activity in human glioma by inducing cellular migration and MMP-2. *Cancer Science, 102*(3), 540–548.

Kandoth, C., et al. (2013). Mutational landscape and significance across 12 major cancer types. *Nature, 502*(7471), 333–339.

Karpel-Massler, G., & Halatsch, M.-E. (2014). *Erlotinib in glioblastoma—a current clinical perspective. Tumors of the central nervous system-primary and secondary.* IntechOpen,.

Khang, M., et al. (2023). Intrathecal delivery of nanoparticle PARP inhibitor to the cerebrospinal fluid for the treatment of metastatic medulloblastoma. *Science Translational Medicine, 15*(720), https://doi.org/10.1126/scitranslmed.adi1617.

Kilic, T., et al. (2000). Intracranial inhibition of platelet-derived growth factor-mediated glioblastoma cell growth by an orally active kinase inhibitor of the 2-phenylaminopyrimidine class. *Cancer Research, 60*(18), ⟨https://pubmed.ncbi.nlm.nih.gov/11016641/⟩.

Kim, S. S., et al. (2015). A tumor-targeting p53 nanodelivery system limits chemoresistance to temozolomide prolonging survival in a mouse model of glioblastoma multiforme. *Nanomedicine: Nanotechnology, Biology, and Medicine, 11*(2), 301–311. https://doi.org/10.1016/j.nano.2014.09.005.

Kim, S. H., et al. (2016). Serine/threonine kinase MLK4 determines mesenchymal identity in glioma stem cells in an NF-κB-dependent manner. *Cancer Cell, 29*(2), https://doi.org/10.1016/j.ccell.2016.01.005.

Kondo, E., et al. (2008). Potent synergy of dual antitumor peptides for growth suppression of human glioblastoma cell lines. *Molecular Cancer Therapeutics, 7*(6), 1461–1471.

Kwatra, M., et al. (2017). Exth-46. A precision medicine approach to target egfrviii in gbm: Osimertinib (azd9291) inhibits the growth of egfrviii-positive glioblastoma stem cells and increases survival of mice bearing intracranial egfrviii-positive gbm. *Neuro-Oncology, 19*(Suppl 6), vi82.

Lam, K., et al. (2022). Prognostic value of O6-methylguanine-DNA methyltransferase methylation in isocitrate dehydrogenase mutant gliomas. *Neuro-Oncology Advances, 4*(1), vdac030.

Lamalice, L., Le Boeuf, F., & Huot, J. (2007). Endothelial cell migration during angiogenesis. *Circulation Research.* https://doi.org/10.1161/01.RES.0000259593.07661.1e.

Lambiv, W. L., et al. (2011). The Wnt inhibitory factor 1 (WIF1) is targeted in glioblastoma and has a tumor suppressing function potentially by induction of senescence. *Neuro-Oncology, 13*(7), 736–747.

Lang, F. F., et al. (2003). Phase I trial of adenovirus-mediated p53 gene therapy for recurrent glioma: Biological and clinical results. *Journal of Clinical Oncology: Official Journal of the American Society of Clinical Oncology, 21*(13), 2508–2518.

Laplante, M., & Sabatini, D. M. (2012). mTOR signaling in growth control and disease. *Cell, 149*(2), 274–293.

Laribee, R. N., et al. (2023). The STAT3-regulated autophagy pathway in glioblastoma. *Pharmaceuticals, 16*(5), https://doi.org/10.3390/ph16050671.

Lavon, I., et al. (2007). Novel mechanism whereby nuclear factor kappaB mediates DNA damage repair through regulation of O(6)-methylguanine-DNA-methyltransferase. *Cancer Research, 67*(18), https://doi.org/10.1158/0008-5472.CAN-06-3820.

Lee, E. Q., et al. (2015). A multicenter, phase II, randomized, noncomparative clinical trial of radiation and temozolomide with or without vandetanib in newly diagnosed glioblastoma patients. *Clinical Cancer Research: An Official Journal of the American Association for Cancer Research, 21*(16), 3610.

Lee, Y., et al. (2016). WNT signaling in glioblastoma and therapeutic opportunities. *Laboratory Investigation; A Journal of Technical Methods and Pathology, 96*(2), 137–150.

Li, G. H., et al. (2009). STAT3 silencing with lentivirus inhibits growth and induces apoptosis and differentiation of U251 cells. *Journal of Neuro-oncology, 91*(2), https://doi.org/10.1007/s11060-008-9696-0.

Li, H., et al. (2015). Tyrphostin AG 1296 induces glioblastoma cell apoptosis and. *Oncology Letters, 10*(6), 3429–3433.

Li, L., et al. (2007). Transfection with anti-p65 intrabody suppresses invasion and angiogenesis in glioma cells by blocking nuclear factor-kappaB transcriptional activity. *Clinical Cancer Research: An Official Journal of the American Association for Cancer Research, 13*(7), https://doi.org/10.1158/1078-0432.CCR-06-1711.

Li, M. Y., et al. (2016). DW10075, a novel selective and small-molecule inhibitor of VEGFR, exhibits antitumor activities both in vitro and in vivo. *Acta Pharmacologica Sinica, 37*(3), https://doi.org/10.1038/aps.2015.117.

Liu, T., et al. (2018). PDGF-mediated mesenchymal transformation renders endothelial resistance to anti-VEGF treatment in glioblastoma. *Nature Communications, 9*(1), 1–13.

Liu, T.-J., et al. (2009). NVP-BEZ235, a novel dual phosphatidylinositol 3-kinase/mammalian target of rapamycin inhibitor, elicits multifaceted antitumor activities in human gliomas. *Molecular Cancer Therapeutics, 8*(8), 2204–2210.

Liu, X., et al. (2019). The third-generation EGFR inhibitor AZD9291 overcomes primary resistance by continuously blocking ERK signaling in glioblastoma. *Journal of Experimental & Clinical Cancer Research: CR, 38*(1), 1–14.

Louis, D. N., et al. (2016). The 2016 world health organization classification of tumors of the central nervous system: A summary. *Acta Neuropathologica, 131*(6), 803–820.

Louis, D. N., et al. (2021). The 2021 WHO classification of tumors of the central nervous system: A summary. *Neuro-Oncology, 23*(8), 1231.

Lucifero, A. G., & Luzzi, S. (2022). Immune landscape in PTEN-related glioma microenvironment: A bioinformatic analysis. *Brain Sciences, 12*(4), https://doi.org/10.3390/brainsci12040501.

Maehama, T., & Dixon, J. E. (1998). The tumor suppressor, PTEN/MMAC1, dephosphorylates the lipid second messenger, phosphatidylinositol 3,4,5-trisphosphate. *The Journal of Biological Chemistry, 273*(22), 13375–13378.

Mao, H., et al. (2012). Deregulated signaling pathways in glioblastoma multiforme: Molecular mechanisms and therapeutic targets. *Cancer Investigation.* https://doi.org/10.3109/07357907.2011.630050.

Matsuura, I., et al. (2004). Cyclin-dependent kinases regulate the antiproliferative function of Smads. *Nature, 430*(6996), 226–231.

McFarland, B. C., et al. (2013). Activation of the NF-κB pathway by the STAT3 inhibitor JSI-124 in human glioblastoma cells. *Molecular Cancer Research: MCR, 11*(5), 494–505.

Merzak, A., et al. (1994). Control of human glioma cell growth, migration and invasion in vitro by transforming growth factor β1. *British Journal of Cancer, 70*(2), 199–203.

Mizobuchi, Y., et al. (2008). REIC/Dkk-3 induces cell death in human malignant glioma. *Neuro-Oncology, 10*(3), 244–253.

Momota, H., Nerio, E., & Holland, E. C. (2005). Perifosine inhibits multiple signaling pathways in glial progenitors and cooperates with temozolomide to arrest cell proliferation in gliomas in vivo. *Cancer Research, 65*(16), 7429–7435.

Moodie, S. A., et al. (1993). Complexes of Ras.GTP with Raf-1 and mitogen-activated protein kinase kinase. *Science (New York, N. Y.), 260*(5114), 1658–1661.

Nakamura, M., et al. (2001). p14ARF deletion and methylation in genetic pathways to glioblastomas. *Brain Pathology, 11*(2), 159–168.

Nazarenko, I., et al. (2012). PDGF and PDGF receptors in glioma. *Upsala Journal of Medical Sciences, 117*(2), https://doi.org/10.3109/03009734.2012.665097.

Nogueira, L., et al. (2011). The NFκB pathway: A therapeutic target in glioblastoma (Available at:) *Oncotarget, 2*(8), https://doi.org/10.18632/oncotarget.322.

Noman, M. Z., et al. (2014). PD-L1 is a novel direct target of HIF-1α, and its blockade under hypoxia enhanced MDSC-mediated T cell activation. *The Journal of Experimental Medicine, 211*(5), https://doi.org/10.1084/jem.20131916.

Nørøxe, D. S., Poulsen, H. S., & Lassen, U. (2016). Hallmarks of glioblastoma: A systematic review. *ESMO Open, 1*(6), e000144.

Noushmehr, H., et al. (2010). Identification of a CpG island methylator phenotype that defines a distinct subgroup of glioma. *Cancer Cell, 17*(5), 510–522.

Ohgaki, H., et al. (2004). Genetic pathways to glioblastoma: A population-based study. *Cancer Research, 64*(19), 6892–6899.

Ohtsu, N., et al. (2016). Eva1 maintains the stem-like character of glioblastoma-initiating cells by activating the noncanonical NF-κB signaling pathway. *Cancer Research, 76*(1), https://doi.org/10.1158/0008-5472.CAN-15-0884.

Omuro, A., & DeAngelis, L. M. (2013). Glioblastoma and other malignant gliomas: A clinical review. *JAMA: The Journal of the American Medical Association, 310*(17), 1842–1850.

Ott, M., et al. (2020). Radiation with STAT3 blockade triggers dendritic cell-T cell interactions in the glioma microenvironment and therapeutic efficacy. *Clinical Cancer Research: An Official Journal of the American Association for Cancer Research, 26*(18), https://doi.org/10.1158/1078-0432.CCR-19-4092.

Pan, H., et al. (2013). The involvement of Nrf2–ARE pathway in regulation of apoptosis in human glioblastoma cell U251. *Neurological Research.* ⟨https://www.tandfonline.com/doi/abs/10.1179/1743132812Y.0000000094⟩.

Pan, H., et al. (2017). VPA and MEL induce apoptosis by inhibiting the Nrf2-ARE signaling pathway in TMZ-resistant U251 cells. *Molecular Medicine Reports, 16*(1), 908–914.

Paolillo, M., Boselli, C., & Schinelli, S. (2018). Glioblastoma under siege: An overview of current therapeutic strategies. *Brain Sciences, 8*(1), 15.

Park, A. K., et al. (2018). Subtype-specific signaling pathways and genomic aberrations associated with prognosis of glioblastoma. *Neuro-Oncology, 21*(1), 59–70.

Parsons, D. W., et al. (2008). An integrated genomic analysis of human glioblastoma multiforme. *Science (New York, N. Y.), 321*(5897), 1807–1812.

Paulus, W., et al. (1995). Effects of transforming growth factor-β1 on collagen synthesis, integrin expression, adhesion and invasion of glioma cells. *Journal of Neuropathology and Experimental Neurology, 54*(2), 236–244.

Pearson, J. R. D., & Regad, T. (2017). Targeting cellular pathways in glioblastoma multiforme. *Signal Transduction and Targeted Therapy, 2*(1), 1–11.

Pellot Ortiz, K. I., et al. (2023). MDM2 inhibition in the treatment of glioblastoma: From concept to clinical investigation. *Biomedicines, 11*(7), 1879.

Perdrix, A., et al. (2017). PRIMA-1 and PRIMA-1Met (APR-246): From mutant/wild type p53 reactivation to unexpected mechanisms underlying their potent anti-tumor effect in combinatorial therapies. *Cancers, 9*(12), 172.

Piperi, C., Papavassiliou, K. A., & Papavassiliou, A. G. (2019). Pivotal role of STAT3 in shaping glioblastoma immune microenvironment. *Cells, 8*(11), https://doi.org/10.3390/cells8111398.

Pölönen, P., et al. (2019). Nrf2 and SQSTM1/p62 jointly contribute to mesenchymal transition and invasion in glioblastoma. *Oncogene, 38*(50), 7473–7490.

Popescu, A. M., et al. (2015). Targeting the VEGF and PDGF signaling pathway in glioblastoma treatment. *International Journal of Clinical and Experimental Pathology, 8*(7), 7825.

Porru, M., et al. (2014). Medical treatment of orthotopic glioblastoma with transferrin-conjugated nanoparticles encapsulating zoledronic acid. *Oncotarget, 5*(21), 10446–10459.

Premkumar, D. R., et al. (2018). Mitochondrial dysfunction RAD51, and Ku80 proteolysis promote apoptotic effects of Dinaciclib in Bcl-xL silenced cells. *Molecular Carcinogenesis, 57*(4), 469–482.

Qiu, X. Y., et al. (2018). PD-L1 confers glioblastoma multiforme malignancy via Ras binding and Ras/Erk/EMT activation. *Biochimica et Biophysica Acta (BBA)—Molecular Basis of Disease, 1864*(5), 1754–1769.

Rahaman, S. O., Vogelbaum, M. A., & Haque, S. J. (2005). Aberrant Stat3 Signaling by Interleukin-4 in Malignant Glioma Cells: Involvement of IL-13Rα2. *Cancer Research, 65*(7), 2956–2963.

Reardon, D. A., et al. (2011). Effect of CYP3A-inducing anti-epileptics on sorafenib exposure: Results of a phase II study of sorafenib plus daily temozolomide in adults with recurrent glioblastoma. *Journal of Neuro-oncology, 101*(1), 57.

Reifenberger, J., et al. (1996). Analysis of p53 mutation and epidermal growth factor receptor amplification in recurrent gliomas with malignant progression. *Journal of Neuropathology and Experimental Neurology, 55*(7), 822–831.

Richardson, T. P., et al. (2001). Polymeric system for dual growth factor delivery. *Nature Biotechnology, 19*(11), 1029–1034.

Rinkenbaugh, A. L., et al. (2016). IKK/NF-κB signaling contributes to glioblastoma stem cell maintenance. *Oncotarget, 7*(43), https://doi.org/10.18632/oncotarget.12507.

Rivera, V. M., et al. (2011). Ridaforolimus (AP23573; MK-8669), a potent mTOR inhibitor, has broad antitumor activity and can be optimally administered using intermittent dosing regimens. *Molecular Cancer Therapeutics, 10*(6), 1059–1071.

Rockwell, N., et al. (2021). p53 mutations exhibit sex specific gain-of-function activity in gliomagenesis. *bioRxiv.* https://doi.org/10.1101/2021.06.11.448124.

Rodriguez, S. M. B., et al. (2023). An Overview of EGFR mechanisms and their implications in targeted therapies for glioblastoma. *International Journal of Molecular Sciences, 24*(13), https://doi.org/10.3390/ijms241311110.

Roth, W., et al. (2000). Secreted frizzled-related proteins inhibit motility and promote growth of human malignant glioma cells. *Oncogene, 19*(37), 4210–4220.

Salzano, G., et al. (2016). Transferrin-targeted nanoparticles containing zoledronic acid as a potential tool to inhibit glioblastoma growth. *Journal of Biomedical Nanotechnology, 12*(4), 811–830. https://doi.org/10.1166/jbn.2016.2214.

Seo, Y. E., et al. (2019). Nanoparticle-mediated intratumoral inhibition of miR-21 for improved survival in glioblastoma. *Biomaterials, 201.* https://doi.org/10.1016/j.biomaterials.2019.02.016.

Shergalis, A., et al. (2018). Current challenges and opportunities in treating glioblastoma. *Pharmacological Reviews, 70*(3), 412–445.

Sherr, C. J. (2001). The INK4a/ARF network in tumour suppression. *Nature Reviews. Molecular Cell Biology, 2*(10), 731–737.

Shinojima, N., et al. (2003). Prognostic value of epidermal growth factor receptor in patients with glioblastoma multiforme. *Cancer Research, 63*(20), ⟨https://pubmed.ncbi.nlm.nih.gov/14583498/⟩.

Shukla, S., et al. (2013). A DNA methylation prognostic signature of glioblastoma: Identification of NPTX2-PTEN-NF-κB nexus. *Cancer Research, 73*(22), https://doi.org/10.1158/0008-5472.CAN-13-0298.

Siegelin, M. D., et al. (2010). Sorafenib exerts anti-glioma activity in vitro and in vivo. *Neuroscience Letters, 478*(3), 165.

Soubannier, V., & Stifani, S. (2017). NF-κB signalling in glioblastoma. *Biomedicines, 5*(2), https://doi.org/10.3390/biomedicines5020029.

Stambolic, V., et al. (1998). Negative regulation of PKB/Akt-dependent cell survival by the tumor suppressor PTEN. *Cell, 95*(1), 29–39.

Stott, F. J., et al. (1998). The alternative product from the human CDKN2A locus, p14(ARF), participates in a regulatory feedback loop with p53 and MDM2. *The EMBO Journal, 17*(17), 5001–5014.

Struve, N., et al. (2020). EGFRvIII upregulates DNA mismatch repair resulting in increased temozolomide sensitivity of MGMT promoter methylated glioblastoma. *Oncogene, 39*(15), 3041–3055.

Stupp, R., et al. (2005). Radiotherapy plus concomitant and adjuvant temozolomide for glioblastoma. *New England Journal of Medicine, 352*(10), 987–996. https://doi.org/10.1056/NEJMoa043330.

Su, Y., et al. (2008). JSI-124 inhibits glioblastoma multiforme cell proliferation through G2/M cell cycle arrest and apoptosis augmentation. *Cancer Biology & Therapy.* https://doi.org/10.4161/cbt.7.8.6263 [Preprint].

Sun, W., et al. (2020). Inhibition of Nrf2 might enhance the anti-tumor effect of temozolomide in glioma cells via inhibition of Ras/Raf/MEK signaling pathway. *The International Journal of Neuroscience.* https://doi.org/10.1080/00207454.2020.1766458.

Suwala, A. K., et al. (2018). Inhibition of Wnt/beta-catenin signaling downregulates expression of aldehyde dehydrogenase isoform 3A1 (ALDH3A1) to reduce resistance against temozolomide in glioblastoma. *Oncotarget, 9*(32), 22703–22716.

Szopa, W., et al. (2017). Diagnostic and therapeutic biomarkers in glioblastoma: Current status and future perspectives. *BioMed Research International, 2017.* https://doi.org/10.1155/2017/8013575.

Tan, A. C., et al. (2020). Management of glioblastoma: State of the art and future directions. *CA: A Cancer Journal for Clinicians, 70*(4), 299–312.

Tan, M. S. Y., et al. (2019). A STAT3-based gene signature stratifies glioma patients for targeted therapy. *Nature Communications, 10*(1), 1–15.

Tchoghandjian, A., et al. (2013). Identification of non-canonical NF-κB signaling as a critical mediator of Smac mimetic-stimulated migration and invasion of glioblastoma cells. *Cell Death & Disease, 4*(3), https://doi.org/10.1038/cddis.2013.70.

Thomas, S. M., et al. (1992). Ras is essential for nerve growth factor- and phorbol ester-induced tyrosine phosphorylation of MAP kinases. *Cell, 68*(6), 1031–1040.

Tompa, M., et al. (2019). Wnt pathway markers in molecular subgroups of glioblastoma. *Brain Research, 1718*, 114–125.

Torp, S. H., Solheim, O., & Skjulsvik, A. J. (2022). The WHO 2021 classification of central nervous system tumours: A practical update on what neurosurgeons need to know—A minireview. *Acta Neurochirurgica, 164*(9), 2453.

Tran, N. L., et al. (2006). Increased fibroblast growth factor-inducible 14 expression levels promote glioma cell invasion via Rac1 and nuclear factor-kappaB and correlate with poor patient outcome. *Cancer Research, 66*(19), https://doi.org/10.1158/0008-5472.CAN-06-0418.

Vassallo, I., et al. (2015). WIF1 re-expression in glioblastoma inhibits migration through attenuation of non-canonical WNT signaling by downregulating the lncRNA MALAT1. *Oncogene, 35*(1), 12–21.

Velpula, K. K., et al. (2012). EGFR and c-Met cross talk in glioblastoma and its regulation by human cord blood stem cells. *Translational Oncology, 5*(5), 379–392.

Vengoji, R., et al. (2019). Afatinib and temozolomide combination inhibits tumorigenesis by targeting EGFRvIII-cMet signaling in glioblastoma cells. *Journal of Experimental & Clinical Cancer Research: CR, 38*(1), 1–13.

Verhaak, R. G. W., et al. (2010). Integrated genomic analysis identifies clinically relevant subtypes of glioblastoma characterized by abnormalities in PDGFRA, IDH1, EGFR, and NF1. *Cancer Cell, 17*(1), 98–110.

Verreault, M., et al. (2016). Preclinical efficacy of the MDM2 inhibitor RG7112 in MDM2-amplified and TP53 wild-type glioblastomas. *Clinical Cancer Research: An Official Journal of the American Association for Cancer Research, 22*(5), 1185–1196.

Wang, J., et al. (2018). Chrysin suppresses proliferation, migration, and invasion in glioblastoma cell lines via mediating the ERK/Nrf2 signaling pathway. *Drug Design, Development and Therapy, 12*, 721–733.

Wang, Y., et al. (2019). Apatinib plus temozolomide for recurrent glioblastoma: An uncontrolled, open-label study. *OncoTargets and Therapy, 12*, 10579–10585.

Wang, Y., et al. (2021). The new PI3K/mTOR inhibitor GNE-477 inhibits the malignant behavior of human glioblastoma cells. *Frontiers in Pharmacology, 12*, 659511.

Wang, Y., et al. (2022). SS-4 is a highly selective small molecule inhibitor of STAT3 tyrosine phosphorylation that potently inhibits GBM tumorigenesis in vitro and in vivo. *Cancer Letters, 533.* https://doi.org/10.1016/j.canlet.2022.215614.

Wang, Y., et al. (2023). Anti-seed PNAs targeting multiple oncomiRs for brain tumor therapy. *Science Advances, 9*(6), https://doi.org/10.1126/sciadv.abq7459.

Warner, L. C., et al. (1993). RAS is required for epidermal growth factor-stimulated arachidonic acid release in rat-1 fibroblasts. *Oncogene, 8*(12), 3249–3255.

Watanabe, K., et al. (1996). Overexpression of the EGF receptor and p53 mutations are mutually exclusive in the evolution of primary and secondary glioblastomas. *Brain Pathology, 6*(3), 217–223.

Wesolowska, A., et al. (2008). Microglia-derived TGF-beta as an important regulator of glioblastoma invasion–an inhibition of TGF-beta-dependent effects by shRNA against human TGF-beta type II receptor. *Oncogene, 27*(7), https://doi.org/10.1038/sj.onc.1210683.

Wick, A., et al. (2020). Superiority of temozolomide over radiotherapy for elderly patients with RTK II methylation class, MGMT promoter methylated malignant astrocytoma. *Neuro-Oncology, 22*(8), https://doi.org/10.1093/neuonc/noaa033.

Wong, A. J., et al. (1992). Structural alterations of the epidermal growth factor receptor gene in human gliomas. *Proceedings of the National Academy of Sciences of the United States of America, 89*(7), 2965–2969.

Wood, E. R., et al. (2004). A unique structure for epidermal growth factor receptor bound to GW572016 (Lapatinib): Relationships among protein conformation, inhibitor off-rate, and receptor activity in tumor cells. *Cancer Research, 64*(18), 6652–6659.

Xie, C., et al. (2018). Preclinical characterization of anlotinib, a highly potent and selective vascular endothelial growth factor receptor-2 inhibitor. *Cancer Science, 109*(4), 1207.

Xu, R. X., et al. (2015). DNA damage-induced NF-κB activation in human glioblastoma cells promotes miR-181b expression and cell proliferation. *Cellular Physiology and Biochemistry: International Journal of Experimental Cellular Physiology, Biochemistry, and Pharmacology, 35*(3), https://doi.org/10.1159/000369748.

Yamada, N., et al. (1995). Enhanced expression of transforming growth factor-β and its type-I and type-II receptors in human glioblastoma. *International Journal of Cancer, 62*(4), 386–392.

Yan, H., et al. (2009). IDH1 and IDH2 mutations in gliomas. *The New England Journal of Medicine, 360*(8), 765–773.

Yan, T., et al. (2022). TGF-β induces GBM mesenchymal transition through upregulation of CLDN4 and nuclear translocation to activate TNF-α/NF-κB signal pathway. *Cell Death & Disease, 13*(4), 1–11.

Yang, L., et al. (2022). VPS9D1-AS1 overexpression amplifies intratumoral TGF-β signaling and promotes tumor cell escape from CD8+ T cell killing in colorectal cancer. *eLife, 11*. https://doi.org/10.7554/eLife.79811.

Yu, C.-Y., et al. (2013). Lgr4 promotes glioma cell proliferation through activation of Wnt signaling. *Asian Pacific Journal of Cancer Prevention: APJCP, 14*(8), 4907–4911.

Yu, J. M., et al. (2006). Increase in proliferation and differentiation of neural progenitor cells isolated from postnatal and adult mice brain by Wnt-3a and Wnt-5a. *Molecular and Cellular Biochemistry, 288*(1), 17–28.

Yu, Z., et al. (2015). NVP-BEZ235, a novel dual PI3K–mTOR inhibitor displays anti-glioma activity and reduces chemoresistance to temozolomide in human glioma cells. *Cancer Letters, 367*(1), 58–68.

Zahonero, C., et al. (2015). Preclinical test of dacomitinib, an irreversible EGFR inhibitor, confirms its effectiveness for glioblastoma. *Molecular Cancer Therapeutics, 14*(7), 1548–1558.

Zanotto-Filho, A., et al. (2011). NFκB inhibitors induce cell death in glioblastomas. *Biochemical Pharmacology, 81*(3), https://doi.org/10.1016/j.bcp.2010.10.014.

Zhang, K., et al. (2012). Wnt/beta-Catenin Signaling in Glioma. *Journal of Neuroimmune Pharmacology: The Official Journal of the Society on NeuroImmune Pharmacology, 7*(4), 740–749.

Zhang, L., & Wang, H. (2017). FTY720 inhibits the Nrf2/ARE pathway in human glioblastoma cell lines and sensitizes glioblastoma cells to temozolomide. *Pharmacological Reports: PR, 69*(6), 1186–1193.

Zhang, P., et al. (2020). Current opinion on molecular characterization for GBM classification in guiding clinical diagnosis, prognosis, and therapy. *Frontiers in Molecular Biosciences, 7*, 562798.

Zhao, H.-F., et al. (2017). Recent advances in the use of PI3K inhibitors for glioblastoma multiforme: Current preclinical and clinical development. *Molecular Cancer, 16*(1), 100.

Zhao, W., et al. (2023). The CDK inhibitor AT7519 inhibits human glioblastoma cell growth by inducing apoptosis, pyroptosis and cell cycle arrest. *Cell Death & Disease, 14*(1), 11.

Zheng, H., et al. (2008). Pten and p53 converge on c-Myc to control differentiation, self-renewal, and transformation of normal and neoplastic stem cells in glioblastoma. *Cold Spring Harbor Symposia on Quantitative Biology, 73*, 427–437.

Zheng, H., et al. (2010). PLAGL2 regulates Wnt signaling to impede differentiation in neural stem cells and gliomas. *Cancer Cell, 17*(5), 497–509.

Zhou, W., et al. (2017). Targeting glioma stem cell-derived pericytes disrupts the blood-tumor barrier and improves chemotherapeutic efficacy. *Cell Stem Cell, 21*(5), 591–603.e4.

Zhu, J., et al. (2013). Nrf2 is required to maintain the self-renewal of glioma stem cells. *BMC Cancer, 13*(1), 1–11.

Zhu, J., et al. (2014). Differential Nrf2 expression between glioma stem cells and non-stem-like cells in glioblastoma. *Oncology letters, 7*(3), 693–698.

Zhu, Y., & Shah, K. (2014). Multiple lesions in receptor tyrosine kinase pathway determine glioblastoma response to pan-ERBB inhibitor PF-00299804 and PI3K/mTOR dual inhibitor PF-05212384. *Cancer Biology & Therapy, 15*(6), 815–822.

# CHAPTER TWO

# Therapeutic potential of hedgehog signaling in advanced cancer types

## Richa Singh[a,*] and Anindita Ray[b]
[a]Department of Pathology and Laboratory Medicine, Weill Cornell Medicine, New York, NY, United States
[b]Neurodegenerative Diseases Research Unit, National Institute of Neurological Disorders and Stroke, Bethesda, MD, United States
*Corresponding author. e-mail address: ris4001@med.cornell.edu

## Contents

| | |
|---|---|
| 1. Overview of hedgehog signaling | 51 |
|    1.1 Hedgehog protein | 51 |
|    1.2 Components of hedgehog signaling | 51 |
|    1.3 Canonical hedgehog signaling | 52 |
|    1.4 Non-canonical hedgehog signaling | 53 |
| 2. Hedgehog signaling in cancer | 55 |
|    2.1 Mechanisms of hedgehog signaling in cancer | 55 |
|    2.2 Hedgehog signaling in metastatic tumors | 59 |
|    2.3 Relevance of hedgehog signaling in neuroendocrine tumors | 62 |
|    2.4 Clinical prevalence of hedgehog signaling in different cancers | 63 |
| 3. Therapeutic relevance of hedgehog signaling in cancer | 64 |
|    3.1 Targeting hedgehog signaling components | 64 |
|    3.2 Combination with epigenetic inhibitors | 69 |
| 4. Perspective | 70 |
| References | 72 |

## Abstract

In this chapter, we have made an attempt to elucidate the relevance of hedgehog signaling pathway in tumorigenesis. Here, we have described different types of hedgehog signaling (canonical and non-canonical) with emphasis on the different mechanisms (mutation-driven, autocrine, paracrine and reverse paracrine) it adopts during tumorigenesis. We have discussed the role of hedgehog signaling in regulating cell proliferation, invasion and epithelial-to-mesenchymal transition in both local and advanced cancer types, as reported in different studies based on preclinical and clinical models. We have specifically addressed the role of hedgehog signaling in aggressive neuroendocrine tumors as well. We have also elaborated on the studies showing therapeutic relevance of the inhibitors of hedgehog signaling in cancer.

Evidence of the crosstalk of hedgehog signaling components with other signaling pathways and treatment resistance due to tumor heterogeneity have also been briefly discussed. Together, we have tried to put forward a compilation of the studies on therapeutic potential of hedgehog signaling in various cancers, specifically aggressive tumor types with a perspective into what is lacking and demands further investigation.

## Abbreviations

| | |
|---|---|
| Hh | Hedgehog |
| SHH | sonic hedgehog |
| IHH | indian hedgehog |
| DHH | desert hedgehog |
| PKA | protein kinase a |
| CKI | casein kinase i |
| GSK | glycogen synthase kinase 3 |
| GLIA | active gli |
| GLI-FL | gli full-length |
| GLIR | repressive gli |
| BCC | basal cell carcinoma |
| RMS | rhabdomyosarcoma |
| MB | Medulloblastoma |
| VEGF | vascular endothelial growth factor |
| IGF | insulin-like growth factor |
| IL-6 | interleukin-6 |
| PDGF | platelet-derived growth factor |
| BMP | bone morphogenetic proteins |
| PDAC | pancreatic ductal adenocarcinoma |
| DGT | Degalactotigonin |
| HCC | hepatocellular carcinoma |
| ALK | activin receptor-like kinase |
| OPN | osteopontin |
| HIP | hedgehog-interacting protein |
| HTPC | hormone treated prostate cancer |
| HNPC | hormone naïve prostate cancer |
| HRPC | hormone refractory prostate cancer |
| NET | neuroendocrine tumor |
| hASH1 | human achaete-scute homologue gene 1 |
| TRKB | tropomyosin-related kinase b |
| LCNEC | large cell neuroendocrine carcinoma |
| PNET | pancreatic neuroendocrine tumor |
| aPKC | atypical protein kinase c |
| BET | bromodomain and extra-terminal domain |
| AML | acute myeloid leukemia |
| MDS | myelodysplastic syndrome |
| LMS | Leiomyosarcoma |

# 1. Overview of hedgehog signaling
## 1.1 Hedgehog protein

Hedgehog (Hh) gene was discovered as a regulatory gene in the development of embryo in Drosophila by Nusslein-Volhard and Wieschaus in 1980 (Nüsslein-Volhard & Wieschaus, 1980). It was identified as a secretory segment polarity gene which was shown to be essential for the embryonic cuticle pattern in Drosophila (Lee, Von Kessler, Parks, & Beachy, 1992). Later, it was identified as an evolutionarily conserved protein with three mammalian orthologs, Sonic hedgehog (SHH), Indian hedgehog (IHH), and Desert hedgehog (DHH). In vertebrates, hedgehog proteins function as dose-dependent morphogens or mitogens controlling cell fate, germ cell differentiation and body organization including growth, patterning and morphogenesis (Briscoe & Thérond, 2013; Ingham & Mcmahon, 2001). Specifically, SHH protein is known to be involved in the early development of motor neurons, notochord development and central nervous system polarity (Echelard et al., 1993; Jessell, 2000; Tanabe, Roelink, & Jessell, 1995). IHH is essential for bone formation and regulates chondrocyte proliferation and differentiation (St-Jacques, Hammerschmidt, & Mcmahon, 1999). DHH regulates germ cell differentiation in testis (Mäkelä et al., 2011).

## 1.2 Components of hedgehog signaling

Apart from the three hedgehog ligands, hedgehog signaling in vertebrates comprises of two 12-pass transmembrane receptors, PTCH1 and PTCH2, a G-protein coupled receptor-like 7-pass transmembrane receptor, SMO and three transcription factors, GLI1, GLI2 and GLI3 (Mäkelä et al., 2011; Rahnama, Toftgård, & Zaphiropoulos, 2004; Wu, Zhang, Sun, Mcmahon, & Wang, 2017). Although both PTCH receptors can bind to the three hedgehog ligands (Teglund & Toftgård, 2010), PTCH1 was reported to be expressed in mesenchymal cells in embryos while PTCH2 was shown to express in skin cells and spermatocytes. PTCH1 was shown to interact mostly with the SHH ligand whereas PTCH2 most probably interacted with DHH in germ cells (Carpenter et al., 1998; Rahnama et al., 2004). PTCH upon binding with hedgehog ligands, undergo a conformational change and release SMO receptor, which is a positive regulator of hedgehog signaling (Chen & Struhl, 1998). SMO is known to physically interact with PTCH but not hedgehog ligands (Stone et al., 1996). Additionally, downstream activity of SMO is known to require phosphorylation by Protein kinase A (PKA) and

Casein kinase I (CK1). Blocking PKA or CK1 results in attenuated hedgehog signaling (Jia, Tong, Wang, Luo, & Jiang, 2004).

GLI proteins, identified first in glioblastoma have highly conserved zinc-finger DNA binding domains (Kinzler et al., 1987). Interestingly, GLI1 has only the activator domain (C-terminal) whereas GLI2 and GLI3 both have activator and repressor domains (C-terminal and N-terminal, respectively) and can undergo post-translational modifications (Abbasi, Goode, Amir, & Grzeschik, 2009). GLI1 functions as an activator of hedgehog signaling regulating G1 cell cycle and tumor formation (Santoni et al., 2013). GLI2 primarily act as an activator in regulating the patterning of ventral regions in spinal cord whereas GLI3 functions as transcriptional repressor in regulating patterning intermediate regions of spinal cord (Persson et al., 2002).

## 1.3 Canonical hedgehog signaling

When no hedgehog ligand is present, PTCH is bound to SMO restricting its accumulation in cilia and inhibiting its activity. This is followed by phosphorylation of C-terminal of GLI proteins by PKA, CK1 and GSK (glycogen synthase kinase 3) which are then bound to ubiquitin-ligase complex followed by proteasomal degradation. The truncated GLI proteins with only N-terminal, GLIR are then translocated in the nucleus where they repress the downstream targets (Aberger & Ruiz I Altaba, 2014; Hui & Angers, 2011; Pan, Bai, Joyner, & Wang, 2006).

When hedgehog ligands are present, PTCH receptors are bound to them and their conformational change releases the SMO receptor. This sets the activation of canonical hedgehog signaling. Hedgehog-PTCH complex is eventually internalized followed by its degradation (Mastronardi, Dimitroulakos, Kamel-Reid, & Manoukian, 2000). SMO accumulation further causes the release of GLI protein from SUFU (suppressor of fused) protein which otherwise retrains the GLI translocation in nucleus (Humke, Dorn, Milenkovic, Scott, & Rohatgi, 2010; Rohatgi, Milenkovic, & Scott, 2007). Similarly, a kinesin protein (KIF7) physically interacts with GLI and in the absence of hedgehog signaling causes its degradation. However, in the presence of active hedgehog signal, KIF7 releases GLI and prevents its truncation to GLIR form (Cheung et al., 2009; Endoh-Yamagami et al., 2009; Liem, He, Ocbina, & Anderson, 2009). GLIFL is eventually converted to GLIA (active form) which then translocate to nucleus and activates the transcription of hedgehog signaling targets including PTCH (Fig. 1) (Cohen et al., 2015).

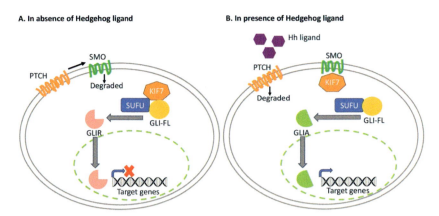

**Fig. 1** Canonical Hedgehog signaling pathway. (A) In the absence of Hh ligand, PTCH degrades SMO and GLI-FL is cleaved to repressive GLI (GLIR) which shuts down the GLI-mediated transcription. (B) In the presence of Hh ligand, PTCH binds to it and gets degraded. SMO binds to KIF7 and GLI is cleaved to its active form, GLIA, switching on GLI-mediated transcription.

This canonical hedgehog signaling is known to induce genes such as cyclin D1 and NMyc and regulate cell proliferation (Kenney, Cole, & Rowitch, 2003).

## 1.4 Non-canonical hedgehog signaling

Unlike canonical hedgehog signaling which involves GLI mediated transcriptional activation/repression, non-canonical hedgehog signaling is GLI-independent (Robbins, Fei, & Riobo, 2012). Mainly two such mechanisms of non-canonical hedgehog signaling have been elucidated.

**A.** SMO-independent but via PTCH
**B.** Mediated by small GTPases downstream of SMO

SMO-independent or Type I non-canonical hedgehog signaling can induce pro-angiogenic response via PTCH in endothelial cells (Chinchilla, Xiao, Kazanietz, & Riobo, 2010; Thibert et al., 2003). PTCH1, in the absence of hedgehog ligand, undergo a cleavage by caspases at a conserved aspartic acid (Asp1392) C-terminal motif, releasing its proapoptotic domain. PTCH1 then associates with a multiprotein complex that induced cell death (Kagawa et al., 2011; Mille et al., 2009). However, in the presence of hedgehog ligand, this association of PTCH1 with proapoptotic complex is inhibited. This was further validated when C-terminal truncated PTCH1 protein induced cell death even in the presence of SHH

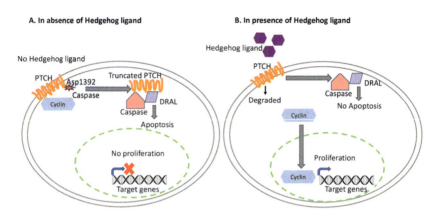

**Fig. 2** Type I non-canonical hedgehog signaling. (A) In absence of Hh ligand, PTCH is cleaved at Asp1392 motif by caspase and inhibits nuclear translocation of Cyclin B1. The truncated PTCH binds to multiprotein cell death complex and induces apoptosis. (B) In presence of Hh ligand, PTCH1 is degraded and cyclin B1 is released and translocated to nucleus.

while a mutation in PTCH1 at the cleavage site of caspase, blocks apoptosis. Further, all three hedgehog ligands (SHH, IHH and DHH) were able to induce PTCH1-mediated antiapoptotic, which could not be reversed by SMO blockers (Chinchilla et al., 2010). Unlike canonical hedgehog signaling, PTCH-mediated non-canonical signaling regulate cell cycle at G2/M checkpoint by binding to phosphorylated cyclin B1 and inhibited its nuclear localization (Barnes, Kong, Ollendorff, & Donoghue, 2001). Upon SHH binding to PTCH1, a conformational change of PTCH1 occurs that increase its affinity for GRK2 and replaces cyclin B1 (Jiang, Yang, & Ma, 2009). Cyclin B1 can then relocate to nucleus and mitosis continues independent of SMO involvement (Fig. 2).

Type II non-canonical hedgehog pathway mediated by small Rho family GTPases is known to regulate actin cytoskeleton in endothelial cells (Renault et al., 2010). Hedgehog ligands are known to induce the formation of actin fiber and tubulogenesis by activating Rho in SMO-dependent manner. Rho and Rac1 are known to be activated by SMO in presence of SHH which is essential for migration in fibroblasts (Polizio et al., 2011). In another study, it was reported that in the absence of hedgehog ligand, inactive SMO interacts with TIAM1 (T-lymphoma invasion and metastasis 1) inactivating Rac1. Upon activation by SHH, SMO-TIAM1 complex dissociates and TIAM1, a Rac guanine nucleotide exchange factor, activates Rac1 (Sasaki, Kurisu, & Kengaku, 2010). Additionally, SHH is known to acts as a chemoattractant

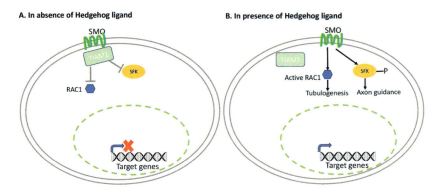

**Fig. 3** Type II non-canonical hedgehog signaling. (A) In absence of Hh-ligand, SMO-TIAM1 complex inactivates Rac1 and SFK. (B) In presence of Hh-ligand, SMO-TIAM1 complex dissociates activating Rac1 and phosphohorylating SFK. These in turn, promote tubulogenesis of actin cytoskeleton and axon guidance, respectively.

serving as an axon guidance cue. Independent of GLI-mediated transcription regulation, SHH can locally stimulate the phosphorylation of SFK kinases in an SMO-dependent manner inducing axon-turning (Charron & Tessier-Lavigne, 2005). Another evidence of non-GLI mediated hedgehog signaling was shown in neural tubes where SHH regulated dose dependent Ca$^{2+}$ spike (Belgacem & Borodinsky, 2011). This opens the possibility of a wide range of calcium-dependent physiological processes to be regulated by non-canonical hedgehog signaling in neuronal cells, for instance, differentiation and migration (Fig. 3).

## 2. Hedgehog signaling in cancer
### 2.1 Mechanisms of hedgehog signaling in cancer

Although hedgehog signaling has been widely studied in embryonic development, it has wide implications in stem cell biology and cancer. SHH is also known to induce stem cell and proliferation related genes such as MYC, cyclin D1, IGF2 and BMI1 during cerebellum development (Ng & Curran, 2011; Wechsler-Reya & Scott, 1999). Hedgehog signaling also maintains stem cells in adult brain, hair follicles and involved in the injury-dependent regeneration of organs (Ahn & Joyner, 2005; Beachy, Karhadkar, & Berman, 2004; Lai, Kaspar, Gage, & Schaffer, 2003). Importance of hedgehog signaling in stem cell and tissue regeneration hints on its relevance in cancer biology. In tumorigenesis, mainly four models of hedgehog signaling has been elucidated (Fig. 4; Table 1) (Sari et al., 2018; Wang et al., 2023a).

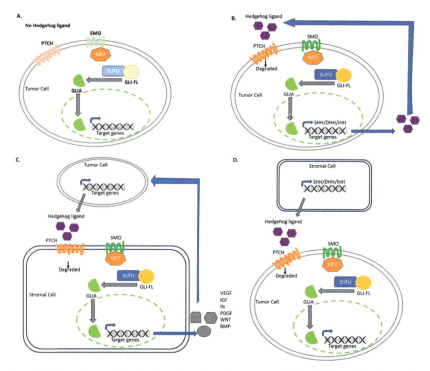

**Fig. 4** Mechanisms of hedgehog signaling in cancer. (A) In Hh ligand-independent activation, loss-of-function mutations in PTCH and SUFU, gain-of-function mutations in SMO and gene amplifications in GLIs lead to an uncontrolled activation of the signaling pathway in absence of Hh ligands. (B) In Hh ligand-dependent autocrine activation, hedgehog ligands are overexpressed by tumor cells and released which are then taken up by the same or neighboring tumor cells resulting in positive feedback loop. (C) In Hh ligand-dependent paracrine activation, Hh ligands secreted by tumor cells activate the stromal cells, which then release chemokines and other small molecules. These are taken up by tumor cells and activate cancer related pathways in them. (D) In Hh ligand-dependent reverse paracrine activation, Hh ligands secreted by stromal cells bind to PTCH in tumor cells and activate the hedgehog signaling in them.

**A.** Hh ligand–independent activation
**B.** Hh ligand–dependent autocrine activation
**C.** Hh ligand–dependent paracrine activation
**D.** Hh ligand–dependent reverse paracrine activation

Hh ligand–independent mechanism in tumors is mutation driven, such as germline loss-of-function mutation in PTCH1 and SUFU, gain-of-function mutation in SMO or gene amplification in GLI1 and GLI2. This

**Table 1** Mechanisms of hedgehog signaling in cancer.

| Model | Drivers | Cancer types |
|---|---|---|
| Hh ligand-independent activation | Inactivating mutation in PTCH and SUFU; Activating mutation in SMO; Amplification in GLI | Basal cell carcinoma; medulloblastoma; rhabdomyosarcoma; invasive transitional cell carcinoma of the bladder; esophageal squamous cell carcinoma; trichoepitheliomas |
| Hh ligand-dependent autocrine activation | Overexpression of SHH and IHH | Small cell lung tumors; pancreatic tumor; breast cancer; colon cancer; bladder cancer; gastric cancer |
| Hh ligand-dependent paracrine activation | Growth factors (VEGF, IGF, IL6, PDGF and BMP) from stroma; SHH from tumor cells; GLI from stroma | Pancreatic cancer; colorectal cancer; esophageal cancer; prostate cancer |
| Hh ligand-dependent reverse paracrine activation | Hh ligands from stroma | B-cell and plasma cell malignancies; gliomas |

was found to be common in basal cell carcinoma (BCC), medulloblastoma (MB) and rhabdomyosarcoma (RMS) resulting in uncontrolled hedgehog signaling activation even in the absence of Hh ligand and induce tumorigenesis. Inactivating mutation in PTCH1 or activating mutation in SMO were reported in sporadic BCCs, while MB and RMS have inactivating mutation in PTCH and SUFU (Taylor et al., 2002; Tostar et al., 2006; Xie et al., 1998). In rare cases of sporadic BCC, mutations in SUFU were also reported (Reifenberger et al., 2005). Additionally, overexpression of GLI1 in epidermis or heterozygous mutation in SUFU resembles BCC syndrome (also known as Gorlin syndrome). In children (more than 3 years old) with MB were reported to have GLI2 amplification (Kool et al., 2014). In invasive transitional cell carcinoma of the bladder (Mcgarvey, Maruta, Tomaszewski, Linnenbach, & Malkowicz, 1998), esophageal squamous cell carcinoma (Maesawa et al., 1998) and trichoepitheliomas (Vorechovský, Undén, Sandstedt, Toftgård, & Ståhle-Bäckdahl, 1997), loss-of-heterozygosity and somatic mutations were identified in PTCH.

In Hh ligand-dependent autocrine signaling, an overexpression of Hh ligand occurs which when released is taken up by same or surrounding tumor cells. Unlike ligand-independent hedgehog signaling, these tumors are highly dependent on Hh ligand and do not have mutation in the components of hedgehog pathway. This was proven using Hh ligand antibody and exogenous Hh ligand (Berman et al., 2003). Further, cyclopamine, an SMO inhibitor could enhance the apoptosis and inhibit tumor cell proliferation. As Hh ligand is mostly expressed in foregut endoderm and mammalian lung, the ligand dependent tumors were also identified in epithelial progenitors and small cell lung tumors (Szczepny et al., 2017). Studies have also shown SHH and IHH ligands to be highly expressed in various other tumors, such as pancreatic (Niyaz, Khan, Wani, Shah, & Mudassar, 2020), colon (Geyer & Gerling, 2021), breast (Goel et al., 2013), bladder (Shin et al., 2014) and gastric (Ertao et al., 2016). Additionally, PTCH1 and GLI were also found to be highly expressed in some tumors.

In another Hh-ligand dependent signaling, tumor microenvironment plays a major role. Here, Hh pathway is activated in a paracrine mechanism where released Hh-ligand from tumor cells can bind to PTCH of stromal cells and activate their hedgehog pathway. The activated stromal cells then secrete growth factors such as vascular endothelial growth factor (VEGF), insulin-like growth factor (IGF), interleukin-6 (IL-6), Wnt, Platelet-derived growth factor (PDGF) and Bone morphogenetic proteins (BMP) which promote tumor growth (Amakye, Jagani, & Dorsch, 2013). This type of Hh- signaling activation is observed in many cancer types such as pancreatic (Tian et al., 2009), colorectal (Gerling et al., 2016) and some esophageal cancers (Ma et al., 2006). In prostate cancer, tumor epithelial cells were observed to produce SHH ligand while GLI1 was observed to be expressed in stromal cells supporting the paracrine mechanism of activation of Hh target genes and promote tumor growth (Fan et al., 2004). Similarly in esophageal cancer, SHH was expressed in tumor cells while PTCH1 and GLI1 were expressed in both tumor and stroma (Ma et al., 2006). Interestingly, in some paradoxical reports on pancreatic ductal adenocarcinoma (PDAC), inhibiting Hh ligand increases tumor progression. It was also observed that deleting SHH in pancreatic epithelium or SMO deletion/pharmacological inhibition in fibroblasts leads to increase in PDAC progression and decrease survival suggesting involvement of fibroblast heterogeneity in tumor microenvironment (Lee et al., 2014; Liu et al., 2016; Rhim et al., 2014).

Lastly, in another mechanism of Hh-ligand dependent reverse paracrine signaling, it was shown the Hh-ligand could be released by stromal cells, which activate the tumor cells promoting their growth. This was observed in B-cell and plasma cell malignancies, where Hh-ligands from bone marrow and splenic stroma could activate Hh signaling in tumor cells (Blotta et al., 2012; Dierks et al., 2007). This was also observed in gliomas where SHH produced from microenvironments or neurospheres were required for GLI activation in tumors (Becher et al., 2008). Similar evidence was observed in colorectal cancer stem cells where macrophages conditioned media could promote tumor growth by activating Hh-signaling in tumors (Fan et al., 2018).

## 2.2 Hedgehog signaling in metastatic tumors

Being involved in developmental pathways including cell proliferation and cellular differentiation, role of hedgehog signaling in regulating epithelial-mesenchymal transition in tumors is expected. Numerous studies have shown hedgehog signaling to be implicated in tumorigenesis. Some have also shown its association with poor prognosis in breast (Bièche et al., 2004; Jeng et al., 2013), colon (Xu, Li, Liu, Leng, & Zhang, 2012), glioma (Li et al., 2011), bladder (Mohd Ariffin et al., 2021) and prostate cancers (Kim, Lee, Hwang, Kang, & Choi, 2011). For instance, SHH expression was correlated with clinically aggressive stage and metastasis in retinoblastoma (Choe et al., 2015). Importance of Hh-signaling in tumor invasiveness and migration was also evident from studies on human renal cell carcinoma xenograft model, where blocking the pathway by SMO inhibitor, NVP-LDE225 (Erismodegib), inhibited cell proliferation, migration and metastasis to lung tissues (D'amato et al., 2014).

Patients with triple negative breast cancer showed significant association of activated hedgehog pathway with advanced tumor stages and poor prognosis. SHH, DHH and GLI were also reported to be over-expressed in triple-negative and luminal B subtype of breast cancer patients with increase in tumor size, nodal spread and distant metastasis. Aforementioned components of Hh-signaling positively correlated with Ki67 suggesting their role in cell proliferation in breast cancer. Another component of Hh-signaling, SMO showed limited clinical relevance in the same cohort with no significant association with distant metastasis. On the contrary, GLI1 inhibitor, GANT61 showed reduced proliferation in both MCF-7 (ER +ve) and MDA-MB-231 (ER −ve) cell lines and decreased transcriptional activation of both PTCH1 and SHH alongside GLI1. GANT61 was also able to reduce tumor cell invasion and motility (Benvenuto et al., 2016; Riaz et al., 2018). In basal-like breast cancer, FOXC1 was found to induce SMO-independent non-canonical Hh signaling by activating GLI2 activation that induced stemness in tumors (Han, Qu, Yu-Rice, Johnson, & Cui, 2016). In another report on ER +ve breast cancer, ER expression correlated with Gli1 and ALDH1. Estrogen could induce Gli1 expression and promoted cancer stem cell renewal, invasiveness and EMT in ER +/Gli1+ cells but not in Gli1-knockdown cells (Sun et al., 2014).

In glioblastoma multiforme, increased mTORC2 activity increased the expression of Hh-pathway components (GLI1, GLI2 and PTCH1) alongwith the expression of its target genes (Cyclin D1, Cyclin D2, Cyclin E, Snail, Slug and VEGF). This was corroborated with increase in increased metastasis, angiogenesis, cellular proliferation and stem cell regeneration. Inhibiting mTORC2 could inactivate Hh-signaling and the associated processes including cancer cell stemness suggesting their crosstalk during cancer progression. mTORC2 also inactivated GSK3β which inhibited GLI2 ubiquitinylation, promoting its stability and nuclear translocation (Maiti, Mondal, Satyavarapu, & Mandal, 2017).

Hh pathway is also known to regulate progression in osteosarcoma where it is involved in the metastasis of tumor cells to lung tissues. A target gene of GLI2 and a marker of invasive osteosarcoma, RPS3 (ribosomal protein 3) was found to be overexpressed in lung metastatic tumors of osteosarcoma compared to non-metastatic tumors (Nagao-Kitamoto et al., 2015). Degalactotigonin (DGT) is also reported to reduce cell proliferation and metastasis in osteosarcoma by inhibiting GLI1-mediated Hh signaling by inactivating GSK3β (Zhao et al., 2018). Genes involved in hedgehog signaling pathway and Wnt signaling pathway, such as IHH, WNT10B and TCF7 were significantly upregulated three metastatic osteosarcoma cell lines compared to the parental cell lines (Muff et al., 2015; Yao et al., 2018).

In hepatocellular carcinomas (HCC), expression of GLI1 correlated with not only with tumor grade, but also with tumor invasion, metastatic potential and expression of p-ERK1/2 and MMP9. Blocking Hh-signaling with cyclopamine could inhibit cell invasion and migration and downregulate the expression of GLI1, MMP9 and pERK1/2 proteins in Bel-7402, a hepatoma cell line. MAPK pathway inhibitors, U0126 and PD98059 was also had a similar effect on Bel-7402 and could inhibit SHH induced invasion and metastasis, although GLI1 expression remained unaltered (Lu, Zhao, He, & Wei, 2012). In an EMT network on HCC, SHH and WNT signaling pathways were shown to be activated by TGF-beta (Steinway et al., 2014).

In gastric cancer cells, recombinant SHH increased cell motility and invasiveness which was stalled upon treatment with SMO-inhibitor, cyclopamine or anti-SHH antibody. This SHH-mediated invasiveness was also reduced upon treatment with anti-TGF-beta antibody. SHH could also increase TGF-beta1 secretion, TGF-beta-mediated expression of activin receptor-like kinase (ALK) 5 protein and phosphorylation of Smad 3. ALK5 inhibitor could reverse the phosphorylation of Smad3, activity of MMP and cell motility and invasion induced by SHH. Similar findings were observed using Smad3 or ALK siRNAs suggesting that SHH induces cell motility and invasiveness in gastric cancer (Yoo, Kang, Kim, & Oh, 2008).

GLI1 and osteopontin (OPN), a direct transcriptional target of GLI1, expression was also observed to increase during the progression of melanoma from primary cutaneous cancer to metastatic melanoma. OPN expression was further decreased upon treatment with cyclopamine with inhibited cell proliferation, migration, and invasion of cancer cells in vitro. It also inhibited with tumor growth and metastasis in xenografts. These were rescued by overexpressing OPN in the GLI1-silenced cells (Das et al., 2009).

About 70% of prostate cancer with Gleason score 8–10, high levels of PTCH1 and hedgehog-interacting protein (HIP) was observed. This was also validated in 4 prostate metastatic tumors whereas SUFU, a negative regulator of Hh-signaling was not expressed in them. Hedgehog target genes were also expressed in prostate cancer cell lines (TSU, DU145, LN-Cap and PC3). Further, cyclopamine was able to inhibit cell invasiveness and induce apoptosis in these models (Karhadkar et al., 2004; Sheng et al., 2004). Additionally, long-term anti-androgen therapy in prostate cancer have shown to upregulate Hh signaling. SHH and DHH expression was increased in the epithelium of hormone treated prostate cancer (HTPC) compared to hormone naïve prostate cancer (HNPC). Similar results were observed in LNCaP, an androgen sensitive cell line, upon maintaining them in androgen deprivation conditions. DHH expression in epithelium was also found to be increased in hormone refractory prostate cancer (HRPC) compared to HNPC, while its expression in stoma was downregulated. In HNPC, SHH expression in the epithelial cells was significantly associated with high Gleason scores ($p = 0.03$), metastatic lymph nodes ($p = 0.004$) while DHH epithelium expression was found to be associated with high pT stages ($p = 0.003$), seminal vesicle invasion ($p = 0.03$) and bladder neck invasion ($p = 0.0008$). Absence of SHH in stromal cells was associated with high Gleason scores ($p = 0.015$), high pT stages ($p = 0.01$) and bladder neck invasion ($p = 0.04$) (Azoulay et al., 2008).

## 2.3 Relevance of hedgehog signaling in neuroendocrine tumors

Evidence in favor of hedgehog signaling involvement in various neuroendocrine tumors were reported in many studies. In a study on neuroendocrine tumors (NETs) of the ileum, Snail gene expression was reported in 59% of the invasive tumor samples and in 86% of liver metastatic tumors. Snail is a transcription factor known to repress E-cadherin and induces epithelial–mesenchymal transition. E-cadherin was further downregulated in Snail expressing cells. Interestingly, SHH was also reported to be co-expressed with Snail in 53% of primary NETS suggesting its role in invasion and metastasis (Fendrich et al., 2007). Hh signaling was also found to be upregulated and GLI1-dependent in gastrointestinal neuroendocrine carcinomas (NECs). GLI1 expression was found to be most in NECs compared to carcinoid tumors, adenocarcinomas or normal tissues. This was also validated NEC cell lines which showed high expression of GLI1. Blocking Hh signaling with cyclopamine lead to downregulation of GLI1, PTCH1, Snail and hASH1 (human achaete-scute homologue gene 1) and upregulated of E-cadherin in NECs but not in GLI1-negative adenocarcinomas. Cyclopamine suppressed both cell proliferation and invasion and induced apoptosis, suggesting its therapeutic potential in NECs (Shida et al., 2006).

Expression of INSM1, a neuroendocrine tumor-specific marker was shown to be upregulated by SHH signaling via NMyc and ASCL1 in small cell lung cancer. NMyc stability was also enhanced by INSM1 via PI3K/AKT and MEK/ERK pathway in neuroendocrine lung cancer. A combination of SHH signaling inhibitors that target INSM1 and NMyc was able to inhibit lung cancer cell growth (Chen, Breslin, & Lan, 2018). In another small cell lung carcinoma, SBC-5, GLI1 siRNA transfection led to a decrease in tropomyosin-related kinase B (TRKB). A combined knockdown of GLI1 and TRKB significantly decreased invasiveness, cell proliferation and migration in SBC-5 cells than when just one was inhibited. The results suggested that inhibiting Hh-signaling increases TRKB expression that counter tumor suppression activity in these cells (Onishi et al., 2017). In another study on large cell neuroendocrine carcinoma (LCNEC) of the lung, therapeutic potential of Hh signaling were assessed. Amongst the other components of signaling pathway, SMO inhibitor, BMS-833923 and GLI inhibitor, GANT61 showed reduced cell viability using MTT assay in three LCNEC lines, H460, H1299 and H810.

Reduced cell growth was also observed upon treating the three cell lines with GL1 and/or GLI2 siRNAs. Suppression of GLI1/2 made the cells more sensitive to cisplatin and increased apoptosis (Ishiwata et al., 2018).

In small intestine neuroendocrine xenograft model (GOT1), in vivo treatment with SMO inhibitor, Sonidegib (also known as erismodegib) resulted in reduced tumor growth while a significant decrease in tumor volume was observed when treated with $^{177}$Lu-octreotate alone or combined with Sonidegib. However, tumor progression was decreased in combination treatment compared to monotherapy. The results also showed several cancer-related signaling pathways (i.e., Wnt/β-catenin, PI3K/AKT/mTOR, G-protein coupled receptor, and Notch) to be affected in combination therapy (Spetz et al., 2017).

Hh signaling was also known to be relevant in pancreatic neuroendocrine tumors (PNETs) which can arise both sporadically and in association with multiple endocrine neoplasia type 1 (MEN-1), a hereditary syndrome. In both these cases, Hh signaling was found to be upregulated and inhibiting the pathway reduced tumor proliferation in MEN-1. PTCH1 showed positive IHC staining in 55% of PNETs and in 80% of MEN-1 patients. Patients with metastasis showed increased staining of PTCH1 than with localized disease (Gurung et al., 2015).

## 2.4 Clinical prevalence of hedgehog signaling in different cancers

Hedgehog signaling components have been found to have mutation, structural variation and copy number alteration in multiple cancer types. Upon querying the clinical samples from three major publicly available cohorts in cBioportal, Pan-cancer analysis of whole genomes (ICGC/TCGA, Nature 2020), Metastatic Solid Cancers (UMich, Nature 2017) and MSK-IMPACT Clinical Sequencing Cohort (MSK, Nat Med 2017), 9% of overall clinical samples (n = 1312) were found to have altered hedgehog signaling components. These samples were queried for mutation, structural variation and copy number alteration data in SHH, IHH, DHH, PTCH1, PTCH2, GLI1, GLI2, GLI3, SMO and SUFU genes.

Specifically, most alterations were found in melanoma (18.33%; n = 22) where 15.83% of samples (n = 19) had gene amplification while the rest 2.5% had multiple alterations in at least one of the hedgehog signaling components. Interestingly, these genes mostly had amplification alterations across most cancer types whereas deep deletions or mutations were observed in < 4% of samples for only few such as colorectal cancer, non-

Table 2 Alterations of hedgehog signaling components in clinical samples.

| Gene symbol | Number of altered samples | % altered samples |
|---|---|---|
| GLI3 | 246 | 8 |
| SHH | 183 | 6 |
| GLI1 | 397 | 4 |
| GLI2 | 110 | 3 |
| SMO | 354 | 3 |
| PTCH1 | 354 | 3 |
| DHH | 73 | 2 |
| PTCH2 | 69 | 2 |
| IHH | 44 | 1 |
| SUFU | 125 | <1 |

small cell lung cancer, soft tissue sarcoma and bladder cancer. Other cancer types such as ovarian cancer (12.2%; n = 15), lung cancer (10.53%; n = 4) and renal cell carcinoma (9.79%; n = 19) were found to have only gene amplifications of hedgehog signaling components.

The most altered gene of the hedgehog pathway was shown to be GLI3 with collective alterations in 8% of the samples (n = 246) across all cancers. This was followed by SHH, altered in 6% (n = 183) and GLI1 altered in 4% (n = 397) of all cancers (Table 2).

## 3. Therapeutic relevance of hedgehog signaling in cancer

### 3.1 Targeting hedgehog signaling components

Targeting hedgehog signaling pathway has been considered as a treatment for multiple cancer types. Hedgehog signaling pathway inhibitors can target at different levels and are broadly classified as Hh ligand inhibitors, SMO antagonists, GLI inhibitors, inhibitors of bromodomain and extra-terminal domain (BET) family of proteins, atypical protein kinase C (aPKC) inhibitors, and phosphodiesterase inhibitors (Xu, Song, Wang, & Ajani, 2019) (Table 3). Hedgehog ligand inhibitors such as neutralizing antibodies (5E1)

Table 3 Inhibitors of hedgehog signaling pathway and its therapeutic relevance in cancer.

| Target | Inhibitor | Cancer | References | Clinical Status & organization |
|---|---|---|---|---|
| Hh ligand | 5E1 | Gastric Cancer | Maun et al. (2010) | |
| SMO | Vismodegib (GDC-0449) | Basal cell carcinoma | O'Kane et al. (2014) | NCT01201915 (Recruiting)- Phase 2; Genentech, Inc.NCT00833417 (completed)- Phase 2; Genentech, Inc. |
| | | Medulloblastoma | Kian et al. (2020) | NCT01878617 (Recruiting)- Phase 2; St. Jude Children's Research Hospital |
| | | Castration-resistant prostate cancer | Ishii et al. (2020) | NCT02115828 (Completed)- Early Phase 1; Sidney Kimmel Comprehensive Cancer Center at Johns Hopkins |
| | | Hepatocellular carcinoma | Pinter et al. (2013) | |
| | | Gastric cancer | Yoon et al. (2014) | NCT00982592 (Completed)- Phase 2; National Cancer Institute |
| | Cyclopamine | Gastric cancer | Jo, Kim, Yoo, and Lee (2018), Yoo et al. (2008) | |
| | | Hepatocellular carcinoma | Steinway et al. (2014) | |
| | | Melanoma | Das et al. (2009) | |

*(continued)*

**Table 3** Inhibitors of hedgehog signaling pathway and its therapeutic relevance in cancer. (cont'd)

| Target | Inhibitor | Cancer | References | Clinical Status & organization |
|---|---|---|---|---|
| | | Prostate cancer | Sheng et al. (2004); Karhadkar et al. (2004) | |
| | | Small cell lung cancer | Szczepny et al. (2017) | |
| | | Gastrointestinal neuroendocrine cancer | Shida et al. (2006) | |
| | Sonidegib (LDE225, erismodegib) | Gastrointestinal neuroendocrine cancer | Spetz et al. (2017) | NCT04007744 (Recruiting)- Phase 1; Mayo Clinic |
| | | Renal cell carcinoma | D'Amato et al. (2014) | |
| | IPI-926 (saridegib) | Gastric cancer | Ma et al. (2017) | |
| | | Pancreatic cancer | Olive et al. (2009) | NCT01130142 (Completed)-Phase 1; Infinity Pharmaceuticals, Inc. |
| | | Head and neck cancer | Keysar et al. (2013) | NCT01255800 (Completed)-Phase 1; University of Colorado, Denver |

| | | | |
|---|---|---|---|
| GLI1/2 | GANT61 | Breast cancer | Riaz et al. (2018), Sun et al. (2014), Benvenuto et al. (2016) |
| | | Large cell lung neuroendocrine | Ishiwata et al. (2018) |
| | | Gastric cancer | Xu et al. (2019), Yan et al. (2013) |
| BRD4 | JQ1 | Medulloblastoma | Tang et al. (2014) |

can inhibit the binding of Hh ligand to PTCH receptor by binding at the SHH pseudo-active site groove (Maun et al., 2010). Inhibiting Hh signaling using 5E1 can abrogate the proliferative response of mesenchymal stem cells to cytokine interferon-gamma which would otherwise, be recruited from bone marrow and promote tumor nice and gastric cancer progression (Donnelly, Chawla, Houghton, & Zavros, 2013).

One of the commonly used SMO inhibitors, cyclopamine is a teratogenic plant-derived alkaloid, extracted from corn lilies (Cooper, Porter, Young, & Beachy, 1998). Cyclopamine binds to heptahelical bundle of SMO and inactivates it abrogating Hh-signaling and tumor growth both in vitro and in vivo (Chen, Taipale, Cooper, & Beachy, 2002; Lu et al., 2012; Szczepny et al., 2017; Yoo et al., 2008). However, its off-target effects and toxicity make it unsuitable for therapeutic use (Lipinski et al., 2008). IPI-926 (saridegib), a derivative of cyclopamine, have shown better results in addressing chemoresistance and inhibiting tumor growth. It has shown been included in phase I and phase II clinical trials for few cancer types (Keysar et al., 2013; Ma, Tian, & Yu, 2017; Olive et al., 2009). Another inhibitor of SMO, GDC-0449 (vismodegib), suppresses Hh-signaling by binding to the extracellular domain of SMO. It has shown improved results in suppressing tumor growth both in vitro and in vivo and has been the first FDA approved Hh signaling inhibitor for local and metastatic basal cell carcinoma (O'kane et al., 2014). GDC-0049 have also been used in combination with chemotherapy in phase II trial for gastric cancer (NCT00982592). Although no significant difference was observed in survival for gastric cancer patients treated with GDC-0449 in combination with chemotherapy, patients with higher CD44 expression showed improved survival upon combination therapy than with just chemotherapy (Yoon et al., 2014). In endocrine resistant breast cancer cells, treatment with PI3K inhibitor showed decreased expression of SMO and GLI1 which could be rescued by blocking GSK3β mediated proteasomal degradation. Additionally, targeting Hh signaling by GDC-0449 in tamoxifen-resistant breast cancer xenografts reduced tumor growth, suggesting a possible link between PI3K/Akt and Hh signaling that leads to tamoxifen resistance (Ramaswamy et al., 2012). GDC-0449, has also shown to significantly reduce cell proliferation and promote apoptosis in all castration-resistant prostate cancer cell lines in vitro ($p < 0.05$). It also significantly inhibited EMT in these cells and reduced tumor growth in C4–2B xenograft model in vivo ($p < 0.05$) (Ishii et al., 2020).

Other SMO inhibitors such as vitamin D3 and Itraconazole have shown to regulate cell viability and induced apoptosis and cell cycle arrest in gastric cancer cells by inhibiting Hh signaling (Baek et al., 2011;

Hu, Hou, Huang, Fang, & Xiong, 2017). Another SMO inhibitor, Sonidegib (also known as LDE225 or erismodegib), as mentioned before, have shown reduced tumor growth in small intestine neuroendocrine tumor and renal cell carcinoma (D'amato et al., 2014; Spetz et al., 2017).

Some GLI inhibitors (GANT58 and GANT61) have also shown therapeutic potential in few cancer types. GANT61, known to bind zinc fingers 2 and 3 in GLI1 can inhibit GLI1 binding to DNA (Agyeman, Jha, Mazumdar, & Houghton, 2014). Combining cyclopamine with GANT61 have shown to repress cell growth in gastric cancer cells (Yan et al., 2013). GANT61 also showed increased apoptosis in CD44 positive gastric cancer cells (Xu et al., 2015). As previously described, GANT61 can reduce cell motility and alter EMT in breast cancer (Benvenuto et al., 2016; Riaz et al., 2018; Sun et al., 2014). In combination with SMO inhibitor, BMS-833923 and GLI inhibitor, GANT61 showed reduced cell viability in lung large cell neuroendocrine carcinoma (Ishiwata et al., 2018).

## 3.2 Combination with epigenetic inhibitors

Hedgehog pathway has been reported to reprogram histone methylation in triple negative breast cancer, mediated through overexpression of ZIC1 which promotes Hh-activation and facilitates the switch from H3K27me3 to H3K27ac (Wang et al., 2023b). In another study, it was shown that Shh-target genes were poised by bivalent H3K27me3 and H3K4me3 marks. Shh activation leads to a Gli-dependent epigenetic switch from PRC2 to Jmjd3-Kdm6b complex, activating the Shh target genes. Further, Jmjd3 inactivation showed reduced growth in Shh-subtype dependent medulloblastoma (Shi et al., 2014).

There also exist indirect inhibitors of GLI that show reduced growth in Hh-dependent tumors. One such example is BET (bromodomain and extra-terminal domain) inhibitor, JQ1. BET proteins regulate transcription by binding to chromatin or forming complexes with transcriptional regulators including RNA polymerase II and elongation factor b (Jang et al., 2005; Tang et al., 2014). It was also shown that a BET family member, BRD4 could also bind to the promoter of GLI1 and GLI2 and regulate their transcription (Tang et al., 2014). Therefore, JQ1 could abrogate Hh signaling by inhibiting BRD4 from binding to GLI1/2. These evidences together suggest possible therapeutic vulnerabilities as combination therapies with epigenetic drugs.

Interestingly, a few studies on combination therapy with epigenetic inhibitors and Hh-inhibitors have been conducted. In one such study, erismodegib combined with 5-Aza, DNMT inhibitor showed synergistic effect in acute myeloid leukemia (AML) and myelodysplastic syndrome (MDS) with reduced clonogenic growth than either of the single agent. The study also showed concurrent dosing to have a stronger synergy than sequential dosing of both agents (Tibes et al., 2015). Based on this phase 1 trial, NCT02129101 was conducted.

Another study reported decrease in DNMT1, DNMT3A and DNMT3B expression upon treatment with Hh inhibitors in leiomyosarcoma cells (LMS). Combination treatment using 5-Aza and GANT61 also showed synergy by inhibiting Hh-pathway and affecting the cell proliferation and migration in LMS cells (Garcia, Al-Hendy, Baracat, Carvalho, & Yang, 2021). Similarly in liver cancer, combined treatment with SMO and histone deacetylase (HDAC) inhibitors showed reduced cell viability, colony formation and increased apoptosis (Li et al., 2019). Increased cell cycle arrest and delayed in vivo tumor growth was also observed in aerodigestive cancer treated with combined SMO and HDAC inhibitors than single agent (Chun et al., 2015).

## 4. Perspective

Both canonical and non-canonical hedgehog signaling pathway have been proven to regulate many cellular processes such as embryonic development, cell differentiation, proliferation, regeneration and maintenance of stem cells. Relevance of hedgehog signaling in tumorigenesis have also been elucidated in many in vitro and in vivo studies. These studies highlight on how hedgehog signaling pathway adopts different mechanisms (mutation-driven, autocrine, paracrine and reverse paracrine) to regulate tumor growth and metastasis in both local and advanced cancer types. Inhibiting hedgehog signaling by Hh-ligand, SMO or GLI antagonists showed reduced cell proliferation, viability, invasion and migration in various cancer cell lines and preclinical models. This suggests the therapeutic potential of blocking hedgehog signaling pathway in these cancers. However, use of these inhibitors in clinical trials have not shown as much promising results. This may be due to unexplained underlying factors which lead to treatment resistance.

One of the limitations of using hedgehog signaling inhibitors as therapeutic strategies in cancer is associated with their drug resistance. Although, Smo inhibitors vismodegib and sonidegib have been FDA approved, increased resistance to vismodegib has been a major clinical concern (Danial, Sarin, Oro, & Chang, 2016). In a prior study, 50% of the patients were reported to show primary resistance while 20% showed secondary resistance where recurrence occurred after initial treatment (Chang & Oro, 2012). SMO mutation, G497W have been associated with primary resistance while SMO D473Y and PTCH1 mutations have been shown to be linked with secondary resistance. To overcome these resistance, taladegib, LEQ-506 and TAK-441 have been now developed and needs further clinical validation. Arsenic trioxide has also been shown to bypass resistance due to SMO mutation. Apart from SMO mutations, the gain-of-function in GLI and loss-of-function in SUFU may also lead to the acquired resistance (Wu et al., 2017).

Another such factor can be the crosstalk between other signaling pathways which allow the tumor to bypass growth suppression including upregulation of PI3K-mTOR signaling, aPKC-ι/λ activation, BRD4 activation, and PDE4 activation (Atwood, Li, Lee, Tang, & Oro, 2013; Buonamici et al., 2010). Combined treatment of sonidegib with buparlisib or dactolisib could overcome the upregulation of PI3K-mTOR signaling and was further clinically evaluated in NCT01576666 and NCT02303041 trials. Hedgehog signaling components has been shown to interact with many other pathways that can regulate proliferation and confer resistance such as PI3K/AKT signaling, MAPK signaling and Wnt signaling. Additionally, combination therapy with more than one hedgehog inhibitor or a hedgehog inhibitor with inhibitors of other pathways have shown better results. Some clinical trials also focus on combining epigenetic inhibitors, chemotherapy or radiation therapy with Hh signaling blockers. These evidences together demand further understanding of the crosstalk between different signaling pathways to address treatment resistance.

Another such unexplored factor can be tumor heterogeneity which is not well studied in relation to Hh signaling. A previous study has shown how combining chemotherapy with Hh-signaling blocker did not show expected results in clinical trial. However, stratifying the patient cohort based on CD44 expression showed difference in survival in the combination group. Overall, this chapter highlights the potential avenues which can be explored to better target hedgehog signaling pathways in both localized and advanced cancers.

## References

Abbasi, A. A., Goode, D. K., Amir, S., & Grzeschik, K. H. (2009). Evolution and functional diversification of the GLI family of transcription factors in vertebrates. *Evolutionary Bioinformatics Online*, 5, 5–13.

Aberger, F., & Ruiz I Altaba, A. (2014). Context-dependent signal integration by the GLI code: The oncogenic load, pathways, modifiers and implications for cancer therapy. *Seminars in Cell & Developmental Biology, 33*, 93–104.
Agyeman, A., Jha, B. K., Mazumdar, T., & Houghton, J. A. (2014). Mode and specificity of binding of the small molecule GANT61 to GLI determines inhibition of GLI-DNA binding. *Oncotarget, 5*, 4492–4503.
Ahn, S., & Joyner, A. L. (2005). In vivo analysis of quiescent adult neural stem cells responding to Sonic hedgehog. *Nature, 437*, 894–897.
Amakye, D., Jagani, Z., & Dorsch, M. (2013). Unraveling the therapeutic potential of the Hedgehog pathway in cancer. *Nature Medicine, 19*, 1410–1422.
Atwood, S. X., Li, M. C., Lee, A., Tang, J. Y., & Oro, A. E. (2013). GLI activation by atypical protein kinase C $\iota/\lambda$ regulates the growth of basal cell carcinomas. *Nature, 494*, 484–488.
Azoulay, S., Terry, S., Chimingqi, M., Sirab, N., Faucon, H., Diez De Medina, G. I. L., ... Allory, Y. (2008). Comparative expression of Hedgehog ligands at different stages of prostate carcinoma progression. *The Journal of Pathology, 216*, 460–470.
Baek, S., Lee, Y. S., Shim, H. E., Yoon, S., Baek, S. Y., Kim, B. S., & Oh, S. O. (2011). Vitamin D3 regulates cell viability in gastric cancer and cholangiocarcinoma. *Anatomy & Cell Biology, 44*, 204–209.
Barnes, E. A., Kong, M., Ollendorff, V., & Donoghue, D. J. (2001). Patched1 interacts with cyclin B1 to regulate cell cycle progression. *The EMBO Journal, 20*, 2214–2223.
Beachy, P. A., Karhadkar, S. S., & Berman, D. M. (2004). Tissue repair and stem cell renewal in carcinogenesis. *Nature, 432*, 324–331.
Becher, O. J., Hambardzumyan, D., Fomchenko, E. I., Momota, H., Mainwaring, L., Bleau, A. M., ... Holland, E. C. (2008). Gli activity correlates with tumor grade in platelet-derived growth factor-induced gliomas. *Cancer Research, 68*, 2241–2249.
Belgacem, Y. H., & Borodinsky, L. N. (2011). Sonic hedgehog signaling is decoded by calcium spike activity in the developing spinal cord. *Proceedings of the National Academy of Sciences of the United States of America, 108*, 4482–4487.
Benvenuto, M., Masuelli, L., Smaele, D. E., Fantini, E., Mattera, M., Cucchi, R., ... Bei, R. (2016). In vitro and in vivo inhibition of breast cancer cell growth by targeting the Hedgehog/GLI pathway with SMO (GDC-0449) or GLI (GANT-61) inhibitors. *Oncotarget, 7*, 9250–9270.
Berman, D. M., Karhadkar, S. S., Maitra, A., Montes De Oca, R., Gerstenblith, M. R., Briggs, K., ... Beachy, P. A. (2003). Widespread requirement for Hedgehog ligand stimulation in growth of digestive tract tumours. *Nature, 425*, 846–851.
Bièche, I., Lerebours, F., Tozlu, S., Espie, M., Marty, M., & Lidereau, R. (2004). Molecular profiling of inflammatory breast cancer: Identification of a poor-prognosis gene expression signature. *Clinical Cancer Research: An Official Journal of the American Association for Cancer Research, 10*, 6789–6795.
Blotta, S., Jakubikova, J., Calimeri, T., Roccaro, A. M., Amodio, N., Azab, A. K., ... Munshi, N. C. (2012). Canonical and noncanonical Hedgehog pathway in the pathogenesis of multiple myeloma. *Blood, 120*, 5002–5013.
Briscoe, J., & Thérond, P. P. (2013). The mechanisms of Hedgehog signalling and its roles in development and disease. *Nature Reviews. Molecular Cell Biology, 14*, 416–429.
Buonamici, S., Williams, J., Morrissey, M., Wang, A. L., Guo, R. B., Vattay, A., ... Dorsch, M. (2010). Interfering with resistance to smoothened antagonists by inhibition of the PI3K pathway in medulloblastoma. *Science Translational Medicine, 2*.
Carpenter, D., Stone, D. M., Brush, J., Ryan, A., Armanini, M., Frantz, G., ... De Sauvage, F. J. (1998). Characterization of two patched receptors for the vertebrate hedgehog protein family. *Proceedings of the National Academy of Sciences of the United States of America, 95*, 13630–13634.

Chang & Oro. (2012). Initial assessment of tumor regrowth after vismodegib in advanced basal cell carcinoma. *Archives of Dermatology, 148*, 1376.

Charron, F., & Tessier-Lavigne, M. (2005). Novel brain wiring functions for classical morphogens: A role as graded positional cues in axon guidance. *Development (Cambridge, England), 132*, 2251–2262.

Chen, C., Breslin, M. B., & Lan, M. S. (2018). Sonic hedgehog signaling pathway promotes INSM1 transcription factor in neuroendocrine lung cancer. *Cellular Signalling, 46*, 83–91.

Chen, J. K., Taipale, J., Cooper, M. K., & Beachy, P. A. (2002). Inhibition of hedgehog signaling by direct binding of cyclopamine to smoothened. *Genes & Development, 16*, 2743–2748.

Chen, Y., & Struhl, G. (1998). In vivo evidence that patched and smoothened constitute distinct binding and transducing components of a hedgehog receptor complex. *Development (Cambridge, England), 125*, 4943–4948.

Cheung, H. O., Zhang, X., Ribeiro, A., Mo, R., Makino, S., Puviindran, V., ... Hui, C. C. (2009). The kinesin protein Kif7 is a critical regulator of Gli transcription factors in mammalian hedgehog signaling. *Science Signaling, 2*, ra29.

Chinchilla, P., Xiao, L., Kazanietz, M. G., & Riobo, N. A. (2010). Hedgehog proteins activate pro-angiogenic responses in endothelial cells through non-canonical signaling pathways. *Cell Cycle (Georgetown, Tex.), 9*, 570–579.

Choe, J. Y., Yun, J. Y., Jeon, Y. K., Kim, S. H., Choung, H. K., Oh, S., ... Kim, J. E. (2015). Sonic hedgehog signalling proteins are frequently expressed in retinoblastoma and are associated with aggressive clinicopathological features. *Journal of Clinical Pathology, 68*, 6–11.

Chun, S. G., Park, H., Pandita, R. K., Horikoshi, N., Pandita, T. K., Schwartz, D. L., & Yordy, J. S. (2015). Targeted inhibition of histone deacetylases and hedgehog signaling suppress tumor growth and homologous recombination in aerodigestive cancers. *American Journal of Cancer Research, 5*, 1337–1352.

Cohen, M., Kicheva, A., Ribeiro, A., Blassberg, R., Page, K. M., Barnes, C. P., & Briscoe, J. (2015). Ptch1 and Gli regulate Shh signalling dynamics via multiple mechanisms. *Nature Communications, 6*, 6709.

Cooper, M. K., Porter, J. A., Young, K. E., & Beachy, P. A. (1998). Teratogen-mediated inhibition of target tissue response to Shh signaling. *Science (New York, N. Y.), 280*, 1603–1607.

D'amato, C., Rosa, R., Marciano, R., D'amato, V., Formisano, L., Nappi, L., ... Bianco, R. (2014). Inhibition of Hedgehog signalling by NVP-LDE225 (Erismodegib) interferes with growth and invasion of human renal cell carcinoma cells. *British Journal of Cancer, 111*, 1168–1179.

Danial, C., Sarin, K. Y., Oro, A. E., & Chang, A. L. S. (2016). An investigator-initiated open-label trial of sonidegib in advanced basal cell carcinoma patients resistant to vismodegib. *Clinical Cancer Research, 22*, 1325–1329.

Das, S., Harris, L. G., Metge, B. J., Liu, S., Riker, A. I., Samant, R. S., & Shevde, L. A. (2009). The hedgehog pathway transcription factor GLI1 promotes malignant behavior of cancer cells by up-regulating osteopontin. *The Journal of Biological Chemistry, 284*, 22888–22897.

Dierks, C., Grbic, J., Zirlik, K., Beigi, R., Englund, N. P., Guo, G. R., ... Warmuth, M. (2007). Essential role of stromally induced hedgehog signaling in B-cell malignancies. *Nature Medicine, 13*, 944–951.

Donnelly, J. M., Chawla, A., Houghton, J., & Zavros, Y. (2013). Sonic hedgehog mediates the proliferation and recruitment of transformed mesenchymal stem cells to the stomach. *PLoS One, 8*, e75225.

Echelard, Y., Epstein, D. J., St-Jacques, B., Shen, L., Mohler, J., Mcmahon, J. A., & Mcmahon, A. P. (1993). Sonic hedgehog, a member of a family of putative signaling molecules, is implicated in the regulation of CNS polarity. *Cell, 75*, 1417–1430.

Endoh-Yamagami, S., Evangelista, M., Wilson, D., Wen, X., Theunissen, J. W., Phamluong, K., ... Peterson, A. S. (2009). The mammalian Cos2 homolog Kif7 plays an essential role in modulating Hh signal transduction during development. *Current Biology: CB, 19*, 1320–1326.

Ertao, Z., Jianhui, C., Chuangqi, C., Changjiang, Q., Sile, C., Yulong, H., ... Shirong, C. (2016). Autocrine Sonic hedgehog signaling promotes gastric cancer proliferation through induction of phospholipase C$\gamma$1 and the ERK1/2 pathway. *Journal of Experimental & Clinical Cancer Research: CR, 35*, 63.

Fan, F., Wang, R., Boulbes, D. R., Zhang, H., Watowich, S. S., Xia, L., ... Ellis, L. M. (2018). Macrophage conditioned medium promotes colorectal cancer stem cell phenotype via the hedgehog signaling pathway. *PLoS One, 13*, e0190070.

Fan, L., Pepicelli, C. V., Dibble, C. C., Catbagan, W., Zarycki, J. L., Laciak, R., ... Bushman, W. (2004). Hedgehog signaling promotes prostate xenograft tumor growth. *Endocrinology, 145*, 3961–3970.

Fendrich, V., Waldmann, J., Esni, F., Ramaswamy, A., Mullendore, M., Buchholz, M., ... Feldmann, G. (2007). Snail and Sonic Hedgehog activation in neuroendocrine tumors of the ileum. *Endocrine-Related Cancer, 14*, 865–874.

Garcia, N., Al-Hendy, A., Baracat, E. C., Carvalho, K. C., & Yang, Q. W. (2021). Targeting hedgehog pathway and DNA methyltransferases in uterine leiomyosarcoma cells. *Cells, 10*.

Gerling, M., Büller, N. V., Kirn, L. M., Joost, S., Frings, O., Englert, B., ... Toftgård, R. (2016). Stromal Hedgehog signalling is downregulated in colon cancer and its restoration restrains tumour growth. *Nature Communications, 7*, 12321.

Geyer, N., & Gerling, M. (2021). Hedgehog signaling in colorectal cancer: All in the stroma? *International Journal of Molecular Sciences, 22*.

Goel, H. L., Pursell, B., Chang, C., Shaw, L. M., Mao, J., Simin, K., ... Mercurio, A. M. (2013). GLI1 regulates a novel neuropilin-2/$\alpha$6$\beta$1 integrin based autocrine pathway that contributes to breast cancer initiation. *EMBO Molecular Medicine, 5*, 488–508.

Gurung, B., Hua, Y., Runske, M., Bennett, B., Livolsi, V., Roses, R., ... Metz, D. C. (2015). PTCH 1 staining of pancreatic neuroendocrine tumor (PNET) samples from patients with and without multiple endocrine neoplasia (MEN-1) syndrome reveals a potential therapeutic target. *Cancer Biology & Therapy, 16*, 219–224.

Han, B., Qu, Y., Yu-Rice, Y., Johnson, J., & Cui, X. (2016). FOXC1-induced Gli2 activation: A non-canonical pathway contributing to stemness and anti-Hedgehog resistance in basal-like breast cancer. *Molecular & Cellular Oncology, 3*, e1131668.

Hu, Q., Hou, Y. C., Huang, J., Fang, J. Y., & Xiong, H. (2017). Itraconazole induces apoptosis and cell cycle arrest via inhibiting Hedgehog signaling in gastric cancer cells. *Journal of Experimental & Clinical Cancer Research: CR, 36*, 50.

Hui, C. C., & Angers, S. (2011). Gli proteins in development and disease. *Annual Review of Cell and Developmental Biology, 27*, 513–537.

Humke, E. W., Dorn, K. V., Milenkovic, L., Scott, M. P., & Rohatgi, R. (2010). The output of Hedgehog signaling is controlled by the dynamic association between suppressor of fused and the gli proteins. *Genes & Development, 24*, 670–682.

Ingham, P. W., & Mcmahon, A. P. (2001). Hedgehog signaling in animal development: Paradigms and principles. *Genes & Development, 15*, 3059–3087.

Ishii, A., Shigemura, K., Kitagawa, K., Sung, S. Y., Chen, K. C., Yi-Te, C., ... Fujisawa, M. (2020). Anti-tumor effect of hedgehog signaling inhibitor, vismodegib, on castration-resistant prostate cancer. *Anticancer Research, 40*, 5107–5114.

Ishiwata, T., Iwasawa, S., Ebata, T., Fan, M., Tada, Y., Tatsumi, K., & Takiguchi, Y. (2018). Inhibition of Gli leads to antitumor growth and enhancement of cisplatin-induced cytotoxicity in large cell neuroendocrine carcinoma of the lung. *Oncology Reports, 39*, 1148–1154.

Jang, M. K., Mochizuki, K., Zhou, M., Jeong, H. S., Brady, J. N., & Ozato, K. (2005). The bromodomain protein Brd4 is a positive regulatory component of P-TEFb and stimulates RNA polymerase II-dependent transcription. *Molecular Cell, 19*, 523–534.

Jeng, K. S., Sheen, I. S., Jeng, W. J., Yu, M. C., Hsiau, H. I., & Chang, F. Y. (2013). High expression of Sonic Hedgehog signaling pathway genes indicates a risk of recurrence of breast carcinoma. *OncoTargets and Therapy, 7*, 79–86.

Jessell, T. M. (2000). Neuronal specification in the spinal cord: Inductive signals and transcriptional codes. *Nature Reviews. Genetics, 1*, 20–29.

Jia, J., Tong, C., Wang, B., Luo, L., & Jiang, J. (2004). Hedgehog signalling activity of Smoothened requires phosphorylation by protein kinase A and casein kinase I. *Nature, 432*, 1045–1050.

Jiang, X., Yang, P., & Ma, L. (2009). Kinase activity-independent regulation of cyclin pathway by GRK2 is essential for zebrafish early development. *Proceedings of the National Academy of Sciences of the United States of America, 106*, 10183–10188.

Jo, Y. S., Kim, M. S., Yoo, N. J., & Lee, S. H. (2018). Inactivating frameshift mutations of HACD4 and TCP10L tumor suppressor genes in colorectal and gastric cancers. *Pathology Oncology Research: POR*.

Kagawa, H., Shino, Y., Kobayashi, D., Demizu, S., Shimada, M., Ariga, H., & Kawahara, H. (2011). A novel signaling pathway mediated by the nuclear targeting of C-terminal fragments of mammalian Patched 1. *PLoS One, 6*, e18638.

Karhadkar, S. S., Bova, G. S., Abdallah, N., Dhara, S., Gardner, D., Maitra, A., ... Beachy, P. A. (2004). Hedgehog signalling in prostate regeneration, neoplasia and metastasis. *Nature, 431*, 707–712.

Kenney, A. M., Cole, M. D., & Rowitch, D. H. (2003). Nmyc upregulation by sonic hedgehog signaling promotes proliferation in developing cerebellar granule neuron precursors. *Development (Cambridge, England), 130*, 15–28.

Keysar, S. B., Le, P. N., Anderson, R. T., Morton, J. J., Bowles, D. W., Paylor, J. J., ... Jimeno, A. (2013). Hedgehog signaling alters reliance on EGF receptor signaling and mediates anti-EGFR therapeutic resistance in head and neck cancer. *Cancer Research, 73*, 3381–3392.

Kian, W., Roisman, L. C., Goldstein, I. M., Abo-Quider, A., Samueli, B., Wallach, N., ... Yakobson, A. (2020). Vismodegib as first-line treatment of mutated sonic hedgehog pathway in adult medulloblastoma. *JCO Precision Oncology, 4*.

Kim, T. J., Lee, J. Y., Hwang, T. K., Kang, C. S., & Choi, Y. J. (2011). Hedgehog signaling protein expression and its association with prognostic parameters in prostate cancer: A retrospective study from the view point of new 2010 anatomic stage/prognostic groups. *Journal of Surgical Oncology, 104*, 472–479.

Kinzler, K. W., Bigner, S. H., Bigner, D. D., Trent, J. M., Law, M. L., O'brien, S. J., ... Vogelstein, B. (1987). Identification of an amplified, highly expressed gene in a human glioma. *Science (New York, N. Y.), 236*, 70–73.

Kool, M., Jones, D. T., Jäger, N., Northcott, P. A., Pugh, T. J., Hovestadt, V., & Project, I. P. T. (2014). Genome sequencing of SHH medulloblastoma predicts genotype-related response to smoothened inhibition. *Cancer Cell, 25*, 393–405.

Lai, K., Kaspar, B. K., Gage, F. H., & Schaffer, D. V. (2003). Sonic hedgehog regulates adult neural progenitor proliferation in vitro and in vivo. *Nature Neuroscience, 6*, 21–27.

Lee, J. J., Perera, R. M., Wang, H., Wu, D. C., Liu, X. S., Han, S., ... Beachy, P. A. (2014). Stromal response to Hedgehog signaling restrains pancreatic cancer progression. *Proceedings of the National Academy of Sciences of the United States of America, 111*, E3091–E3100.

Lee, J. J., Von Kessler, D. P., Parks, S., & Beachy, P. A. (1992). Secretion and localized transcription suggest a role in positional signaling for products of the segmentation gene hedgehog. *Cell, 71*, 33–50.

Li, J. M., Cai, H., Li, H. X., Liu, Y. G., Wang, Y. R., Shi, Y., ... Wang, D. (2019). Combined inhibition of sonic Hedgehog signaling and histone deacetylase is an effective treatment for liver cancer. *Oncology Reports, 41*, 1991–1997.

Li, Q., Zhang, Y., Zhan, H., Yuan, Z., Lu, P., Zhan, L., & Xu, W. (2011). The Hedgehog signalling pathway and its prognostic impact in human gliomas. *ANZ Journal of Surgery, 81*, 440–445.

Liem, K. F., He, M., Ocbina, P. J., & Anderson, K. V. (2009). Mouse Kif7/Costal2 is a cilia-associated protein that regulates Sonic hedgehog signaling. *Proceedings of the National Academy of Sciences of the United States of America, 106*, 13377–13382.

Lipinski, R. J., Hutson, P. R., Hannam, P. W., Nydza, R. J., Washington, I. M., Moore, R. W., ... Bushman, W. (2008). Dose- and route-dependent teratogenicity, toxicity, and pharmacokinetic profiles of the hedgehog signaling antagonist cyclopamine in the mouse. *Toxicological Sciences: An Official Journal of the Society of Toxicology, 104*, 189–197.

Liu, X., Pitarresi, J. R., Cuitiño, M. C., Kladney, R. D., Woelke, S. A., Sizemore, G. M., ... Ostrowski, M. C. (2016). Genetic ablation of Smoothened in pancreatic fibroblasts increases acinar-ductal metaplasia. *Genes & Development, 30*, 1943–1955.

Lu, J. T., Zhao, W. D., He, W., & Wei, W. (2012). Hedgehog signaling pathway mediates invasion and metastasis of hepatocellular carcinoma via ERK pathway. *Acta Pharmacologica Sinica, 33*, 691–700.

Ma, H., Tian, Y., & Yu, X. (2017). Targeting smoothened sensitizes gastric cancer to chemotherapy in experimental models. *Medical Science Monitor: International Medical Journal of Experimental and Clinical Research, 23*, 1493–1500.

Ma, X., Sheng, T., Zhang, Y., Zhang, X., He, J., Huang, S., ... Xie, J. (2006). Hedgehog signaling is activated in subsets of esophageal cancers. *International Journal of Cancer. Journal International du Cancer, 118*, 139–148.

Maesawa, C., Tamura, G., Iwaya, T., Ogasawara, S., Ishida, K., Sato, N., ... Satodate, R. (1998). Mutations in the human homologue of the Drosophila patched gene in esophageal squamous cell carcinoma. *Genes, Chromosomes & Cancer, 21*, 276–279.

Maiti, S., Mondal, S., Satyavarapu, E. M., & Mandal, C. (2017). mTORC2 regulates hedgehog pathway activity by promoting stability to Gli2 protein and its nuclear translocation. *Cell Death & Disease, 8*, e2926.

Mäkelä, J. A., Saario, V., Bourguiba-Hachemi, S., Nurmio, M., Jahnukainen, K., Parvinen, M., & Toppari, J. (2011). Hedgehog signalling promotes germ cell survival in the rat testis. *Reproduction (Cambridge, England), 142*, 711–721.

Mastronardi, F. G., Dimitroulakos, J., Kamel-Reid, S., & Manoukian, A. S. (2000). Co-localization of patched and activated sonic hedgehog to lysosomes in neurons. *Neuroreport, 11*, 581–585.

Maun, H. R., Wen, X., Lingel, A., De Sauvage, F. J., Lazarus, R. A., Scales, S. J., & Hymowitz, S. G. (2010). Hedgehog pathway antagonist 5E1 binds hedgehog at the pseudo-active site. *The Journal of Biological Chemistry, 285*, 26570–26580.

Mcgarvey, T. W., Maruta, Y., Tomaszewski, J. E., Linnenbach, A. J., & Malkowicz, S. B. (1998). PTCH gene mutations in invasive transitional cell carcinoma of the bladder. *Oncogene, 17*, 1167–1172.

Mille, F., Thibert, C., Fombonne, J., Rama, N., Guix, C., Hayashi, H., ... Mehlen, P. (2009). The Patched dependence receptor triggers apoptosis through a DRAL-caspase-9 complex. *Nature Cell Biology, 11*, 739–746.

Mohd Ariffin, K., Abd Ghani, F., Hussin, H., Md Said, S., Yunus, R., Veerakumarasivam, A., & Abdullah, M. A. (2021). Hedgehog signalling molecule, SMO is a poor prognostic marker in bladder cancer. *The Malaysian Journal of Pathology, 43*, 49–54.

Muff, R., Rath, P., Ram Kumar, R. M., Husmann, K., Born, W., Baudis, M., & Fuchs, B. (2015). Genomic instability of osteosarcoma cell lines in culture: Impact on the prediction of metastasis relevant genes. *PLoS One, 10*, e0125611.

Nagao-Kitamoto, H., Setoguchi, T., Kitamoto, S., Nakamura, S., Tsuru, A., Nagata, M., ... Komiya, S. (2015). Ribosomal protein S3 regulates GLI2-mediated osteosarcoma invasion. *Cancer Letters, 356*, 855–861.

Ng, J. M., & Curran, T. (2011). The Hedgehog's tale: Developing strategies for targeting cancer. *Nature Reviews. Cancer, 11*, 493–501.

Niyaz, M., Khan, M. S., Wani, R. A., Shah, O. J., & Mudassar, S. (2020). Sonic hedgehog protein is frequently up-regulated in pancreatic cancer compared to colorectal cancer. *Pathology Oncology Research: POR, 26*, 551–557.

Nüsslein-Volhard, C., & Wieschaus, E. (1980). Mutations affecting segment number and polarity in Drosophila. *Nature, 287*, 795–801.

O'kane, G. M., Lyons, T., Mcdonald, I., Mulligan, N., Moloney, F. J., Murray, D., & Kelly, C. M. (2014). Vismodegib in the treatment of advanced BCC. *Irish Medical Journal, 107*, 215–216.

Olive, K. P., Jacobetz, M. A., Davidson, C. J., Gopinathan, A., Mcintyre, D., Honess, D., ... Tuveson, D. A. (2009). Inhibition of Hedgehog signaling enhances delivery of chemotherapy in a mouse model of pancreatic cancer. *Science (New York, N. Y.), 324*, 1457–1461.

Onishi, H., Nakamura, K., Nagai, S., Yanai, K., Yamasaki, A., Kawamoto, M., ... Morisaki, T. (2017). Hedgehog inhibition upregulates TRK expression to antagonize tumor suppression in small cell lung cancer cells. *Anticancer Research, 37*, 4987–4992.

Pan, Y., Bai, C. B., Joyner, A. L., & Wang, B. (2006). Sonic hedgehog signaling regulates Gli2 transcriptional activity by suppressing its processing and degradation. *Molecular and Cellular Biology, 26*, 3365–3377.

Persson, M., Stamataki, D., Te Welscher, P., Andersson, E., Böse, J., Rüther, U., ... Briscoe, J. (2002). Dorsal-ventral patterning of the spinal cord requires Gli3 transcriptional repressor activity. *Genes & Development, 16*, 2865–2878.

Pinter, M., Sieghart, W., Schmid, M., Dauser, B., Prager, G., Dienes, H. P., ... Peck-Radosavljevic, M. (2013). Hedgehog inhibition reduces angiogenesis by downregulation of tumoral VEGF-A expression in hepatocellular carcinoma. *United European Gastroenterologisches Journal: Organ der Gesellschaft für Gastroenterologie der DDR, 1*, 265–275.

Polizio, A. H., Chinchilla, P., Chen, X., Kim, S., Manning, D. R., & Riobo, N. A. (2011). Heterotrimeric Gi proteins link Hedgehog signaling to activation of Rho small GTPases to promote fibroblast migration. *The Journal of Biological Chemistry, 286*, 19589–19596.

Rahnama, F., Toftgård, R., & Zaphiropoulos, P. G. (2004). Distinct roles of PTCH2 splice variants in Hedgehog signalling. *The Biochemical Journal, 378*, 325–334.

Ramaswamy, B., Lu, Y., Teng, K. Y., Nuovo, G., Li, X., Shapiro, C. L., & Majumder, S. (2012). Hedgehog signaling is a novel therapeutic target in tamoxifen-resistant breast cancer aberrantly activated by PI3K/AKT pathway. *Cancer Research, 72*, 5048–5059.

Reifenberger, J., Wolter, M., Knobbe, C. B., Köhler, B., Schönicke, A., Scharwächter, C., ... Reifenberger, G. (2005). Somatic mutations in the PTCH, SMOH, SUFUH and TP53 genes in sporadic basal cell carcinomas. *The British Journal of Dermatology, 152*, 43–51.

Renault, M. A., Roncalli, J., Tongers, J., Thorne, T., Klyachko, E., Misener, S., ... Losordo, D. W. (2010). Sonic hedgehog induces angiogenesis via Rho kinase-dependent signaling in endothelial cells. *Journal of Molecular and Cellular Cardiology, 49*, 490–498.

Rhim, A. D., Oberstein, P. E., Thomas, D. H., Mirek, E. T., Palermo, C. F., Sastra, S. A., & Stanger, B. Z. (2014). Stromal elements act to restrain, rather than support, pancreatic ductal adenocarcinoma. *Cancer Cell, 25*, 735–747.

Riaz, S. K., Khan, J. S., Shah, S. T. A., Wang, F., Ye, L., Jiang, W. G., & Malik, M. F. A. (2018). Involvement of hedgehog pathway in early onset, aggressive molecular subtypes and metastatic potential of breast cancer. *Cell Communication and Signaling: CCS, 16*, 3.

Robbins, D. J., Fei, D. L., & Riobo, N. A. (2012). The Hedgehog signal transduction network. *Science Signaling, 5*, re6.

Rohatgi, R., Milenkovic, L., & Scott, M. P. (2007). Patched1 regulates hedgehog signaling at the primary cilium. *Science (New York, N. Y.), 317*, 372–376.

Santoni, M., Burattini, L., Nabissi, M., Morelli, M. B., Berardi, R., Santoni, G., & Cascinu, S. (2013). Essential role of Gli proteins in glioblastoma multiforme. *Current Protein & Peptide Science, 14*, 133–140.

Sari, I. N., Phi, L. T. H., Jun, N., Wijaya, Y. T., Lee, S., & Kwon, H. Y. (2018). Hedgehog signaling in cancer: A prospective therapeutic target for eradicating cancer stem cells. *Cells, 7*.

Sasaki, N., Kurisu, J., & Kengaku, M. (2010). Sonic hedgehog signaling regulates actin cytoskeleton via Tiam1-Rac1 cascade during spine formation. *Molecular and Cellular Neurosciences, 45*, 335–344.

Sheng, T., Li, C., Zhang, X., Chi, S., He, N., Chen, K., ... Xie, J. (2004). Activation of the hedgehog pathway in advanced prostate cancer. *Molecular Cancer, 3*, 29.

Shi, X., Zhang, Z., Zhan, X., Cao, M., Satoh, T., Akira, S., ... Wu, J. (2014). An epigenetic switch induced by Shh signalling regulates gene activation during development and medulloblastoma growth. *Nature Communications, 5*, 5425.

Shida, T., Furuya, M., Nikaido, T., Hasegawa, M., Koda, K., Oda, K., ... Ishikura, H. (2006). Sonic Hedgehog-Gli1 signaling pathway might become an effective therapeutic target in gastrointestinal neuroendocrine carcinomas. *Cancer Biology & Therapy, 5*, 1530–1538.

Shin, K., Lim, A., Odegaard, J. I., Honeycutt, J. D., Kawano, S., Hsieh, M. H., & Beachy, P. A. (2014). Cellular origin of bladder neoplasia and tissue dynamics of its progression to invasive carcinoma. *Nature Cell Biology, 16*, 469–478.

Spetz, J., Langen, B., Rudqvist, N., Parris, T. Z., Helou, K., Nilsson, O., & Forssell-Aronsson, E. (2017). Hedgehog inhibitor sonidegib potentiates. *BMC Cancer, 17*, 528.

St-Jacques, B., Hammerschmidt, M., & Mcmahon, A. P. (1999). Indian hedgehog signaling regulates proliferation and differentiation of chondrocytes and is essential for bone formation. *Genes & Development, 13*, 2072–2086.

Steinway, S. N., Zañudo, J. G., Ding, W., Rountree, C. B., Feith, D. J., Loughran, T. P., & Albert, R. (2014). Network modeling of TGFβ signaling in hepatocellular carcinoma epithelial-to-mesenchymal transition reveals joint sonic hedgehog and Wnt pathway activation. *Cancer Research, 74*, 5963–5977.

Stone, D. M., Hynes, M., Armanini, M., Swanson, T. A., Gu, Q., Johnson, R. L., ... Rosenthal, A. (1996). The tumour-suppressor gene patched encodes a candidate receptor for Sonic hedgehog. *Nature, 384*, 129–134.

Sun, Y., Wang, Y., Fan, C., Gao, P., Wang, X., Wei, G., & Wei, J. (2014). Estrogen promotes stemness and invasiveness of ER-positive breast cancer cells through Gli1 activation. *Molecular Cancer, 13*, 137.

Szczepny, A., Rogers, S., Jayasekara, W. S. N., Park, K., Mccloy, R. A., Cochrane, C. R., ... Watkins, D. N. (2017). The role of canonical and non-canonical Hedgehog signaling in tumor progression in a mouse model of small cell lung cancer. *Oncogene, 36*, 5544–5550.

Tanabe, Y., Roelink, H., & Jessell, T. M. (1995). Induction of motor neurons by Sonic hedgehog is independent of floor plate differentiation. *Current Biology: CB, 5*, 651–658.

Tang, Y., Gholamin, S., Schubert, S., Willardson, M. I., Lee, A., Bandopadhayay, P., ... Cho, Y. J. (2014). Epigenetic targeting of Hedgehog pathway transcriptional output through BET bromodomain inhibition. *Nature Medicine, 20*, 732–740.

Taylor, M. D., Liu, L., Raffel, C., Hui, C. C., Mainprize, T. G., Zhang, X., ... Hogg, D. (2002). Mutations in SUFU predispose to medulloblastoma. *Nature Genetics, 31*, 306–310.

Teglund, S., & Toftgård, R. (2010). Hedgehog beyond medulloblastoma and basal cell carcinoma. *Biochimica et Biophysica Acta, 1805*, 181–208.

Thibert, C., Teillet, M. A., Lapointe, F., Mazelin, L., Le Douarin, N. M., & Mehlen, P. (2003). Inhibition of neuroepithelial patched-induced apoptosis by sonic hedgehog. *Science (New York, N. Y.), 301*, 843–846.

Tian, H., Callahan, C. A., Dupree, K. J., Darbonne, W. C., Ahn, C. P., Scales, S. J., & Sauvage, F. J, D. E. (2009). Hedgehog signaling is restricted to the stromal compartment during pancreatic carcinogenesis. *Proceedings of the National Academy of Sciences of the United States of America, 106*, 4254–4259.

Tibes, R., Al-Kali, A., Oliver, G. R., Delman, D. H., Hansen, N., Bhagavatula, K., ... Bogenberger, J. M. (2015). The Hedgehog pathway as targetable vulnerability with 5-azacytidine in myelodysplastic syndrome and acute myeloid leukemia. *Journal of Hematology & Oncology, 8*.

Tostar, U., Malm, C. J., Meis-Kindblom, J. M., Kindblom, L. G., Toftgård, R., & Undén, A. B. (2006). Deregulation of the hedgehog signalling pathway: A possible role for the PTCH and SUFU genes in human rhabdomyoma and rhabdomyosarcoma development. *The Journal of Pathology, 208*, 17–25.

Vorechovský, I., Undén, A. B., Sandstedt, B., Toftgård, R., & Ståhle-Bäckdahl, M. (1997). Trichoepitheliomas contain somatic mutations in the overexpressed PTCH gene: Support for a gatekeeper mechanism in skin tumorigenesis. *Cancer Research, 57*, 4677–4681.

Wang, J., Cui, B., Li, X., Zhao, X., Huang, T., & Ding, X. (2023a). The emerging roles of Hedgehog signaling in tumor immune microenvironment. *Frontiers in Oncology, 13*, 1171418.

Wang, X., Xu, J., Sun, Y., Cao, S., Zeng, H., Jin, N., ... Huang, M. (2023b). Hedgehog pathway orchestrates the interplay of histone modifications and tailors combination epigenetic therapies in breast cancer. *Acta Pharmaceutica Sinica B, 13*, 2601–2612.

Wechsler-Reya, R. J., & Scott, M. P. (1999). Control of neuronal precursor proliferation in the cerebellum by Sonic Hedgehog. *Neuron, 22*, 103–114.

Wu, F., Zhang, Y., Sun, B., Mcmahon, A. P., & Wang, Y. (2017). Hedgehog signaling: From basic biology to cancer therapy. *Cell Chemistry & Biology, 24*, 252–280.

Xie, J., Murone, M., Luoh, S. M., Ryan, A., Gu, Q., Zhang, C., ... Sauvage, F. J, D. E. (1998). Activating Smoothened mutations in sporadic basal-cell carcinoma. *Nature, 391*, 90–92.

Xu, M., Gong, A., Yang, H., George, S. K., Jiao, Z., Huang, H., ... Zhang, Y. (2015). Sonic hedgehog-glioma associated oncogene homolog 1 signaling enhances drug resistance in CD44(+)/Musashi-1(+) gastric cancer stem cells. *Cancer Letters, 369*, 124–133.

Xu, M., Li, X., Liu, T., Leng, A., & Zhang, G. (2012). Prognostic value of hedgehog signaling pathway in patients with colon cancer. *Medical Oncology (Northwood, London, England), 29*, 1010–1016.

Xu, Y., Song, S., Wang, Z., & Ajani, J. A. (2019). The role of hedgehog signaling in gastric cancer: Molecular mechanisms, clinical potential, and perspective. *Cell Communication and Signaling: CCS, 17*, 157.

Yan, R., Peng, X., Yuan, X., Huang, D., Chen, J., Lu, Q., ... Luo, S. (2013). Suppression of growth and migration by blocking the Hedgehog signaling pathway in gastric cancer cells. *Cellular Oncology (Dordr), 36*, 421–435.

Yao, Z., Han, L., Chen, Y., He, F., Sun, B., Kamar, S., ... Yang, Z. (2018). Hedgehog signalling in the tumourigenesis and metastasis of osteosarcoma, and its potential value in the clinical therapy of osteosarcoma. *Cell Death & Disease, 9*, 701.

Yoo, Y. A., Kang, M. H., Kim, J. S., & Oh, S. C. (2008). Sonic hedgehog signaling promotes motility and invasiveness of gastric cancer cells through TGF-beta-mediated activation of the ALK5-Smad 3 pathway. *Carcinogenesis, 29*, 480–490.

Yoon, C., Park, D. J., Schmidt, B., Thomas, N. J., Lee, H. J., Kim, T. S., ... Yoon, S. S. (2014). CD44 expression denotes a subpopulation of gastric cancer cells in which Hedgehog signaling promotes chemotherapy resistance. *Clinical Cancer Research: An Official Journal of the American Association for Cancer Research, 20*, 3974–3988.

Zhao, Z., Jia, Q., Wu, M. S., Xie, X., Wang, Y., Song, G., ... Shen, J. (2018). Degalactotigonin, a natural compound from. *Clinical Cancer Research: An Official Journal of the American Association for Cancer Research, 24*, 130–144.

# CHAPTER THREE

# An Overview of the Unfolded Protein Response (UPR) and Autophagy Pathways in Human Viral Oncogenesis

Shovan Dutta[a,1], Anirban Ganguly[b,1], and Sounak Ghosh Roy[c,*]

[a]Center for Immunotherapy & Precision Immuno-Oncology (CITI), Lerner Research Institute, Cleveland Clinic, Cleveland, OH, United States
[b]Department of Biochemistry, All India Institute of Medical Sciences, Deoghar, Jharkhand, India
[c]Henry M Jackson for the Advancement of Military Medicine, Naval Medical Research Command, Silver Spring, MD, United States
*Corresponding author. e-mail address: sounak87@gmail.com

## Contents

| | |
|---|---|
| 1. Cross talk between autophagy and UPR | 82 |
| 2. Autophagy and UPR in oncogenic viruses | 84 |
| 3. Epstein-Barr virus (EBV) and its relationship to cancer | 84 |
|    3.1 ERS/UPR and autophagy in EBV-induced solid tumors | 85 |
| 4. Human papillomavirus (HPV) and its relationship to cancer | 87 |
|    4.1 ERS/UPR and autophagy in HPV-induced solid tumors | 89 |
| 5. Human immunodeficiency virus (HIV) and its relationship to cancer | 90 |
|    5.1 HIV-mediated non-Hodgkin lymphoma (NHL) | 91 |
|    5.2 ERS/UPR and autophagy in non-Hodgkin lymphoma (NHL) | 91 |
| 6. Human herpes virus-8 (HHV-8) and its relationship to cancer | 92 |
|    6.1 Important oncoproteins in HHV-8 | 93 |
|    6.2 Autophagy in KSHV-induced solid tumors | 94 |
|    6.3 ERS/UPR in KSHV-induced solid tumors | 95 |
| 7. HTLV-1 (human T-cell lymphotropic virus type 1) and its relationship to cancer | 97 |
|    7.1 Important oncoproteins in HTLV-1 | 98 |
|    7.2 Autophagy in HTLV-1 induced solid tumors | 100 |
|    7.3 ERS/UPR in HTLV-1 induced solid tumors | 101 |
| 8. Hepatitis B (HBV) and its relationship to cancer | 102 |
|    8.1 ERS/UPR in HBV-induced solid tumors | 103 |
|    8.2 Autophagy in HBV-induced solid tumors | 107 |
| 9. Discussion | 117 |
| References | 118 |

[1] Equal contribution.

## Abstract

Autophagy and Unfolded Protein Response (UPR) can be regarded as the safe keepers of cells exposed to intense stress. Autophagy maintains cellular homeostasis, ensuring the removal of foreign particles and misfolded macromolecules from the cytoplasm and facilitating the return of the building blocks into the system. On the other hand, UPR serves as a shock response to prolonged stress, especially Endoplasmic Reticulum Stress (ERS), which also includes the accumulation of misfolded proteins in the ER. Since one of the many effects of viral infection on the host cell machinery is the hijacking of the host translational system, which leaves in its wake a plethora of misfolded proteins in the ER, it is perhaps not surprising that UPR and autophagy are common occurrences in infected cells, tissues, and patient samples. In this book chapter, we try to emphasize how UPR, and autophagy are significant in infections caused by six major oncolytic viruses—Epstein-Barr (EBV), Human Papilloma Virus (HPV), Human Immunodeficiency Virus (HIV), Human Herpesvirus-8 (HHV-8), Human T-cell Lymphotropic Virus (HTLV-1), and Hepatitis B Virus (HBV). Here, we document how whole-virus infection or overexpression of individual viral proteins in vitro and in vivo models can regulate the different branches of UPR and the various stages of macro autophagy. As is true with other viral infections, the relationship is complicated because the same virus (or the viral protein) exerts different effects on UPR and Autophagy. The nature of this response is determined by the cell types, or in some cases, the presence of diverse extracellular stimuli. The vice versa is equally valid, i.e., UPR and autophagy exhibit both anti-tumor and pro-tumor properties based on the cell type and other factors like concentrations of different metabolites. Thus, we have tried to coherently summarize the existing knowledge, the crux of which can hopefully be harnessed to design vaccines and therapies targeted at viral carcinogenesis.

## 1. Cross talk between autophagy and UPR

Several well-orchestrated processes involved with maintaining cellular homeostasis or committing to cell death are activated by cellular stress which in turn is induced by external or internal triggers. The UPR, hypoxia, autophagy, and mitochondrial function represent such processes which form an integral part of global endoplasmic reticulum (ER) stress response (ERS). Pathological conditions causing subsequent impairment of ERS elements may lead to perturbation of overall cellular homeostasis. Moreover, changes in mitochondrial function or autophagy may be triggered via activation of the UPR leading to modulating effects on the UPR, demonstrating presence of crosstalk processes. A significant role is played by the established UPR process which restores homeostasis subsequent to accumulation of toxic misfolded proteins (Hetz, 2012a; Rutkowski & Hegde, 2010). Regulation of UPR is modulated by the coordinated action of three ER membrane-embedded sensors namely the activating transcription factor 6 (ATF6), double-stranded RNA-activated protein kinase (PKR)-like ER kinase (PERK), and inositol-requiring enzyme 1 (IRE1) which are in turn activated by perturbed ER homeostasis. All these sensors act via specialized transcriptional programs mediated by distinct transducers namely spliced X-box binding protein 1 (sXBP1) (for IRE1), ATF4 (for PERK), and cleaved ATF6 (for ATF6). The functions of these factors include direct activation of the transcription of proteins or chaperones involved in redox homeostasis, lipid biosynthesis, secretion of protein, or cell death programs (Senft & Ronai, 2015).

Autophagy may be defined as an important catabolic process specialized to provide proteins, organelles, and cytoplasmic components to lysosomes for carrying out degradation and recycling. A suitably coordinated program comprising over 30 autophagy-related (ATG) genes regulates autophagy, which can be positively regulated by nutrient starvation followed by subsequent inhibition of mechanistic target of rapamycin (mTOR) signaling (Mizushima & Komatsu, 2011) or by the UPR with the accumulation of aggregated misfolded proteins (Deegan et al., 2013). The crosstalk between autophagy and the UPR can be further demonstrated by the fact that PERK–eukaryotic translation initiation factor 2 alpha (eIF2a) pathway is essential for induction of autophagy after ERS (known to be activated by tunicamycin treatment) (B'Chir et al., 2013). To be more specific, the proteins ATF4 and C/EBP homologous protein (CHOP) (which is a known transcription factor induced by ATF4) were demonstrated to transcriptionally regulate about a dozen ATG genes (B'Chir et al., 2013). Along with this, IRE1 is also implicated in the process of autophagy activation. It has been shown that tumor necrosis factor receptor-associated factor 2 (TRAF2) dependent activation of IRE1 and c-Jun N-terminal kinase (JNK) both lead to Bcl-2 phosphorylation, facilitating dissociation of Beclin-1 (known autophagy regulatory protein), phosphoinositide-3-kinase (PI3K) complex activation and autophagy (Deegan et al., 2013). Noteworthy to mention here that in presence of oxidative stress, the regulatory control of autophagy via JNK can be IRE1 independent (Haberzettl & Hill, 2013). Studies have shown that within intestinal epithelial cells of ATG-knockout mice there was significantly raised IRE1 activity along with IRE1-dependent inflammation which implied that dysregulated autophagy leads to IRE1 activity triggering associated with concomitant activation of UPR via the sXBP arm which in turn reflects the presence of a possible feedback mechanism in the regulation of UPR signaling (Adolph et al., 2013; Deegan et al., 2013). These molecular mechanisms linking ERS and the UPR to autophagy (Deegan et al., 2013), highlight the bidirectional interaction between autophagy and ERS.

## 2. Autophagy and UPR in oncogenic viruses

Oncogenic viruses affect several molecular mechanisms of their hosts which lead to inhibition of cell death, uncontrolled proliferation as well as chronic inflammation and this modulation in the long term directly or indirectly promotes tumorigenesis (Vescovo et al., 2020a). Researchers have also found an innate ability of these oncoviruses to alter the pathways linked to autophagy during immune response to infections which ultimately triggers onset of malignant transformation. It has also been found out that autophagy can have a role in both tumor suppression as well as tumor progression and the development of chemoresistance (Yun & Lee, 2018). Faulty autophagic pathways result in lesser control of cell quality attributed to genomic damage, metabolic stress, and tumor formation (Lorin et al., 2013). On the other hand, upregulated autophagy pathways promote the progression of malignancy by guiding cancer cells to tackle metabolic stress and overcome chemotherapeutic cytotoxicity (Kögel, 2012). Autophagy has also been shown to play an important role in maintaining the stemness of cancerous stem cells and affecting tumor recurrence and chemoresistance (Nazio et al., 2019; Vescovo et al., 2020b).

## 3. Epstein-Barr virus (EBV) and its relationship to cancer

EBV (formerly known as human herpesvirus-4) is a type of virus that remains dormant in most people. EBV causes infectious mononucleosis and has been associated with certain cancers, including **Burkitt lymphoma, immunoblastic lymphoma, nasopharyngeal cancer, and stomach (gastric) cancer** (Baumforth et al., 1999a). It has been studied that approximately 20% of all cancers are associated with infectious agents, 12% of all cancers are caused by oncoviruses, and 80% of viral cancers occur in the developing world (Bouvard et al., 2009; De Martel et al., 2012; Mui, Haley, & Tyring, 2017a; Zur Hausen, 2009).

In 1958, Denis Burkitt, an English surgeon working in Uganda, first described a common cancer affecting children in regions of equatorial Africa and named as Burkitt's lymphoma. Burkitt discovers that because of the climatic and geographical distribution, there is a vector-borne virus responsible for Burkitt's lymphoma. In 1964, Epstein and Barr identified herpesvirus-like particles under the electron microscope in a cell line established in cell culture from Burkitt's lymphoma biopsy. Additionally, it has also been observed that Burkitt's lymphoma had much higher antibody titers to EBV antigens compared to the controls. In 1973, it was revealed that EBV DNA could be found in the Burkitt's lymphoma and nasopharyngeal carcinoma tumor cells from the marmosets and owl monkeys, which suggested that had oncogenic potential in both human and non-human primates (Baumforth et al., 1999b). Two types of EBV can be found in humans. EBV-1 and EBV-2 (initially known as EBV-A and EBV-B). Both types have homologies except the region that encodes EBV nuclear antigens (EBNAs; EBNA1, EBNA2, EBNA3A, EBNA3B, EBNA3C, EBNA leader protein) and the Epstein-Barr early RNAs (EBERs; EBER1 and EBER2). EBV-1 shows higher efficiency of transforming B cells, than EBV-2 (Baumforth et al., 1999b; Mosier et al., 1988; Young et al., 1987).

There is no effective antiviral therapy available for EBV infectious mononucleosis in immunocompetent persons. Acyclovir and ganciclovir may reduce EBV shedding but are ineffective clinically. Ganciclovir is a derivative of acyclovir. They are nucleoside antiviral drugs that have the function of inhibiting the synthesis of EBV-DNA. The anti-EBV effect of them is stronger than other drugs. Treatment of immunocompromised patients with EBV lymphoproliferative disease is controversial. In the treatment of EBV-IM, the therapeutic effect of ganciclovir is superior compared to acyclovir. Ganciclovir can eliminate the symptoms of angina, fever, enlarged lymph nodes, and other signs in children, can improve abnormal blood indicators, and has a higher negative conversion rate of EBV and fewer adverse reactions. On the other hand, the acyclovir has not been proven to be beneficial (Ishii et al., 2019; Coşkun et al., 2020; Zhang et al., 2001).

## 3.1 ERS/UPR and autophagy in EBV-induced solid tumors

The ER is the cell organelle that maintains cellular homeostasis and contributes to lipid synthesis, protein folding, translocation, and post-translational modifications (Greenblatt & Wygnanski, 2003; Walter & Ron, 2011; Asha & Sharma-Walia, 2018). ER stress can be activated by various pathological and physiological conditions including the UPR to restore homeostasis. There are several stress factors like hypoxia, starvation, change in pH, calcium depletion, and viral infection which may be involved in changing or modifying the ER environment. These changes or modifications are responsible for the process of proper protein folding within the ER, which eventually leads to the accumulation of misfolded or unfolded proteins causing ER stress. This ER stress is responsible for the activation of the UPR, a cellular homeostasis response connecting the ER to the nucleus to restore cellular equilibrium (Zhang et al., 2006; Ron & Walter, 2007). ER-associated degradation (ERAD) is facilitated by UPR-activated apoptosis or degradation of unfolded or misfolded proteins, which cannot enter the secretory pathway. Cancer cells and viruses have adaptive mechanisms to control ER stress-induced apoptosis, which allows them to grow aggressively (Asha & Sharma-Walia, 2018).

A study has demonstrated that the novel protein C7 and iron chelators reactivate EBV lytic cycle by chelating intracellular iron and activating the ERK1/2-autophagy axis. Apart from this iron chelation, this autophagy can be induced by ER stress (Corazzari et al., 2017a; Yiu et al., 2018, 2019). An extensive study recently highlighted the crosstalk between EBV and autophagy in B cells. There are other studies demonstrated that several EBV proteins like EBNA1, EBNA3C, LMP1, LMP2A, and Rta/Zta have been associated with regulating autophagy initiation, progression, and completion for EBV lytic reactivation, viral particle formation and release (Yiu et al., 2019; Chun & Kim, 2018; Das, Shravage, & Baehrecke, 2012; Wang, Ye, & Zhao, 2019a; McKnight & Yue, 2013). Studies have revealed that gene knockdown of various autophagic proteins such as berlin-1, ATG5, ATG12, ATG7, LC3B, ATG10, ATG3, and Rab9, have shown the importance of ATG5 in EBV lytic reactivation (Yiu et al., 2019). For the very first time Yiu et al., 2019, demonstrated that for autophagy initiation, the ATG5 protein is required for EBV lytic reactivation in EBV-infected nasopharyngeal carcinoma (NPC) cells. Additionally, it has also been demonstrated that C7/iron chelators and histone deacetylase inhibitors (HDACi) induce autophagy-dependent and independent mechanisms, respectively, to reactivate the lytic cycle of EBV and impose differential cellular effects. In addition to that, the combination of C7 and suberoylanilide hydroxamic acid (SAHA) at their corresponding reactivation kinetics enhances EBV lytic reactivation (Yiu et al., 2019; Hui et al., 2016; Ramayanti et al., 2018; Lee et al., 2008).

Lee et al. (2009), have described that EBV controls both the UPR and autophagy to nurture the latent phase of its life cycle. The latent membrane protein 1 (LMP-1) oncogene of EBV induces the UPR dose-dependently (Lee & Sugden, 2008; Lee, Lee, & Sugden, 2009a). LMP-1 mimics the

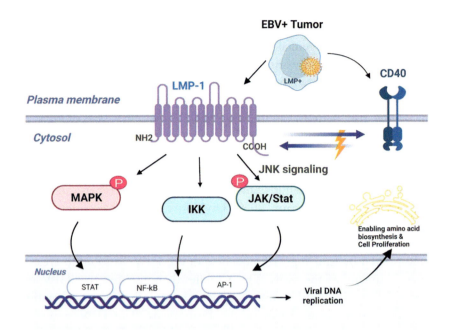

**Fig. 1** UPR and Autophagy in EBV-mediated carcinogenesis.

cellular CD40 receptor and activates NF-κB, AP1, and the Jak/Stat pathways through its carboxy-terminal signaling domain to promote cell proliferation (Brinkmann & Schulz, 2006; Kilger et al., 1998). It uses its amino-terminal, six-membrane-spanning domain to regulate its signaling instead of a ligand, and it is this latter domain that induces the UPR, PERK, ATF6, and IRE-1, and this prolonged activation of PERK eventually leads to apoptosis (Lee & Sugden, 2008; Lee et al., 2009a; Lin et al., 2007; Zinszner et al., 1998) (Fig. 1).

## 4. Human papillomavirus (HPV) and its relationship to cancer

HPVs is a large group of related viruses. Each virus in the group is recognized by a number, which is known as a different HPV type. Most of the HPV infections cause warts on the skin, such as on the arms, chest, hands, or feet. Many other types of HPV are found mainly on the body's mucous membranes specifically on the moist surface layers that line organs and parts of the body that open to the outside, such as the vagina, anus, mouth, and throat. The HPV types found on mucous membranes are sometimes known as genital HPV. They generally do not live on the skin. Some HPVs can cause warts (papillomas) on or around the genitals and anus of both men and women. Women may also have warts on the cervix and in the vagina. These HPV types rarely cause cancer; thus, they are called "low-risk" viruses. On the other hand, some other types of HPV are called "high-risk" because they can cause cancer. Common high-risk HPV types include HPV 16 and 18. In most cases, the body can clear the common HPV infection on its own. But some chronic, or long-lasting infections, especially when it's caused by high-risk HPV types, can lead to cancer over time (Fontham, 2020; Division of STD Prevention, 2022; Society., 2017).

Several pieces of evidence from Spanish, French, and Brazilian scientists demonstrated by studying 118 sequences of the HPV 16 strain to create a genetic timeline of the virus. These studies demonstrated that HPV had been transmitted to modern human species through their ealier and extinct human ancestor Neanderthals or the Denisovans. The HPV strain can hardly be found among the Sub-Saharan Africans, this evidence helps demonstrating that humans whoever moved out from Africa more than 100,000 years ago may have developed the disease from somewhere else in the world. Researchers believe that the human HPV16 strain, that infects 4% percent of Americans and also can lead to cervical cancer, is about 500,00 years old and according to these evidences, researchers can conclude that this strain may have originated in the Neanderthals or the Denisovans lineages (de Sanjosé, Brotons, & Pavón, 2018; Quest, 2016).

According to the CDC, more than 42 million Americans are infected with types of HPV that cause disease. About 13 million Americans, including teens, become infected each year. HPV is spread through intimate skin-to-skin contact. You can get HPV by having vaginal, anal, or oral sex with someone who has the virus, even if they don't have signs or symptoms (CDC, 2023). According to WHO, cervical cancer is the fourth most common cancer among women globally, with an estimated 604,000 new cases and 342,000 deaths in 2020. About 90% of the new cases and deaths worldwide in 2020 occurred in low- and middle-income countries. There are two types of HPV (16 and 18) that are known to be responsible for nearly 50% of high-grade cervical pre-cancers (Rao et al., 2012). HPV is usually transmitted through sexual contact and most people are infected with HPV, shortly after the onset of sexual activity, and almost 90% of the cases clear the infection automatically. It has been reported that woman who is infected by HIV has 6 times more chances of getting HPV infection compared to those who are not infected by HIV. Usually, takes 15–20 years for cervical cancer to develop in women with normal immune systems, but untreated HIV infection patients who have weak immune systems, can trigger the HPV infection within 5–10 years (Stelzle et al., 2021).

## 4.1 ERS/UPR and autophagy in HPV-induced solid tumors

Autophagy regulates the cellular response and is responsible for eliminating damaged organelles, proteins, and cell membranes via a lysosomal pathway and cell stress and diseases can also trigger this process (Surviladze et al., 2013). It has been well described as a self-degradation mechanism associated with tumor progression, including cervical cancer. This process plays a critical role in both promoting and preventing cervical cancer (Li et al., 2015). Another study suggested that HPV-16 E5 suppresses three key players in the ER stress pathway: COX-2, XBP-1, and IRE1a. Many other viral proteins have been reported to cause alterations in the ER stress response pathway, including the canine papillomavirus E5 protein (CfPV2 E5) (Condjella et al., 2009; Tardif et al., 2004; Yu et al., 2006; Su, Liao, & Lin, 2002; Sudarshan, Schlegel, & Liu, 2010).

Studies have suggested that the NF-κB signaling pathway promotes the proliferation, invasion, and metastasis of cervical cancer cells (Pang, Zhang, & Zhang, 2017; Nair et al., 2003; Moore-Carrasco et al., 2007, 2009; Zhu et al., 2017). Additionally, UPR signaling sensors provide a potential link between the activation of the NF-κB pathway, which regulates the expression of various proinflammatory genes and immunomodulatory molecules, and ER stress (Zhu et al., 2017; Hotamisligil, 2010). Study suggested that in primary genital keratinocytes, 16E5 has shown a specific and consistent role in the downregulation of ER stress response gene, at the same time the study also suggested that E5 has a potential role in repressing the cellular ER stress response following HPV infection (Sudarshan et al., 2010).

Studies suggested that environmental risk factors induce the activation of ER stress, as a result, cancer cells learn to adapt and bypass the fate of ER stress-induced apoptosis. Studies indicate that caspase-12 specifically plays an important role in the apoptotic signaling induced by ER stress (Wali, Bachawal, & Sylvester, 2009). A study suggested that caspase-12 and caspase-3 are activated in the apoptotic cell both by extrinsic (death ligand) and intrinsic (mitochondrial) pathways. Inhibitor of apoptosis (IAP) directly regulates apoptosis by preventing the activation of caspase 3. However, the IAP directly regulates apoptosis by preventing the activation of caspase-3 (Lavrik, Golks, & Krammer, 2005). The study also suggested that quinazolinediamine (QNZ), which is an NF-κB inhibitor, decreased the autophagy and apoptosis induced by any external stress, which inhibits the activation of caspase-12 or caspase-3 in cells under ER stress by enhancing the expression of IAP family members. The study indicated that 3-BrPA can enhance therapeutic by improving the efficacy of the CRT/E7 DNA vaccine potency in generating improved antigen-specific immune responses and antitumor effects. This study suggests that the treatment of tumor-bearing mice with 3-BrPA can be considered a potent immune-mediated therapeutic antitumor effect through enhanced tumor-specific immunity, and also through the increased susceptibility of tumor cells to antigen-specific CD8 + T cell-mediated killing (Lee et al., 2019).

## 5. Human immunodeficiency virus (HIV) and its relationship to cancer

Infection with HIV can lead to a high risk of having cancer, although HIV alone does not act directly in the formation of cancer. Over time HIV damages the immune system and due to HIV-related immunosuppression, which impairs control of oncogenic viral infections, results in cancer formation, which are mostly known as cancers are "HIV-associated cancers" (Grulich et al., 2007; Shiels et al., 2009; Shiels & Engels, 2017; NIH-NCI, 2017). Kaposi sarcoma (KS) is a soft tissue tumor that occurs in patients with immunosuppression, such as those with acquired immunodeficiency syndrome (AIDs) or those undergoing immunosuppression due to an organ transplant. For the first time, in 1872, Moritz Kaposi, an Austro-Hungarian dermatologist described "Idiopathisches multiples Pigmentsarkom der Haut", which has become known as KS, demonstrated this disease from the study of 5 patients with the multifocal disease (Stănescu et al., 2007; Bishop & Kaposi Sarcoma, 2022). Later, human herpesvirus/Kaposi sarcoma herpesvirus was discovered as a causative agent of Kaposi sarcoma as the AIDS epidemic progressed in the 1980s (Bishop & Kaposi Sarcoma, 2022; Flore et al., 1998). KS mostly occurs in patients over 50 years old of Eastern European and Mediterranean descent and these patients appear to have a higher risk for secondary malignancies (Fatahzadeh, 2012; Iscovich et al., 2000). Within the United States, incidence has been stable around 1:100,000 since 1997 (Schneider & Dittmer, 2017).

## 5.1 HIV-mediated non-Hodgkin lymphoma (NHL)

NHL is a type of cancer that affects the lymphatic system. The lymphatics are a part of the immune system which helps protect the body from infection and disease. HIV causes these NHL to become a fast-growing type of NHL, which includes diffuse large B-cell lymphoma, Burkitt's lymphoma, and central nervous system (CNS) lymphoma (Armitage & Weisenburger, 1998). Diffuse large B-cell lymphoma (DLBCL) is the most common subtype of NHL worldwide, with >25,000 cases diagnosed in the United States annually and accounting for >10,000 deaths (Best et al., 2019a). NHL is common in ages 65–74, the median age being 67 years (Sant et al., 2010; Smith et al., 2011). Additionally, NHL is 5th most common diagnosis of pediatric cancer in children under the age of 15 years, and approximately 7% of childhood cancers in the developed world. In the United States, approximately 800 new pediatric NHL cases have been reported annually, and an incidence of 10–20 cases per million people per year (Jiang & Li, 2020).

## 5.2 ERS/UPR and autophagy in non-Hodgkin lymphoma (NHL)

The study has revealed that misfolded or unwanted proteins in the ER are translocated to the cytoplasm and ubiquitinated to allow their degradation through the UPS. The adaptive UPR triggers survival mechanisms to reduce protein load by selectively transcribing key protein-folding chaperone elements, attenuating protein synthesis, and blocking the influx of proteins into the ER. Cells unable to restore normal protein balance due to irresolvable ER stress will activate a late-phase UPR, signaling the cell to undergo apoptosis (Best et al., 2019b). UPR regulates protein loads by several mechanisms like selectively transcribing key protein folding chaperone elements, attenuating protein synthesis, and blocking the influx of proteins into the ER. ER stress is responsible for activating late-phase UPR, which results the cells being unable to restore normal protein balance and thus undergo apoptosis. ER stress is also responsible for releasing cellular proteins like; glucose-regulated protein (GRP78), released from major stress sensors by activating UPR, including, PKR- like PERK, inositol-requiring enzyme 1a (IRE1a), and ATF6 (Hetz, 2012b). The study also revealed that upon activation of PERK, phosphorylation of eIF2a occurs which slows down most protein translation except the synthesis of ATF4 (Sano & Reed, 2013). Activated PERK signaling upregulates transcription factor C/EBP homologous protein (CHOP), which induces

GADD34 and proapoptotic BH3-only proteins, and downregulates prosurvival proteins (BCL2). The study also suggested that dimerization and autophosphorylation of IRE1 results in the formation of an active transcription factor, spliced XBP1. Furthermore, the ATF6 translocate to the Golgi apparatus, where the cytosolic fragment of ATF6 (ATF6f) controls ER-associated degradation genes. Together, these processes lead to protein folding in the ER (Best et al., 2019b; Hetz, 2012b; Sano & Reed, 2013).

## 6. Human herpes virus-8 (HHV-8) and its relationship to cancer

HHV-8 belonging to sub-family of *Gammaherpesvirinae* possesses a linear, large double stranded DNA of genome size 165 kb surrounded by icosahedral capsid and an envelope. It is also known as Kaposi's sarcoma -associated herpesvirus (KSHV) due to its significant association with Kaposi's sarcoma (KS). HHV-8 is also implicated with B-cell lymphoproliferative disorders like multicentric Castleman disease (MCD) and primary effusion lymphoma (PEL) as well as inflammatory cytokine syndrome (KICS) (Ethel et al., 2019). Routes of viral transmission primarily include saliva, blood and sexual contact but rarely mother-to-fetus transmission has been seen (De Paoli & Carbone, 2016). Worldwide seroprevalence is estimated around 5% to 20% with variable geographical distribution but despite its high prevalence, only a few patients, especially those who are immunocompromised, go on to develop HHV-8 related diseases (Wakeman, Izumiya, & Speck, 2017; Dow, Cunningham, & Buchanan, 2014). Similar to other herpesviruses, HHV-8 has both latent and lytic phases and after primary infection, it goes to the latent phase as an episome, utilizing host cell machinery for replication (Sychev et al., 2017). Though B lymphocytes serve as primary reservoir in latency followed by endothelial cells, the latter is primarily infected in KS (De Paoli & Carbone, 2016).

Latent proteins involved in the pathogenesis of HHV-8 mediated malignancies include LANA (latency associated nuclear antigen-1), v-FLIP (viral FLICE inhibitory protein), v-cyclin (viral-Cyclin), kaposin and numerous micro-RNAs (Beral et al., 1990). A more favorable microenvironment suitable for tumor initiation and progression is created in primary cells with latent infection of HHV-8 but unlike EBV-infected cells, these cells are not immortalized. Although latent phase is significant for maintenance of viral genome, lytic replication is important in HHV-8 infected cells for development of an inflammatory microenvironment as well as for dissemination, transmission and preservation of the virus (Peterman, Jaffe, & Beral, 1993). Switching between latent and lytic phases is maintained by the open reading frame (ORF) 50 gene which encodes for the replication and transcription activator (RTA) protein (Wakeman et al., 2017).

## 6.1 Important oncoproteins in HHV-8

One of the most consistently expressed oncoproteins in HHV-8 infected tumors is LANA which is a key multifunctional protein in the maintenance of HHV-8 latency and oncogenesis (Kaposi, 1872; Okada, Goto, & Yotsumoto, 2014). LANA plays an inhibitory role in inhibition of signaling pathways such as MAPK, ERK, PI3K/AKT, JAKJ/STAT, Wnt and Notch which aids in escaping immune surveillance and it is also involved in the regulation of proteins such as transcription factors, replication factors and chromatin modifying enzymes involved in maintaining viral latency. Silencing of transcriptional activity of RTA promoter by LANA leads to suppression of lytic reactivation (Okada et al., 2014). LANA inactivates the p53 and Rb tumor suppressor proteins leading to cell cycle progression (Thakker & Verma, 2016; Gantt & Casper, 2011). A cyclin D viral homolog named viral cyclin (v-cyclin) promotes latency by forming a kinase complex with CDK6, thus modulating cell cycle and proliferation Rb protein, CDK inhibitors, p27, histone H1 are all known targets of vCyclin-CDK complex (Okada et al., 2014). Another oncoprotein, V-FLIP stimulates NF-κB signaling pathway, thereby, upregulating anti-apoptotic genes like BCL-2 (Gantt & Casper, 2011; Li et al., 2016) and leading to downstream effects such as increased expression of Notch ligand KJAG1 (Peterman et al., 1993). Kaposin B promotes tumor development in HHV-8 infected cells by increasing cytokine expressions such as interleukins (IL-6 and IL-8), TNF-α, and Macrophage inflammatory protein-1 (MIP-1α) and MIP-1β. IL-8 attaches to HHV-8 encoded viral G protein-coupled receptor (vGPCR) to stimulate the synthesis of angiogenic factors and induced expression of the lytic switch protein ORF 50 RTA (Chen et al., 2017a). HHV-8 lytic genes include v-IL6, vGPCR, v-BCL2, v-MIP, and v-IRF-1 (viral IFN regulatory factor) (Thakker & Verma, 2016). vGPCR promotes the activation of ROS leading to oxidative DNA damage (Gantt & Casper, 2011). Resistance to apoptosis is promoted by vGPCR and v-IRF-1 by activation of NF-κBsignaling pathway and inhibition of pro-apoptotic mediators respectively.

## 6.2 Autophagy in KSHV-induced solid tumors

The unique feature exhibited by lymphotropic viruses such as KSHV (HHV-8) is that they initially promote induction of autophagy followed by inhibition of autophagosome maturation and detrimental decrease in autophagy-mediated clearance: all of which ultimately promotes tumorigenesis (Vescovo et al., 2020a). KSHV genome encodes several oncoproteins promoting both latent and lytic phase proteins which are also involved in autophagy regulation. vCyclin D induces autophagy by AMPK stimulation (Leidal et al., 2012) whereas vFLIP not only limits the autophagy pathways by directly inhibiting ATG3 and LC3 activation (Lee et al., 2009b) but also inhibits caspase-8 and NF-κB to suppress apoptosis (Ganem, 2007).

The oncoprotein LANA inhibits p53 activity which prevents death of KSHV-infected cells and promotes cell proliferation by suppressing pRB expression (Ganem, 2007; Friborg et al., 1999). As the transcription of several autophagic genes is mediated by p53 (White, 2016), the LANA-mediated p53 inhibition should lead to impairment of autophagy during KSHV latency. In addition, during latent infection, studies have shown that KSHV also restricts STAT3 (Roca Suarez et al., 2018) which in turn inhibits p53 and autophagy. One major event found during KSHV infections is highlighted by the process of lytic re-activation after latency, where the induction of autophagy has been shown to play an important role. STAT3 phosporylation and activation is augmented by KSHV which leads to enhancement of autophagy and lytic re-activation (Santarelli et al., 2019). During this crucial step, the vIL6 and viral Kaposin B act as inducers of phosphorylation of STAT3 on Ser727, which in turn leads to augmentation of p53/p21 axis and inhibition of autophagy (Santarelli et al., 2019; Deutsch et al., 2004; Suthaus et al., 2012; King, 2013). In addition, it has been explained by researchers that autophagy inhibition leads to decreased lytic reactivation of KSHV (Wen et al., 2010). Lytic viral proteins like GPCR and Bcl2 inhibit autophagy by the activation of the PI3K/AKT/mTOR pathway and altered activity of BECLIN 1 and regulation of ATG14L stability (Pattingre et al., 2005; Bhatt & Damania, 2013; Zhang et al., 2015). From the above-mentioned findings, it is quite plausible to hypothesize that after an initial activation of autophagy during latent–lytic transition, to avoid viral degradation, autophagy is ultimately inhibited. Strengthening this hypothesis, studies have also described that the KSHV lytic protein K7 promotes the negative interaction between Rubicon and

BECLIN 1 complex II, leading to further blockade to autophagosome maturation (Liang et al., 2013). Taken together, all these evidences point to the fact that although during the periods of de novo infection as well as latency KSHV inhibits autophagy, it again activates it to augment latency–lytic transition, and ultimately blocks it rapidly to complete the KSHV cycle (Cirone, 2018). In the current scenario, it is very important to understand, the extensive process of autophagy regulation by KSHV during viral life cycle to connect the observed inhibition of autophagy, particularly during latency, to oncogenesis.

More recently, HHV-8-encoded viral IFN regulatory factor 1 (vIRF-1) has been shown to localize in part to mitochondria, leading to direct interaction with the mitophagy receptor NIX. This causes subsequent promotion of NIX-mediated mitophagy in HHV-8-infected PEL cells to remove dysfunctional mitochondria during the phase of lytic replication. Induction of DRP1-dependent mitochondrial fragmentation in transfected HeLa cells is also mediated by vIRF-1, which can promote mitophagy (Vo et al., 2019). Nevertheless, though the exact mechanisms for orchestrating mitochondrial dynamics by vIRF-1 to induce mitophagy remain to be established, researchers have claimed that this pro-mitophagy activity of vIRF-1 is associated with the inhibition of apoptosis and enhancement of viral replication (Vo & Choi, 2021).

## 6.3 ERS/UPR in KSHV-induced solid tumors

For avoiding immune surveillance and promote viral replication machinery, herpesviruses modulate cellular stress responses. When the protein load in the ER exceeds folding capacity leading to the accumulation of misfolded proteins there is activation of the conserved stress response popularly termed as the "unfolded protein response (UPR)" which is known to maintain protein homeostasis through translational and transcriptional reprogramming. By chance, if restoration of homeostasis is not feasible, the UPR switches modes from "helper" to "executioner" phases resulting in apoptotic signaling pathways. Rapid burst of HHV glycoprotein synthesis during the phase of lytic replication is responsible for ER stress, and as a result, these viruses devise numerous well-developed mechanisms to modulate UPR signaling for preparing an optimal niche for replication (Johnston & McCormick, 2019).

ER stress has been shown to trigger reactivation of EBV, KSHV and murine gamma herpesvirus 68 (MHV68) (Bhende et al., 2007; Matar et al., 2014; Wilson et al., 2007; Yu et al., 2007). In response to ER stress, lytic reactivation is primarily due to the transcription factor: XBP1s whose target sequences in the promoters of early viral genes function as ER stress-sensing mechanism. In order to demonstrate that UPR transcription might affect KSHV lytic replication, researchers ectopically expressed the spliced isoform of XBP1 which turned out to be a strong inhibitor of virion production in epithelial cells, acting in a dose-dependent manner (Johnston & McCormick, 2019). Hence, although XBP1s plays a critical role in reactivation from latent phase, the virus inhibits its expression in the lytic cycle to avoid deleterious effects during viral replication.

Excessive UPR signaling is known to inhibit KSHV lytic replication. Strong induction of the UPR using pharmacologic agents like 2-deoxyglucose, tunicamycin, brefeldin A, or treatment with proteasome inhibitors known to induce ER stress (e.g. bortezomib, lactacystin, proteasome inhibitor I, MG132) may lead to lytic reactivation, triggering apoptotic pathways and inhibition of virion production (Granato et al., 2017; Saji et al., 2011; Leung et al., 2012; Shigemi et al., 2016).

Modulation of UPR signaling can be done by KSHV protein viral interleukin-6 (vIL-6; also called K2) during the lytic cycle. vIL-6 possesses an amino-terminal signal peptide that directs its translation in the ER and leads to its subsequent secretion. But, a significant fraction of vIL6 remains within the ER where it interacts with hypoxia upregulated 1 protein (HYOU1, also called Grp170) (Giffin et al., 2014). The nucleotide exchange factor, HYOU1, promotes ADP release from binding immunoglobulin protein (BiP), allowing a sustained BiP association with misfolded or unfolded proteins (Behnke, Feige, & Hendershot, 2015). HYOU1-BiP interactions may be inhibited by vIL-6 resulting in UPR activation as well as increased protein misfolding. By binding components of the calnexin cycle, such as glucosidase II (GlucII) and UDP-glucose: glycoprotein glucosyltransferase 1 (UGGT1), researchers have elucidated a probable role of vIL-6 in protein folding (Chen, Xiang, & Nicholas, 2017b), and subtle modulations of UPR signaling by direct control of proteostasis.

Studies have proposed that ER localization of new ORF45 isoforms led to the upregulation of UPR markers. Even though the exact roles of alternative ORF45 isoforms during lytic replication remains obscure, silencing BiP expression reduced production of progeny virions to a great extent, despite having trivial impact on viral gene expression. This implies that, in the ER, BiP may be crucial for viral glycoprotein folding. But it is to be noted that, BiP silencing is also shown to trigger ER stress leading to

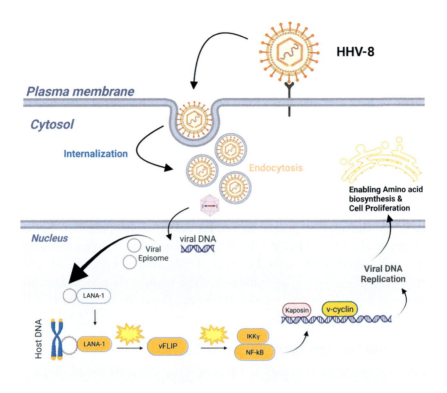

Fig. 2 UPR and Autophagy in HHV8-mediated carcinogenesis.

activation of the UPR, so the decreased virion production may have resulted from increased UPR signaling and not due to the viral protein folding activity of BiP. Hence, further studies and elegant experiments need to be done to study the effects of novel ORF45 protein isoforms on the UPR (Fig. 2).

## 7. HTLV-1 (human T-cell lymphotropic virus type 1) and its relationship to cancer

Isolated in 1979, human T-cell lymphotropic virus type 1 (HTLV-1) is a Delta retrovirus with a positive ssRNA genome of size 9 kb, having seven sub-types, with a tropism for CD4+ T lymphocytes and encoding numerous genes such as gag, env, tax, pol along with two long terminal repeats (Akram et al., 2017). The pX region situated at the 3′ end of the genome encodes accessory genes such as Tax, Rex, HTLV-1 bZIP factor (HBZ), p12, p30 and p13 (Bartoe et al., 2000; Collins et al., 1998; Silverman et al., 2004; Zhao, 2016). Epidemiological studies point that although around 10 million people worldwide have been infected by HTLV-1, it has been found that only about 2–5% have developed adult T-cell leukemia/lymphoma (ATLL), known to be a very aggressive form of leukemia (Zhang et al., 2017). Apart from ATLL, this virus is also shown to cause polymyositis, HTLV-1 associated myelopathy/tropical spastic paraparesis (HAM/TSP) (Osame et al., 1986). The principal routes of HTLV-1 transmission are mainly via sexual route, breast milk and infected blood products (Okochi, Sato, & Hinuma, 1984). Though in animals, retroviruses cause insertional mutations leading to carcinogenesis, studies have pinpointed that HTLV-1 proteins, Tax-1 and HBZ, are the main causative factors for ATLL development in humans (Akram et al., 2017). HTLV-1 has been demonstrated to infect CD4 cells, CD8 cells, and dendritic cells (Lee et al., 1984). During infection, HTLV-1 spreads by cell-to-cell route within the host mainly via two mechanisms: mitosis and viral synapse transfer (Oliveira, Farre, & Bittencourt, 2016; Igakura et al., 2003).

## 7.1 Important oncoproteins in HTLV-1

A trans-activator protein named Tax, coded by the sense strand of the pX provirus region (Mesnard et al., 2015; Nicot, 2015) plays a significant yet complex role in ATLL development, resulting from its interaction with over a hundred cellular proteins to inhibit apoptosis, enhance cell signaling, stimulate dysregulation of cell cycle, interfere with DNA repair and upregulate proto-oncogenes (Boxus et al., 2008; Kannian & Green, 2010; Sun & Yamaoka, 2005). For the induction and repression of transcription of certain genes, Tax interacts with factors like NF-κB, CREB/ATF, AP-1, CBP/p300, and p300/CBP-associated factor (P-CAF) serum responsive factor (SRF) (Sun & Yamaoka, 2005).

The oncogenic function of Tax has been demonstrated *in-vivo* by the generation of different Tax transgenic mouse strains, which develop distinct tumors and/or inflammatory lesions depending on the promoter used to drive Tax expression (Mohanty & Harhaj, 2020; Grossman et al., 1995; Nerenberg et al., 1987; Rauch et al., 2009; Kwon et al., 2005). Tax can transform murine fibroblasts (Tanaka et al., 1990), immortalize primary human CD4 + lymphocytes (Robek & Ratner, 1999) and induce ATLL-like disease in transgenic mice (Hasegawa et al., 2006). Tax interaction with CREB/ATF results in repression of p53, cyclin A, and c-Myb genes (Kibler & Jeang, 2001; Nicot et al., 2000; Fujii et al., 1991). Unregulated lymphocyte development may result from continuous activation of the NF-κB pathway by Tax. Tax positively regulates G0/G1 transition by hyperphosphorylation of hDLG (Drosophila Discs Large) (Boxus et al., 2008) as well as promote G1/S transition by cyclin D/CDK activation (Lee et al., 1984). Apart from modulation of transcription and cell-cycle regulation, Tax also causes genetic damage directly by aneuploidy effect and clastogenic DNA damage and indirectly via hyper activation of the DNA damage response (DDR) pathway (Sun & Yamaoka, 2005). From the above discussion, though Tax appears to be a major factor for initiating ATLL transformation, it is usually found to be repressed after tumor initiation and only about 60% of ATLL circulating cells show detectable levels of Tax (Grassmann, Aboud, & Jeang, 2005; Takeda et al., 2004). Interestingly, the effects of Tax are evident even after its repression (Fujikawa et al., 2016).

HBZ unlike Tax, is present in 100% of ATLL cells (Bazarbachi et al., 2011) and is encoded by the antisense strand of the pX region of HTLV-1 provirus (Bartoe et al., 2000; Collins et al., 1998; Silverman et al., 2004; Zhao, 2016). Unlike Tax, which is highly immunogenic and a target of a cytotoxic T lymphocyte (CTL) response, due to its low immunogenicity, survival of HBZ in ATLL is much more feasible (Zhao, 2016; Mesnard et al., 2015). HBZ is shown to promote T cell proliferation, surprisingly, it is found to be antagonistic to several oncogenic functions of Tax, which includes activation of NF-κB pathway, CREB (cAMP-response element binding), activating protein-1 (AP-1) and NFAT (nuclear factor of activated T cells) pathways (Satou et al., 2006; Mesnard, Barbeau, & Devaux, 2006). Thus, HBZ can repress the actions of Tax protein, facilitating HTLV-1 to evade the immune system for developing ATLL (Lavorgna & Harhaj, 2014). HBZ upregulates E2F1 gene causing increase in lymphocyte proliferation (Satou et al., 2006) through the inhibition of the pro-apoptotic BIM (Bcl-2 Interacting Mediator of cell death) gene, apoptosis is prevented (Zhao, 2016). HBZ counteracts senescence by inhibiting the classical NF-κB pathway (Zhao, 2016; Zhi et al., 2011) and induces several microRNAs which lead to a compromise of integrity of host genome and hence, contributing to ATLL transformation (Watanabe, 2017). All these regulatory mechanisms affected by HBZ demonstrate its importance in oncogenesis and the very fact that its levels correlate with pro viral loads (PVL), the greatest risk factor for ATLL, adds further to its oncogenic potential (Saito et al., 2009; Usui et al., 2008).

## 7.2 Autophagy in HTLV-1 induced solid tumors

HTLV-1, being a lymphotropic oncogenic virus, promotes autophagy induction, as well as blocks autophagosome maturation resulting in an obstructed process. This blockage tends to be more harmful for cells than repression of induction steps, as accumulating autophagosomes have the potential to block multiple cellular processes. Tax-1 positively modulates pro-autophagic protein BECLIN 1 by continuous activation of the IKK complex and, subsequently NF-κB which augments autophagosome formation on lipid raft microdomains (Chen et al., 2015; Ren et al., 2015). In addition, Tax-1 also restricts the fusion of lysosomes and autophagosomes which ultimately results in the accumulation of autophagosomes (Tang et al., 2013). Another important oncoprotein noted in HTLV-1, HBZ is known to suppress both the processes of apoptosis and autophagy, which significantly contributes to HTLV- 1 mediated oncogenesis (Zhang et al., 2017). Interestingly, with the suppression of mTOR inhibitor, GADD34, HBZ restricts autophagy (Mukai & Ohshima, 2014), which additionally leads to cell proliferation and growth.

Autophagy and HTLV-1 infection are intricately linked via the Tax/IKK mediators, and the autophagy proteins regulate bidirectionally. Impairment in cellular growth proportional to decreased NF-κB and STAT3 activity is observed upon silencing of BECN1 in HTLV-1 transformed cells (Chen et al., 2015; Ducasa et al., 2021). In addition, PI3KC3 (Phosphatidylinositol 3-Kinase catalytic Subunit Type 3) or BECN1 depletion markedly retards the proliferation of HTLV-1 infected T cells. Altogether, in the lipid raft domains (LRDs), the oncoprotein Tax, with its interaction with IKKg/NEMO (NF-κB Essential Modulator) recruits the IKK complex, and via its interaction with BECN1, the autophagic PI3KC3 complex is recruited. It seems quite plausible that a positive feedback loop exists between autophagy and NF-κB pathways in HTLV-1 infection (Ren et al., 2015).

Surprisingly, autophagy deregulation by HTLV-1 have been studied to have other goals as well. In an attempt to study IKKg/NEMO interactors, the p47 protein was recently discovered to be highly decreased in cells from HTLV-1 infected as well as from ATLL patients (Shibata et al., 2012). The UBA domain of p47 is linked to degradation of ubiquitinated proteins (Hartmann-Petersen et al., 2004; Wójcik, Yano, & DeMartino, 2004) and p47 is shown to inhibit IKKg/NEMO independent of A20 and CYLD which are two known regulators of the IKK complex (Shibata et al., 2012;

Pujari et al., 2013) leading to antagonizing action of Tax on the NF-kB pathway. The autophagy receptor SQSTM-1/p62 (sequestosome 1 protein) possesses domains for recognition of LC3 (Microtubule-associated protein 1A/1B light chain 3) and ubiquitin chains to create channels for the transport of cargoes for autophagy-mediated degradation (Pankiv et al., 2007; Svenning & Johansen, 2013). It has been shown that SQSTM-1/p62 interacts directly with Tax present in the Tax/IKK complex located in structures associated with Golgi (Schwob et al., 2019).

## 7.3 ERS/UPR in HTLV-1 induced solid tumors

X-box binding protein 1 (XBP1), is a key player in the cellular unfolded protein response (UPR), having two isoforms in cells namely the spliced XBP1S and unspliced XBP1U. Within the LTR region of HTLV-1, XBP1U has been shown to bind with the 21-bp Tax-responsive element *in vitro* leading to transactivation of HTLV-1 transcription. *In-vivo* studies in HTLV-1 show that both XBP1S and XBP1U interact with Tax attach with the LTR region In the HTLV-1-infected C10/MJ and MT2 T-cell lines, It has also been demonstrated that there is elevated mRNA levels of the gene for XBP1 along with detection of several UPR genes highlighting the fact that HTLV-1 infection may trigger the UPR in host cells (Ku et al., 2008). Experiments show the role of MK-2048, a second-generation HIV-1 integrase (IN) inhibitor, as a selective and very potent killer of HTLV-1-infected cells. After treatment with MK-2048, there was significant elevation of the levels of gene expression of UPR- PERK (PKR-like ER kinase) signaling pathway in ATL cell lines, revealed by differential transcriptome profiling. HTLV-1-infected cells have been shown to be hypersensitive to endoplasmic reticulum (ER) stress-mediated apoptosis. In HTLV-1-infected cells, MK-2048 specifically activated the ER stress-related proapoptotic gene, DNA damage-inducible transcript 3 protein (DDIT3), but this activation did not happen in the uninfected cells of HTLV-1-carrier PBMCs (peripheral blood mononuclear cells). The same study showed that MK-2048 selectively promotes apoptosis in HTLV-1-infected cell through the activation of the UPR. Insights into the modulations of this novel mechanism of targeting the PERK-ATF4-CHOP pathway depicted by the HIV IN (inhibitor MK-2048) in HTLV-1-infected cells holds a great promise in the prophylactic and therapeutic management of HTLV-1-related diseases including ATL (Ikebe et al., 2020).

## 8. Hepatitis B (HBV) and its relationship to cancer

Hepatitis B is a non-cytopathic and hepatotropic DNA virus that belongs to the *Hepadnaviridae* family and is considered one of the most feared and frequent oncoviruses. According to official World Health Organization (WHO) estimates from 2019, around 300 million people are at risk of infection, and approximately 820,000 perished, the prevalent mortal factors being liver cirrhosis and hepatocellular carcinoma (HCC). Furthermore, 1.5 million new cases of chronic Hepatitis B (CHB) were identified in 2019 alone, demonstrating the ever-increasing threat posed by this DNA virus. (Jeng, Papatheodoridis, & Lok, 2023; Iannacone & Guidotti, 2022; Locarnini et al., 2015; Yuen et al., 2018; Revill et al., 2019).

Hepatitis B infection, along with Hepatitis C and D infections, alcohol abuse, Non-Alcoholic Fatty Liver Disease (NAFLD)/ Non-Alcoholic Steatohepatitis (NASH), is a crucial risk factor for the progression of HCC, one of the most common cancers worldwide, and hepatic cirrhosis, which is also a risk factor for HCC. HBV infection is linked to a higher incidence of HCC and hepatic cirrhosis in East Asia and Sub-Saharan Africa than in the rest of the world. Furthermore, studies show that HBV infection causes twice as many cirrhosis cases as HCV infection on a global scale and is the most vital force behind HCC, emphasizing the importance of HBV therapy in preventing liver cancer (Llovet et al., 2021; Yang et al., 2019; El-Serag, 2012; Yang & Roberts, 2010; Alberts et al., 2022; Global Burden of Disease Liver Cancer et al., 2017).

Following infection, the HBV genetic material (relaxed circular DNA) is transformed, by host enzymes, into a covalently closed circular (ccc) DNA, an episomal transcriptional template. The integration of viral ccc DNA into the host genome, which results in oncogene activation and enhanced HBV surface antigen (HBsAg) expression, is one of the most thoroughly characterized mechanisms by which HBV infection leads to HCC. Furthermore, the presence of cDNA in the host liver has been linked to the resurgence of HBV replication in immunocompromised patients who had otherwise recovered (Wang et al., 1990; Wooddell et al., 2017; Bousali et al., 2021; Martinez et al., 2021; Raimondo et al., 2019; Seeger & Mason, 2015).

The most used treatments include immunomodulatory drugs such as pegylated interferon alpha and direct-acting antivirals (DAA) such as third generation nucleos(t)ide analogs (NUCs). However, because none of these treatments have been proven to eliminate HBV completely, the emphasis is now on a 'functional' cure—lowering the HBsAg levels in the patient's

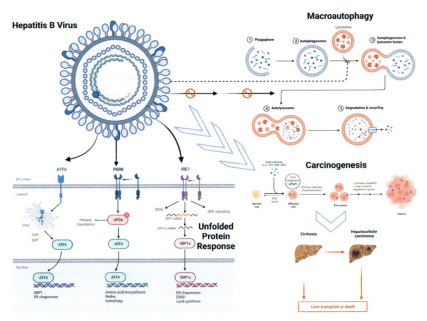

**Fig. 3** The effect of HBV infection on different stages of macro autophagy and the different branches of the Unfolded Protein Response (UPR).

serum (Locarnini et al., 2015; Yuen et al., 2018; Revill et al., 2019; Terrault et al., 2018; European Association for the Study of the Liver, 2017; Idilman, 2017; Levrero, Testoni, & Zoulim, 2016; Fanning et al., 2019; Suzuki et al., 2019; Hou et al., 2020; Marcellin et al., 2019).

### 8.1 ERS/UPR in HBV-induced solid tumors

The involvement of the UPR in human viral infections has grown in importance during the last two decades. UPR has been shown to be regulated in both RNA and DNA viruses, and in numerous cases, a direct link has been shown between UPR activation and other associated processes such as autophagy (explained in detail in this publication), inflammation, and mitochondrial pathways. (Hu et al., 2018; Choi & Song, 2019; Cirone, 2021) HBV, a well-studied member of the *Hepadnaviridae* family (enveloped DNA viruses), is an excellent illustration of how virus infection can affect UPR and vice versa (Fig. 3). For more than a decade, studies have shown that UPR occurs during entire virus infection and in cell cultures when HBV proteins are overexpressed by plasmids (Kim, Kyaw, & Cheong, 2017; Lazar, Uta, & Branza-Nichita, 2014; Lin, Hung, & Huang, 2020a; Hu et al., 2022).

It is worth noting that early research that led to the investigation of UPR in HBV were primarily based on plasmid-mediated overexpression of HBV regulatory protein X (HBx) in human hepatoma cell cultures. These findings not only established the control of numerous UPR pathways in HBx-expressing cells, but also the crosslink between UPR and other pathways in infected cells. One of the first discoveries established that HBx dramatically upregulates two UPR pathways in human hepatoma cell lines (Hep3B, HepG.2.2.15)—Activating Transcription Factor 6 (ATF-6) and Inositol Requiring Enzyme 1 (IRE1) (Li et al., 2007). A subsequent investigation found that the remaining UPR pathway—Protein Kinase R like Endoplasmic Reticulum Kinase (PERK)—was upregulated in HBx transgenic mice and transfected hepatic cell lines (Huh7, HepG2). This study demonstrated a clear relationship between HBx-induced PERK activation, ATP deprivation, and activation of cyclooxygenase-2 (COX-2) mediated inflammation, in addition to demonstrating the overexpression of the PERK pathway in HBx-transfected cells and transgenic mice. It is rather interesting that research linking UPR and metabolic pathways in the setting of HBV infection were accessible quite early on (Cho et al., 2011). Furthermore, a later study found that HBx plays a role in regulating the mitochondrial unfolded protein response (UPRmt) in nutritional restriction, which is a hallmark of cancer. Starvation (induced by Earle's Balanced Salt Solution) induces Parkin translocation to partial mitochondria, facilitating mitophagy in HepG2 and HepG2.2.15 cells expressing the HBx protein. In addition, HBx expression controls the subcellular distribution of the Lon Peptidase I (LONP1) protein, with increased mitochondrial LONP1 promoting UPRmt and decreased cytosolic LONP1 triggering PINK1-Parkin mediated mitophagy. Increased mitophagy was linked to decreased apoptosis, forming a three-pronged mechanism of HBx-mediated tumor development in human hepatocyte cell lines (Huang et al., 2018). In addition, HBXIP has been identified as an oncogene in Tamoxifen-resistant (TmaR) breast cancer cells—and unlike the studies cited above, in this case, HBXIP blocks IRE1α and the downstream ROS production and apoptosis. Moreover, the deletion of HBXIP in these cells reveals that the IRE1α-dependent ROS activates the ATF6 and PERK pathways, thus presenting HBXIP as a suppressor of all three UPR branches in TmaR cells. Since this finding is the exact opposite of the studies described above, it is crucial to consider that HBV proteins exhibit a complex relationship with the UPR pathways, depending upon the cell type and the microenvironment (Zhang et al., 2022).

The number of studies tying HBx to HBV-mediated ER stress and UPR has paved the road for therapeutic research into the viability of targeting HBx to prevent liver fibrosis and HCC. Crysophanol, a traditional Chinese anthraquinone herb, was found to be effective in preventing the spread of HBx-mediated carcinogenesis by activating ferroptosis and ER stress in rat hepatic stellate cell lines (HSC-T6). Unlike several previous research (shown above), in which HBx upregulates UPR, this report shows a Crysophanol-dependent reversal of HBx-mediated repression of both ERS/UPR and Ferroptosis. This study adds to the evidence that the relationship between viral proteins and UPR is dynamic, subject to modification and alteration depending on viral dose and the kind of infected tissues and cells (Kuo et al., 2020). Other extracellular agents like Stearic Acid (SA) and Oleic Acid (OA), common agents to induce Steatosis, are shown to induce the PERK pathway and restrict the release of HBV surface antigen (HBsAg) (Liu et al., 2022).

Like HBx, other components of the HBV genome have been implicated not only in the regulation of the UPR but also in the crosstalk between UPR and other pro-tumor pathways in HBV-induced HCC. A 2016 study, for example, identified canonical NF-κβ signaling as a significant obstacle in the evolution of HCC in HBV Surface Antigen (HBsAg) transgenic mice. According to this study, the Binding Immunoglobulin Protein (BiP)-mediated regulation of UPR is the key mechanism by which NF-κβ inhibits tumor growth in transgenic mice. When HBsAg transgenic mice were bred with ordinary mice with hepatocyte-specific NF- κβ inhibition, the progenies showed enhanced eIF2α activity, CHOP overexpression, and DNA damage, all of which resulted in 100% HCC development (Sunami et al., 2016). HBsAg overexpression has also been linked to the production of ground glass hepatocytes (GGH), a typical characteristic of chronic HBV infection. Infection with entire HBV and overexpression of HBV genes resulted in UPR activation, higher levels of HBsAg in the endoplasmic reticulum, and the development of the GGH phenotype in HepG2-NTCP and primary human hepatocytes (PHH). This reaction is dose and duration dependent; prolonged UPR activates the PERK pathway, which shifts the momentum of cell viability towards apoptosis, or programmed cell death (Li et al., 2019a). The GGH phenotype of HCC is linked to another common component of the HBV genome: the large HBV surface (LHBs) protein, specifically the pre-S domain. Studies have found a link between pre-S domain mutations and increased ERS and GGH development. Pulldown assays in hepatocellular cell lines demonstrate the interaction between the major UPR regulator protein GRP78/BiP/HSPA5 and the pre-S2 domains of the large and middle S proteins. Furthermore, type I and type II GGH are shown to be related to mutations in the pre-S1 (Delta-S1 LHBs) and pre-S2 (Delta-S2 LHBs) domains, respectively, adding to our repertoire of structural knowledge about the HBV genome and its relationship to ER stress and infectivity. In addition to ER stress, these mutations activate associated pathways such as oxidative stress, DNA damage, Cyclooxygenase 2, Cyclin A, and genomic instability, all essential prerequisites for carcinogenesis. Interestingly, Apolipoprotein H (Apo H) has been identified as a unique link between HBsAg and LHBs in infected 293T cells. HBV infection upregulates Apo H expression in 293T cells, which then inhibits HBsAg secretion, while the overexpression of LHBs induces ER stress and CHOP activation, along with ApoH activation. Finally, it has been further demonstrated that all three UPR branches are activated in HCCs after overexpression of the small protein of the Hepatitis B surface antigen—SHBs. Moreover, SHBs-induced UPR activation also converges with the increased expression of the pro-angiogenesis marker VEGFA (Vascular Endothelial Growth Factor A) (Lin et al., 2020a; Fan et al., 2001; Hsieh et al., 2004; Su et al., 2008; Suwanmanee, Wada, & Ueda, 2021; Liu et al., 2021; Wu et al., 2022).

Studies on HBV patients and hepatocellular carcinoma cell lines have also helped to improve our understanding of the cellular interaction between the components of the core UPR machinery (ATF6/ IRE1/ PERK) and additional peripheral pathways of the larger ER stress response and antiviral immune response. A study on HBsAg-transgenic mice and infected humanized-liver mice demonstrate that the interferons (IFNs) α, γ induce cell death and liver injury by the suppression of UPR. In these models, upregulation of UPR by BiP downregulation and Tunicamycin treatment successfully reversed the IFN-induced liver injury (Baudi et al., 2021) An intriguing study on human hepatoma cell lines (Huh7, HepG2, HepG2.2.2.15) emphasizes the complexities of HBV-mediated cellular responses, as in any other infection. HBV infection activates peripheral pathways such as the Endoplasmic Reticulum Associated Degradation (ERAD)—a catabolic process characterized by the degradation of misfolded proteins by ER-degradation enhancing Mannosidase-like (EDEM) proteins. This, in turn, starts a feedback loop in which EDEM-1 (a member of the EDEM family) causes autophagy-mediated destruction of HBV envelope proteins. This is also one of the first papers demonstrating interaction between an ER stress route and autophagy in HBV infection, which is discussed in greater detail in this manuscript (Lazar et al., 2012). A subsequent study, published in 2018, found that the transcription factor XBP1 (IRE1 pathway) is involved in the overexpression of a peripheral ER stress protein called mesencephalic astrocyte-derived neurotrophic factor (MANF) in HBV patients and HCC cell lines. Although MANF relates to neural protection during neurodegenerative illnesses, another important role of MANF is to protect non-neural cells from the degradative and harmful effects of ER stress. According to the findings, XBP-1 binds to two ER-specific elements (ERSE) in the MANF promoter region, promoting MANF overexpression during HBV infection. Furthermore, in the absence of Tunicamycin, a chemical stimulus that activates XBP1, it was demonstrated that MANF directly interacts with XBP1 to enhance its own transcription (Wang et al., 2018).

## 8.2 Autophagy in HBV-induced solid tumors

Autophagy is a homeostatic cellular phenomenon consisting of five distinct stages, beginning with the formation of a double membraned vesicle, followed by the enclosure of this structure, and culminating in the acidic degradation of the contents trapped inside. In the context of HBV infection, the first few steps, until up to the enclosure of the autophagosome, have a preserving effect on viral packaging because it is conducive to perpetual replication and packaging inside the host cell, protected from the antiviral pathways of the host cytoplasm. However, the final steps, namely, the fusion of the autophagosome (containing the viral cargo) and the acidic lysosome, followed by the acidic degradation of this fused autolysosome (autophagic flux), are counterproductive to infection because viral maturation and release are incomplete once viral proteins are degraded. (Fig. 3). Since the initial phase of autophagy starts with borrowed membranes from the Endoplasmic Reticulum, it is perhaps not surprising that investigation into the role of autophagy in HBV replication and infection pathways began almost simultaneously with the first investigations into the control of UPR by HBV infection. In fact, quite a few studies show the crosstalk between autophagy and ERS/UPR in HBV infection during these earliest phases of investigation. An intriguing parallel is that, like with UPR and HBV research, studies on the HBV X protein were among the first to establish the regulation of autophagy in HBV infection. A recent report also links ER Stress (ERS) and Autophagy in infected hepatic cell lines and primary human hepatocytes (PHHs). This study, designed to investigate the effects of a posttranslational modification O-GlcNAcylation, reveals that ERS and autophagosome formation converge in promoting HBV replication. The inhibition of O-GlcNAcylation induces both ER stress and autophagosome formation, accompanied by a block in the autophagosome-lysosome fusion, thus preventing the autophagic degradation of HBV proteins. Such a 'decoupling' between autophagosome formation and autophagic degradation is an important recurring theme in HBV replication, as we shall discuss in greater detail in the following paragraphs. (Tang et al., 2012; Tian, Wang, & Ou, 2015; Abdoli et al., 2019; Wang et al., 2020a; Lin et al., 2020b; Zhang, 2020).

It is interesting to observe the identification of early autophagy pathway components as crucial players in the viral lifecycle long before the rest of the autophagy pathway was linked to viral infection. One of the earliest publications demonstrates the transactivation of the Beclin-1 promoter by the overexpression of HBx protein in hepatoma cell lines (HepG2, HepG2.2.2.15). In addition, HBx overexpression and whole virus infection also induces Beclin-1-dependent autophagy in nutrient-deficit (starvation) scenarios and blocking Beclin-1 activation impedes viral replication. This study further highlights the induction of autophagy vacuole formation in cells with HBx overexpression and HBV genomic DNA expression, and the school of thought put forth in this study is the facilitation of HBV DNA replication by the formation of autophagic vacuoles, rather than the degradation of infectious particles. A contemporary analysis involving HBx overexpression in liver and hepatoma cell lines refines the above finding by illustrating that HBV genotype C triggers a more robust activation of basal autophagy than genotype B. A much more recent report shows direct interaction between HBx and the cell cycle protein Arrestin Beta 1 (ARBB1) as a pro-viral phenomenon that is responsible for an upsurge in autophagosome formation and the concomitant increase in viral replication, and this is confirmed in both mouse models and hepatocytes. In the discussion about the direct interaction between viral proteins and host factors, Interferons (IFNs) and interferon-stimulated genes (ISGs) might hold a significant position, as depicted in a recent document. The ushering in of the "COVID-19 age" has shifted focus toward antiviral immune responses in modulating autophagy during HBV infection, and a recent report shows an antiviral role of type I ISGs in autophagic degradation, the final step of autophagy. Bioluminescence Resonance Energy Transfer (BRET) data reveals a direct interaction between a type I ISG Galectin-9 and the HBV protein HBc, as fundamental to the autophagic degradation of HBc. In addition, another type I ISG, the antiviral Viperin, serves as a third factor in this interaction, which culminates in the HBc proteolysis and restriction of viral infection. Besides HBx and HBc, mutation analysis in hepatoma cells identified the involvement of another HBV protein-small surface protein (SHBs)—in the upregulation of autophagy. Treatment of the infected cells with pharmacological inhibitors (3-MA) and inducers (Rapamycin) demonstrates that autophagy, specifically LC-I lipidation, is crucial for HBV replication. More importantly, this SHBs-mediated autophagy relies on Unfolded Protein Response (UPR), one of the earliest studies illustrating direct crosstalk between UPR and autophagy. Finally, evidence is emerging that the autophagy membranes (phagophores, autophagosomes, amphisomes, autolysosomes) play instrumental roles in different phases of the HBV lifecycle inside the host, and this phenomenon involves the interaction of hyperphosphorylated core proteins and autophagic membranes. While phagophores are linked to the assembly of HBV nucleocapsids, autophagosomes facilitate the trafficking of the HBV proteins (pre-core and core), and amphisomes promote the egress of the mature HBV virions. Moreover, as shall be demonstrated in detail below, the fusion between autophagosome and lysosome, followed by the acidophilic degradation of the autophagolysosome, is a major hindrance to HBV replication and maturation. (Tang et al., 2009; Wang, Shi, & Yang, 2010; Li et al., 2011; Lei et al., 2021; Miyakawa et al., 2022; Chu et al., 2022; Chuang & James Ou, 2023).

A stream of reports soon ensued, identifying more target proteins from the early phase of autophagy that are instrumental in viral infection and favor increased viral replication and increased infectivity. As pointed above, a recurring feature among some of these studies is the 'decoupling' between the upregulation of the early autophagy and the formation of acidic vacuoles in the final step—the 'initiation' stage does not necessarily lead to autophagic vacuole formation and degradation. In a study covering the entire gamut of disease models—hepatoma cell lines, mouse liver tissues, and primary cells—HBx overexpression triggers activation of phosphatidyl-inositol-3-kinase type III (PI3K-III), a downstream target of Beclin-1, and HBV DNA replication is strongly dependent on PI3KIII activation. However, there is no indication of an increased autophagic vacuole formation in these reports, and it underlines the paramount significance of the earliest phase of autophagy induction ("initiation"/ "nucleation"/ "expansion") but not necessarily the autophagic degradation in HBV lifecycle. This phenomenon, involvement of the early autophagy pathway, has been validated by results obtained from one of the earliest transgenic models ever employed in this domain of research—a mouse model with a liver-specific *Atg5* knockout. In addition to a significant reduction in HBV replication in the *Atg5*−/− liver, there is a marked decrease of HBV antigen in the sera collected from the livers of these mice, thus revealing the therapeutic values of these findings. A later study identifies a specific Atg5 polymorphism—rs510432 AA+GA genotype—as being significantly associated with HCC progression. This study reports a significant over expression of this rs510432 AA+GA genotype in HCC and liver cirrhosis patients, as compared to chronic hepatitis patients, thus identifying a novel genetic risk factor. A similar study performed on 551 HBV-infected patients and 247 healthy controls in North India identified a genetic variant of another autophagy-related protein, ATG16, as a potential genetic marker for chronic HBV and hepatic cirrhosis. PCR-RFLP-based analysis revealed that the GG allele of the ATG16L1 variant T300A (rs2241880) was overrepresented in asymptomatic and chronic HBV carriers. In addition to the regulatory proteins depicted above, their downstream and upstream targets are crucial to this network. Among these proteins, notable examples include eIF4E Binding Protein (4EBP1) and P70-S6Kinase (S6K1), downstream targets of the mammalian target of Rapamycin (mTOR), a negative regulator of autophagy. Studies on HepG2.2.15 (HBx overexpression) have shown that S6K1 and 4EBP1 serve different roles in the HBV lifecycle. While S6K1 knockdown significantly downregulates HBx expression and autophagy, 4EBP1 exhibits minimal effect on autophagy and increases HBx expression. Moreover, API-2, an inhibitor of the pro-autophagy Akt protein (upstream regulator of PI3KI/mTOR), restricts viral replication in HepG2 cell lines, thus underlining the importance of Akt in the viral infection lifecycle (Li et al., 2011; Sir et al., 2010; Tian et al., 2011; Li et al., 2019b; Gao et al., 2019; Sharma et al., 2020).

While studying the role of a specific pathway in infection, it is crucial to expand the search for target proteins and biomarkers to pave the way for better therapeutic intervention. While early research in autophagy mostly identified a handful of components (Beclin-1, ATG5, and PI3KIII) from early autophagy, the deployment of more high-throughput technologies is proving indispensable in identifying more target proteins that are deregulated in HBV infection. A study of the peripheral mononuclear blood cells (PBMCs) of eleven CHB patients and nine healthy individuals, pulled together from two families, revealed that eleven genes were downregulated, and seven genes were upregulated in CHB patients compared to their healthy counterparts. It is remarkable that the regulated genes encompass the entire gamut of the autophagy cascade and contain components from each stage of autophagy, beginning with the initiation and ending with the autophagic degradation (Tian et al., 2019). Although the earliest studies pointed at a type of disassociation between the early and later stages of autophagy in HBV infection, it is now well established that the final stages, namely fusion and degradation, are also important. Studies leading to the identification of more autophagic proteins beyond the realm of the conventional trio described above (Beclin-1, ATG5, and PI3KIII) have also paved new avenues in the design of prospective therapies. Several reports illustrate the importance of negative regulators of autophagosome-lysosome fusion, namely, Ras-related GTP binding proteins (Rab7) and associated proteins like the Synaptosome-associated protein 29 (SNAP-29) and VAMP8, in keeping a check on HBV infection. Knockdown and silencing of these anti-fusion factors promote HBV replication and the release of HBV serum antigens (HBsAg), suggesting that these proteins are also antiviral and can thus be used for therapy design. It is vital to recognize that autophagy is such a complicated process that different stages of autophagy serve almost antagonistic requirements for the viral life cycle. As shown above, while the early stages of autophagy are mostly pro-viral and pro-replication in general, the final stages, namely, fusion/degradation have inhibitory effects on the replication and release of HBV components. (Lin et al., 2019a, 2019b).

As research progressed into the later part of last decade, investigators started unearthing more components in the story of HBV infection and its effect on autophagy. Fresh findings did not just uncover new elements of canonical macro-autophagy targeted by viral proteins; in addition, they also revealed non-coding RNAs, exosomal components, non-canonical autophagy, autophagy gene polymorphisms, oxidative stress, glucose metabolism, and extracellular stimuli (like nitric oxide, cisplatin) in the larger canvas of infection and autophagy. (Li et al., 2019b; Gao et al., 2019; Tian et al., 2019; Lin et al., 2019b; Fu et al., 2019; Wang et al., 2019b; Wu, Lan, & Liu, 2019; Chen et al., 2019; Hu et al., 2019; Liu et al., 2019; Zhang et al., 2019).

miRNAs comprise a principal category of non-coding RNAs and thus constitute an important avenue of research in several infections and diseases. HBV is no exception either—multiple studies have linked HBV-mediated autophagy and miRNAs. *In vitro* and *in vivo* data suggests the repression of an anti-autophagy miR-192-3p during HBV infection. Since miR-192-3p exerts an anti-autophagy pathway by binding to the X-linked Inhibitor of Apoptosis (XIAP), suppression of this miRNA activity is crucial to maintaining high basal autophagy and a consequence, high HBV replication. In addition, the same study identifies a direct interaction between HBx and c-Myc as the root cause of miRNA suppression, thus bringing forth a cellular network that can be harnessed for therapeutic purposes. In another study, HBV infection in patients and hepatocytes are linked to an elevation of a pro-autophagy miRNA-miR-146a-5p. In this case, both HBV core (HBc) and x (HBx) proteins are implicated in the upregulation of miR-146a-5p in the serum of CHB patients and HBV-expressing hepatocytes. This further solidifies the notion that miRNAs could serve as potent biomarkers in HBV diagnosis and therapeutic intervention, especially since a whole set of miRNAs on both spectrums of autophagy regulation, i.e., pro-, and anti-autophagy, are available (Fu et al., 2019; Wang et al., 2019b). Besides miRNAs, other non-coding RNAs are also being studied as potential novel biomarkers in HBV-induced HCC. A distinct class of long non-coding RNA (lncRNA), lncRNA activated by TGF-β(LNC-ATB), is found to be closely linked to HBx-mediated autophagy in HepG2 cells and tumor prognosis in HBV patients. Blocking LNC-ATB or TGF-β suppresses both autophagy and the spread of infection, illustrating the efficacy of non-coding RNAs in HBV therapy and diagnosis. In addition, data from HCC patients and cells infected with HBV show that the liver exosome promotes chemoresistance by inducing chaperone-mediated autophagy (LAMP2a-dependent) and the concomitant downregulation of apoptosis. Since exosomes are rich sources of coding and non-coding RNAs, this finding opens new possibilities for unearthing novel biomarkers in HBV research. (Liu et al., 2019; Zhang et al., 2020).

In addition to identifying the core components of autophagy in the HBV lifecycle, investigators also studied the byzantine network maneuvering different branches of cell death. For instance, while Beclin-1 is a prime regulator in HBV-mediated autophagy, studies in human liver cancer cells have also identified Nitric Oxide (NO) as an upstream regulator that affects Beclin-1 to tip the balance of cell death away from autophagy and toward apoptosis. Increased levels of Nitric Oxide Synthase (NOS) trigger NO production in HBV-infected cells, which leads to disruption of the pro-autophagy Beclin-1/Vps34 interaction and promotes the pro-apoptotic Beclin-1/Bcl2 interaction. Interestingly, an antidote to this cellular phenomenon was identified in a study investigating the effect of Lipopolysaccharide (LPS)-induced Toll-Like Receptor (TLR) stimulation of HBx-expressing hepatic cell lines. Following LPS treatment and TLR stimulation, HBx triggers the dissociation of BECLIN-1/BCL-2 and the promotion of TRAF-6/BECLIN-1/VPS34, thus moving the dynamics in favor of autophagy. In addition, reactive oxygen species (ROS) have been linked to autophagy and HBV replication in diverse conditions such as Cisplatin-mediated chemotherapy and the suppression of neutrophil-induced extracellular trap (NET) release during infection. Cisplatin-induced activation of ROS (JNK-dependent) and autophagy (Akt/mTOR) is responsible for HBV reactivation, a common occurrence in patients undergoing chemotherapy. In contrast, studies on HBV-infected C57BL/6 and CBV patients indicate that HBV proteins (HBV C, E, S, X) downregulate the ROS/autophagy nexus, and this plays a significant role in the suppression of neutrophil-induced extracellular trap (NET) release and boosts HBV replication (Chen et al., 2019; Hu et al., 2019; Zhang et al., 2019; Son et al., 2021).

Since autophagy is intricately linked to cellular metabolism, it is perhaps no surprise that glucose metabolism can emerge as a key player in the HBV-autophagy saga. Indeed, just like stress factors described above, glucose metabolism, is crucial in regulating HBV replication. In hepatocytes, a lower glucose concentration (5 mM) promotes HBV replication via the upregulation of the AMPK-ULK1 pathway, whereas a slightly higher glucose dose (10–25 mM) blocks autophagy (via mTOR activation) and restricts HBV replication. Yet a different study shows that interferon alpha 2a (IFNA-2a) enhances autophagy initiation and blocks autophagic degradation in the presence of both high and low glucose concentrations. The induction of autophagy, and the consequential increase in HBV replication, might explain the low efficacy associated with IFNA-2a, a commonly used HBV treatment. Thus, glucose metabolism might determine the response to various immunomodulatory responses designed for HBV prevention. In the same light, autophagy might also be crucial in the high HBV replication frequently linked to Glucosamine, a dietary supplement commonly prescribed for joint pain and osteoarthritis. The monosaccharide Glucosamine promotes HBV replication by activating autophagosome formation (Akt/mTOR pathway) and the siege of autophagosome-lysosomal fusion. It is thus important to note that the balance between the 'protective' (initiation, nucleation, phagophore expansion) and 'destructive' (fusion, degradation) aspects not only dictates the rate of HBV replication and infectivity in untreated conditions but even when treated with suggested remedies like IFNA-2A. (Lin et al., 2020c; Wang et al., 2020b; Li et al., 2022).

Table 1 The crosstalk between Unfolded Protein Response (UPR), Autophagy, and Cancers caused by oncolytic viruses—A tabular summary.

| Cancer type | Oncovirus | Origin/tissue | Role of UPR on tumor progression | Role of autophagy on tumor progression | Infection pattern | Protein activation involve | References |
|---|---|---|---|---|---|---|---|
| Burkitt lymphoma | EBV | Epithelial and Lymphoid | Upregulation | Upregulation | Latent state of EBV infection | LC3-II, BZLF1 and BRLF1 | PMID: 26335716; |
| Immunoblastic lymphoma | EBV | Epithelial and lymphoid | Upregulation | Upregulation | Latent state of EBV infection | EBNA3A, EBNA3B, and EBNA3C | PMID: 32092124; PMID: 30704144 |
| Nasopharyngeal cancer | EBV | Upper mucosal epithelium | Upregulation | Upregulation | Latent state of EBV infection | LMP1, TR-L1, XBP-1, P62, LC3-II | PMID: 19435892; PMID: 30863482 |
| Stomach (gastric) cancer | EBV | Gastrointestinal tissues | Upregulation | Upregulation | Latent state of EBV infection | Sec62, ACSS1, FAM3B, IHH, and TRABD | PMID: 33381453; https://doi.org/10.3389/fonc.2020.583463 |

| Cervical cancer | HPV | Cervix and in the vagina | Upregulation | Upregulation | Upregulation | hr-HPV16 and HPV18, & lr-HPV 6 and HPV11 infection | E5, E6, and E7 oncoproteins, Bcl-1 and LC3-II | PMID: 35456001; https://doi.org/10.3390/cells11081323 |
|---|---|---|---|---|---|---|---|---|
| non-Hodgkin lymphoma (NHL) | HIV | Lymph system | Upregulation | Upregulation | Upregulation | chronic B cell activation | Tat, IL 6 and IL10, pRb2/p130 oncosuppressor protein | PMID: 35517869; https://doi.org/10.2147/BLCTT.S361320 |
| Multicentric Castleman disease (MCD), Primary effusion lymphoma (PEL), & Inflammatory cytokine syndrome (KICS) | HHV-8 | B-cell lymphoproliferative disorders | Upregulation | Upregulation | Upregulation | Lytic replication in epithelial cells, like skin, blood vessels, and organs | LANA, v-FLIP, v-cyclin, kaposin, GPCR and Bcl2, BECLIN 1, ATG14L | PMID: 32310483 |

*(continued)*

Table 1 The crosstalk between Unfolded Protein Response (UPR), Autophagy, and Cancers caused by oncolytic viruses—A tabular summary. (cont'd)

| Cancer type | Oncovirus | Origin/tissue | Role of UPR on tumor progression | Role of autophagy on tumor progression | Infection pattern | Protein activation involve | References |
|---|---|---|---|---|---|---|---|
| Adult T-cell leukemia/ lymphoma (ATLL) | HTLV-1 | Insertional mutations | Upregulation | Upregulation/ downregulation | Transmission of infected lymphocytes | Tax, CREB, AP-1, NFAT, HBZ, XBP-1 | PMID: 32809660 |
| Liver cirrhosis & Hepatocellular carcinoma (HCC) | HBV | Liver infection | Upregulation | Upregulation | Transmitted when blood, semen, or another body fluid | HBsAg, HBeAg, HBcAg, ATF6, IRE1α, PKR, PERK | PMID: 31027244, PMID: 32942717 |

## 9. Discussion

In the preceding passages, we have summarized how six oncolytic viruses—Epstein Barr Virus (EBV) (Fig. 1), Human Papilloma Virus (HPV), Human Immunodeficiency Virus (HIV), Human Herpesvirus 8 (HHV-8) (Fig. 2), Human T-cell Lymphotropic Virus Type 1 (**HTLV-1**), and Hepatitis B virus (HBV) (Fig. 3)—exhibit a close relationship with both the pathways of Unfolded Protein Response (UPR) and Autophagy in infected cells (Table 1). Although studies abound in investigating autophagy and UPR pathways in infection and other diseases separately (Choi & Song, 2019; Klionsky et al., 2021), our review is unique because we try to bring together these closely related and 'linked' processes—UPR and autophagy—in the context of oncolytic tumor formation. This report is also relevant because it delves into two primary aspects of oncoviruses, i.e., infection itself and tumor formation, processes that are heavily dependent upon and frequently regulated by UPR and autophagy. As has been demonstrated in multiple studies, UPR pathways like PERK-eIF2α-CHOP occupies a pivotal position in determining the fate of cell survival and plays crucial roles in deciding if stressed cells commit to pro-survival pathways or pro-apoptotic pathways (Hu et al., 2018; Liu et al., 2015). In a similar fashion, autophagy has significant relevance in both tumor progression and regulation of the spread of viral infection. This book chapter thus aims to summarize different themes in the context of oncolytic viral infection: regulation of virus-induced tumor formation, infectivity of the oncolytic viruses, stress pathways in infected cells, and the status of homeostasis in infected cells (Ma & Hendershot, 2004; Feldman, Chauhan, & Koong, 2005; Corazzari et al., 2017b; Prasad & Greber, 2021; Mehrbod et al., 2019; Chan, 2014; Chiramel, Brady, & Bartenschlager, 2013; Mao et al., 2019; Debnath, Gammoh, & Ryan, 2023; Chen, Gao, & Su, 2021; Wang et al., 2011; Poillet-Perez & White, 2019; Kimmelman & White, 2017; Mowers, Sharifi, & Macleod, 2018). Oncolytic viruses occupy a special relevance in the global health landscape (Mui, Haley, & Tyring, 2017b; Krump & You, 2018). Hence, it is imperative to gain a better understanding of the cellular processes impacted in the course of infection. Like any other virus infection, a robust translational burden is imposed on the endoplasmic reticulum (ER) of the infected host cells, thus triggering multiple waves of UPR pathways. On the other hand, since autophagy is heavily dependent upon the ER as a major source of membranes for the autophagosome, those above translational "overload" has also been shown to induce autophagy via UPR and ER stress. Moreover, several studies have shown the presence of viral particles in autophagosomes, thus highlighting the importance of autophagy in mitigating infection. With the increasing importance of UPR and autophagy, various targeted therapies are also devised to modulate these pathways in cancer and other human diseases (Hetz, Axten, & Patterson, 2019; Rivas, Vidal, & Hetz, 2015; Nagelkerke et al., 2014; Park & Ozcan, 2013; Cirone et al., 2019; Towers & Thorburn, 2016; Cheng et al., 2013; Amaravadi & Thompson, 2007). In this book chapter, the authors have laid out different markers from Unfolded Protein Response and Autophagy that could potentially be harnessed to design similar targeted therapies against the cancers triggered by infection with these oncolytic viruses.

## References

Abdoli, A., et al. (2019). Harmonized autophagy versus full-fledged hepatitis B virus: Victorious or defeated. *Viral Immunology, 32*(8), 322–334.

Adolph, T. E., et al. (2013). Paneth cells as a site of origin for intestinal inflammation. *Nature, 503*(7475), 272–276.

Akram, N., et al. (2017). Oncogenic role of tumor viruses in humans. *Viral Immunology, 30*(1), 20–27.

Alberts, C. J., et al. (2022). Worldwide prevalence of hepatitis B virus and hepatitis C virus among patients with cirrhosis at country, region, and global levels: A systematic review. *The Lancet Gastroenterology and Hepatology, 7*(8), 724–735.

Amaravadi, R. K., & Thompson, C. B. (2007). The roles of therapy-induced autophagy and necrosis in cancer treatment. *Clinical Cancer Research: An Official Journal of the American Association for Cancer Research, 13*(24), 7271–7279.

Armitage, J. O., & Weisenburger, D. D. (1998). New approach to classifying non-Hodgkin's lymphomas: Clinical features of the major histologic subtypes. Non-Hodgkin's Lymphoma Classification Project. *Journal of Clinical Oncology: Official Journal of the American Society of Clinical Oncology, 16*(8), 2780–2795.

Asha, K., & Sharma-Walia, N. (2018). Virus and tumor microenvironment induced ER stress and unfolded protein response: From complexity to therapeutics. *Oncotarget, 9*(61), 31920.

Bartoe, J. T., et al. (2000). Functional role of pX open reading frame II of human T-lymphotropic virus type 1 in maintenance of viral loads in vivo. *Journal of Virology, 74*(3), 1094–1100.

Baudi, I., et al. (2021). Interferon signaling suppresses the unfolded protein response and induces cell death in hepatocytes accumulating hepatitis B surface antigen. *PLoS Pathogens, 17*(5), e1009228.

Baumforth, K., et al. (1999b). The Epstein-Barr virus and its association with human cancers. *Molecular Pathology, 52*(6), 307.

Baumforth, K. R., et al. (1999a). The Epstein-Barr virus and its association with human cancers. *Molecular Pathology: MP, 52*(6), 307–322.

Bazarbachi, A., et al. (2011). How I treat adult T-cell leukemia/lymphoma. *Blood, The Journal of the American Society of Hematology, 118*(7), 1736–1745.

B'Chir, W., et al. (2013). The eIF2alpha/ATF4 pathway is essential for stress-induced autophagy gene expression. *Nucleic Acids Research, 41*(16), 7683–7699.

Behnke, J., Feige, M. J., & Hendershot, L. M. (2015). BiP and its nucleotide exchange factors Grp170 and Sil1: Mechanisms of action and biological functions. *Journal of Molecular Biology, 427*(7), 1589–1608.

Beral, V., et al. (1990). Kaposi's sarcoma among persons with AIDS: A sexually transmitted infection? *The Lancet, 335*(8682), 123–128.

Best, S., et al. (2019a). Targeting ubiquitin-activating enzyme induces ER stress-mediated apoptosis in B-cell lymphoma cells. *Blood Advances, 3*(1), 51–62.

Best, S., et al. (2019b). Targeting ubiquitin-activating enzyme induces ER stress–mediated apoptosis in B-cell lymphoma cells. *Blood Advances, 3*(1), 51–62.

Bhatt, A. P., & Damania, B. (2013). AKTivation of PI3K/AKT/mTOR signaling pathway by KSHV. *Frontiers in Immunology, 3*, 401.

Bhende, P. M., et al. (2007). X-box-binding protein 1 activates lytic Epstein-Barr virus gene expression in combination with protein kinase D. *Journal of Virology, 81*(14), 7363–7370.

Bishop, B.N., & Kaposi Sarcoma, L.D. (2022). StatPearls [Internet]. Treasure Island, FL: StatPearls Publishing.

Bousali, M., et al. (2021). Hepatitis B virus DNA integration, chronic infections and hepatocellular carcinoma. *Microorganisms, 9*(8).

Bouvard, V., et al. (2009). A review of human carcinogens—Part B: Biological agents. *The Lancet Oncology, 10*(4), 321–322.

Boxus, M., et al. (2008). The HTLV-1 tax interactome. *Retrovirology, 5*(1), 76.

Brinkmann, M. M., & Schulz, T. F. (2006). Regulation of intracellular signalling by the terminal membrane proteins of members of the Gammaherpesvirinae. *Journal of General Virology, 87*(5), 1047–1074.

CDC. (2023). *National center for HIV/AIDS, viral hepatitis, STD, and TB prevention.*

Chan, S. W. (2014). The unfolded protein response in virus infections. *Frontiers in Microbiology, 5*, 518.

Chen, C., Gao, H., & Su, X. (2021). Autophagy-related signaling pathways are involved in cancer (Review). *Experimental and Therapeutic Medicine, 22*(1), 710.

Chen, D., Xiang, Q., & Nicholas, J. (2017b). Human herpesvirus 8 interleukin-6 interacts with calnexin cycle components and promotes protein folding. *Journal of Virology, 91*(22) p. 10.1128/jvi. 00965-17.

Chen, H.-S., et al. (2017a). BET-inhibitors disrupt Rad21-dependent conformational control of KSHV latency. *PLoS Pathogens, 13*(1), e1006100.

Chen, L., et al. (2015). The autophagy molecule Beclin 1 maintains persistent activity of NF-κB and Stat3 in HTLV-1-transformed T lymphocytes. *Biochemical and Biophysical Research Communications, 465*(4), 739–745.

Chen, X., et al. (2019). Cisplatin induces autophagy to enhance hepatitis B virus replication via activation of ROS/JNK and inhibition of the Akt/mTOR pathway. *Free Radical Biology & Medicine, 131*, 225–236.

Cheng, Y., et al. (2013). Therapeutic targeting of autophagy in disease: Biology and pharmacology. *Pharmacological Reviews, 65*(4), 1162–1197.

Chiramel, A. I., Brady, N. R., & Bartenschlager, R. (2013). Divergent roles of autophagy in virus infection. *Cells, 2*(1), 83–104.

Cho, H. K., et al. (2011). Endoplasmic reticulum stress induced by hepatitis B virus X protein enhances cyclo-oxygenase 2 expression via activating transcription factor 4. *The Biochemical Journal, 435*(2), 431–439.

Choi, J. A., & Song, C. H. (2019). Insights into the role of endoplasmic reticulum stress in infectious diseases. *Frontiers in Immunology, 10*, 3147.

Chu, J. Y. K., et al. (2022). Autophagic membranes participate in hepatitis B virus nucleocapsid assembly, precore and core protein trafficking, and viral release. *Proceedings of the National Academy of Sciences of the United States of America, 119*(30) e2201927119.

Chuang, Y. C., & James Ou, J. H. (2023). Regulation of hepatitis B virus replication by autophagic membranes. *Autophagy, 19*(4), 1357–1358.

Chun, Y., & Kim, J. (2018). Autophagy: An essential degradation program for cellular homeostasis and life. *Cells, 7*(12), 278.

Cirone, M. (2018). EBV and KSHV infection dysregulates autophagy to optimize viral replication, prevent immune recognition and promote tumorigenesis. *Viruses, 10*(11), 599.

Cirone, M. (2021). ER stress, UPR activation and the inflammatory response to viral infection. *Viruses, 13*(5).

Cirone, M., et al. (2019). Autophagy manipulation as a strategy for efficient anticancer therapies: Possible consequences. *Journal of Experimental & Clinical Cancer Research: CR, 38*(1), 262.

Collins, N. D., et al. (1998). Selective ablation of human T-cell lymphotropic virus type 1 p12I reduces viral infectivity *in vivo*. *Blood, The Journal of the American Society of Hematology, 91*(12), 4701–4707.

Condjella, R., et al. (2009). The canine papillomavirus E5 protein signals from the endoplasmic reticulum. *Journal of Virology, 83*(24), 12833–12841.

Corazzari, M., et al. (2017a). Endoplasmic reticulum stress, unfolded protein response, and cancer cell fate. *Frontiers in Oncology, 7*.

Corazzari, M., et al. (2017b). Endoplasmic reticulum stress, unfolded protein response, and cancer cell fate. *Frontiers in Oncology, 7*, 78.

Coşkun, A., et al. (2020). Investigation of ganciclovir resistance in cytomegalovirus strains obtained from immunocompromised patients. *Mikrobiyoloji Bulteni, 54*(4), 619–628.

Das, G., Shravage, B. V., & Baehrecke, E. H. (2012). Regulation and function of autophagy during cell survival and cell death. *Cold Spring Harbor Perspectives in Biology, 4*(6), a008813.

De Martel, C., et al. (2012). Global burden of cancers attributable to infections in 2008: A review and synthetic analysis. *The Lancet Oncology, 13*(6), 607–615.

De Paoli, P., & Carbone, A. (2016). Kaposi's Sarcoma Herpesvirus: Twenty years after its discovery. *European Review for Medical & Pharmacological Sciences, 20*, 7.

de Sanjosé, S., Brotons, M., & Pavón, M. A. (2018). The natural history of human papillomavirus infection. *Best Practice & Research. Clinical Obstetrics & Gynaecology, 47*, 2–13.

Debnath, J., Gammoh, N., & Ryan, K. M. (2023). Autophagy and autophagy-related pathways in cancer. *Nature Reviews. Molecular Cell Biology, 24*(8), 560–575.

Deegan, S., et al. (2013). Stress-induced self-cannibalism: On the regulation of autophagy by endoplasmic reticulum stress. *Cellular and Molecular Life Sciences: CMLS, 70*(14), 2425–2441.

Deutsch, E., et al. (2004). Role of protein kinase C δ in reactivation of Kaposi's sarcoma-associated herpesvirus. *Journal of Virology, 78*(18), 10187–10192.

Dow, D. E., Cunningham, C. K., & Buchanan, A. M. (2014). A review of human herpesvirus 8, the Kaposi's sarcoma-associated herpesvirus, in the pediatric population. *Journal of the Pediatric Infectious Diseases Society, 3*(1), 66–76.

Ducasa, N., et al. (2021). Autophagy in human T-cell leukemia virus type 1 (HTLV-1) induced leukemia. *Frontiers in Oncology, 11*, 641269.

El-Serag, H. B. (2012). Epidemiology of viral hepatitis and hepatocellular carcinoma. *Gastroenterology, 142*(6), 1264–1273.e1.

Ethel, C., et al. (2019). Kaposi sarcoma (Primer). *Nature Reviews: Disease Primers, 5*(1).

European Association for the Study of the Liver. (2017). Electronic address, e.e.e. and L. European Association for the Study of the, EASL 2017 Clinical Practice Guidelines on the management of hepatitis B virus infection. *Journal of Hepatology, 67*(2), 370–398.

Fan, Y. F., et al. (2001). Prevalence and significance of hepatitis B virus (HBV) pre-S mutants in serum and liver at different replicative stages of chronic HBV infection. *Hepatology (Baltimore, Md.), 33*(1), 277–286.

Fanning, G. C., et al. (2019). Therapeutic strategies for hepatitis B virus infection: Towards a cure. *Nature Reviews. Drug Discovery, 18*(11), 827–844.

Fatahzadeh, M. (2012). Kaposi sarcoma: Review and medical management update. *Oral Surgery, Oral Medicine, Oral Pathology and Oral Radiology, 113*(1), 2–16.

Feldman, D. E., Chauhan, V., & Koong, A. C. (2005). The unfolded protein response: A novel component of the hypoxic stress response in tumors. *Molecular Cancer Research: MCR, 3*(11), 597–605.

Flore, O., et al. (1998). Transformation of primary human endothelial cells by Kaposi's sarcoma-associated herpesvirus. *Nature, 394*(6693), 588–592.

Fontham, E. T., et al. (2020). Cervical cancer screening for individuals at average risk: 2020 guideline update from the American Cancer Society. *CA: A Cancer Journal for Clinicians, 70*(5), 321–346.

Friborg, J. Jr, et al. (1999). p53 inhibition by the LANA protein of KSHV protects against cell death. *Nature, 402*(6764), 889–894.

Fu, L., et al. (2019). miR-146a-5p enhances hepatitis B virus replication through autophagy to promote aggravation of chronic hepatitis B. *IUBMB Life, 71*(9), 1336–1346.

Fujii, M., et al. (1991). HTLV-1 Tax induces expression of various immediate early serum responsive genes. *Oncogene, 6*(6), 1023–1029.

Fujikawa, D., et al. (2016). Polycomb-dependent epigenetic landscape in adult T-cell leukemia. *Blood, The Journal of the American Society of Hematology, 127*(14), 1790–1802.

Ganem, D. (2007). KSHV-induced oncogenesis. *Human Herpesviruses: Biology, Therapy, and Immunoprophylaxis.*

Gantt, S., & Casper, C. (2011). Human herpesvirus 8-associated neoplasms: The roles of viral replication and antiviral treatment. *Current Opinion in Infectious Diseases, 24*(4), 295.

Gao, Q., et al. (2019). Distinct role of 4E-BP1 and S6K1 in regulating autophagy and hepatitis B virus (HBV) replication. *Life Sciences, 220*, 1–7.

Giffin, L., et al. (2014). Modulation of Kaposi's sarcoma-associated herpesvirus interleukin-6 function by hypoxia-upregulated protein 1. *Journal of Virology, 88*(16), 9429–9441.

Global Burden of Disease Liver Cancer, C., et al. (2017). The burden of primary liver cancer and underlying etiologies from 1990 to 2015 at the global, regional, and national level: Results from the global burden of disease study 2015. *JAMA Oncology, 3*(12), 1683–1691.

Granato, M., et al. (2017). Bortezomib promotes KHSV and EBV lytic cycle by activating JNK and autophagy. *Scientific Reports, 7*(1), 13052.

Grassmann, R., Aboud, M., & Jeang, K.-T. (2005). Molecular mechanisms of cellular transformation by HTLV-1 Tax. *Oncogene, 24*(39), 5976–5985.

Greenblatt, D., & Wyganski, I. (2003). Effect of leading-edge curvature on airfoil separation control. *Journal of Aircraft, 40*(3), 473–481.

Grossman, W. J., et al. (1995). Development of leukemia in mice transgenic for the tax gene of human T-cell leukemia virus type I. *Proceedings of the National Academy of Sciences, 92*(4), 1057–1061.

Grulich, A. E., et al. (2007). Incidence of cancers in people with HIV/AIDS compared with immunosuppressed transplant recipients: A meta-analysis. *Lancet, 370*(9581), 59–67.

Haberzettl, P., & Hill, B. G. (2013). Oxidized lipids activate autophagy in a JNK-dependent manner by stimulating the endoplasmic reticulum stress response. *Redox Biology, 1*(1), 56–64.

Hartmann-Petersen, R., et al. (2004). The Ubx2 and Ubx3 cofactors direct Cdc48 activity to proteolytic and nonproteolytic ubiquitin-dependent processes. *Current Biology, 14*(9), 824–828.

Hasegawa, H., et al. (2006). Thymus-derived leukemia-lymphoma in mice transgenic for the Tax gene of human T-lymphotropic virus type I. *Nature Medicine, 12*(4), 466–472.

Hetz, C., Axten, J. M., & Patterson, J. B. (2019). Pharmacological targeting of the unfolded protein response for disease intervention. *Nature Chemical Biology, 15*(8), 764–775.

Hetz, C. (2012a). The unfolded protein response: Controlling cell fate decisions under ER stress and beyond. *Nature Reviews. Molecular Cell Biology, 13*(2), 89–102.

Hetz, C. (2012b). The unfolded protein response: Controlling cell fate decisions under ER stress and beyond. *Nature Reviews. Molecular Cell Biology, 13*(2), 89–102.

Hotamisligil, G. S. (2010). Endoplasmic reticulum stress and the inflammatory basis of metabolic disease. *Cell, 140*(6), 900–917.

Hou, J. L., et al. (2020). Outcomes of long-term treatment of chronic HBV infection with entecavir or other agents from a randomized trial in 24 countries. *Clinical Gastroenterology and Hepatology: The Official Clinical Practice Journal of the American Gastroenterological Association, 18*(2), 457–467.e21.

Hsieh, Y. H., et al. (2004). Pre-S mutant surface antigens in chronic hepatitis B virus infection induce oxidative stress and DNA damage. *Carcinogenesis, 25*(10), 2023–2032.

Hu, H., et al. (2018). The C/EBP homologous protein (CHOP) transcription factor functions in endoplasmic reticulum stress-induced apoptosis and microbial infection. *Frontiers in Immunology, 9*, 3083.

Hu, S., et al. (2019). Hepatitis B virus inhibits neutrophil extracellular trap release by modulating reactive oxygen species production and autophagy. *Journal of Immunology, 202*(3), 805–815.

Hu, T., et al. (2022). Endoplasmic reticulum stress in hepatitis B virus and hepatitis C virus infection. *Viruses, 14*(12).

Huang, X. Y., et al. (2018). Hepatitis B virus X protein elevates Parkin-mediated mitophagy through Lon Peptidase in starvation. *Experimental Cell Research, 368*(1), 75–83.

Hui, K. F., et al. (2016). Inhibition of class I histone deacetylases by romidepsin potently induces Epstein-Barr virus lytic cycle and mediates enhanced cell death with ganciclovir. *International Journal of Cancer, 138*(1), 125–136.

Iannacone, M., & Guidotti, L. G. (2022). Immunobiology and pathogenesis of hepatitis B virus infection. *Nature Reviews. Immunology, 22*(1), 19–32.

Idilman, R. (2017). The summarized of EASL 2017 Clinical Practice Guidelines on the management of hepatitis B virus infection. *The Turkish Journal of Gastroenterology: the Official Journal of Turkish Society of Gastroenterology, 28*(5), 412–416.

Igakura, T., et al. (2003). Spread of HTLV-I between lymphocytes by virus-induced polarization of the cytoskeleton. *Science (New York, N. Y.), 299*(5613), 1713–1716.

Ikebe, E., et al. (2020). Activation of PERK-ATF4-CHOP pathway as a novel therapeutic approach for efficient elimination of HTLV-1–infected cells. *Blood Advances, 4*(9), 1845–1858.

Iscovich, J., et al. (2000). Classic kaposi sarcoma: Epidemiology and risk factors. *Cancer, 88*(3), 500–517.

Ishii, T., et al. (2019). Clinical differentiation of infectious mononucleosis that is caused by Epstein-Barr virus or cytomegalovirus: A single-center case-control study in Japan. *Journal of Infection and Chemotherapy: Official Journal of the Japan Society of Chemotherapy, 25*(6), 431–436.

Jeng, W. J., Papatheodoridis, G. V., & Lok, A. S. F. (2023). Hepatitis B. *Lancet, 401*(10381), 1039–1052.

Jiang, L., & Li, N. (2020). B-cell non-Hodgkin lymphoma: Importance of angiogenesis and antiangiogenic therapy. *Angiogenesis, 23*(4), 515–529.

Johnston, B. P., & McCormick, C. (2019). Herpesviruses and the unfolded protein response. *Viruses, 12*(1), 17.

Kannian, P., & Green, P. L. (2010). Human T lymphotropic virus type 1 (HTLV-1): Molecular biology and oncogenesis. *Viruses, 2*(9), 2037–2077.

Kaposi (1872). Idiopathisches multiples pigmentsarkom der haut. *Archiv für Dermatologie und Syphilis, 4*, 265–273.

Kibler, K. V., & Jeang, K.-T. (2001). CREB/ATF-dependent repression of cyclin a by human T-cell leukemia virus type 1 Tax protein. *Journal of Virology, 75*(5), 2161–2173.

Kilger, E., et al. (1998). Epstein–Barr virus-mediated B-cell proliferation is dependent upon latent membrane protein 1, which simulates an activated CD40 receptor. *The EMBO Journal, 17*(6), 1700–1709.

Kim, S. Y., Kyaw, Y. Y., & Cheong, J. (2017). Functional interaction of endoplasmic reticulum stress and hepatitis B virus in the pathogenesis of liver diseases. *World Journal of Gastroenterology: WJG, 23*(43), 7657–7665.

Kimmelman, A. C., & White, E. (2017). Autophagy and tumor metabolism. *Cell Metabolism, 25*(5), 1037–1043.

King, C. A. (2013). Kaposi's sarcoma-associated herpesvirus kaposin B induces unique monophosphorylation of STAT3 at serine 727 and MK2-mediated inactivation of the STAT3 transcriptional repressor TRIM28. *Journal of Virology, 87*(15), 8779–8791.

Klionsky, D. J., et al. (2021). Autophagy in major human diseases. *The EMBO Journal, 40*(19), e108863.

Kögel, D. (2012). The head of janus: Exploiting autophagy for cancer therapy. *Journal of Developing Drugs, 1*(e109).

Krump, N. A., & You, J. (2018). Molecular mechanisms of viral oncogenesis in humans. *Nature Reviews. Microbiology, 16*(11), 684–698.

Ku, S. C., et al. (2008). XBP-1, a novel human T-lymphotropic virus type 1 (HTLV-1) tax binding protein, activates HTLV-1 basal and tax-activated transcription. *Journal of Virology, 82*(9), 4343–4353.

Kuo, C. Y., et al. (2020). Chrysophanol attenuates hepatitis B virus X protein-induced hepatic stellate cell fibrosis by regulating endoplasmic reticulum stress and ferroptosis. *Journal of Pharmacological Sciences, 144*(3), 172–182.

Kwon, H., et al. (2005). Lethal cutaneous disease in transgenic mice conditionally expressing type I human T cell leukemia virus Tax. *Journal of Biological Chemistry, 280*(42), 35713–35722.

Lavorgna, A., & Harhaj, E. W. (2014). Regulation of HTLV-1 tax stability, cellular trafficking and NF-κB activation by the ubiquitin-proteasome pathway. *Viruses, 6*(10), 3925–3943.

Lavrik, I. N., Golks, A., & Krammer, P. H. (2005). Caspases: Pharmacological manipulation of cell death. *The Journal of Clinical Investigation, 115*(10), 2665–2672.

Lazar, C., et al. (2012). Activation of ERAD pathway by human hepatitis B virus modulates viral and subviral particle production. *PLoS One, 7*(3), e34169.

Lazar, C., Uta, M., & Branza-Nichita, N. (2014). Modulation of the unfolded protein response by the human hepatitis B virus. *Front Microbiol, 5*, 433.

Lee, D. Y., & Sugden, B. (2008). The LMP1 oncogene of EBV activates PERK and the unfolded protein response to drive its own synthesis. *Blood, The Journal of the American Society of Hematology, 111*(4), 2280–2289.

Lee, D. Y., Lee, J., & Sugden, B. (2009a). The unfolded protein response and autophagy: Herpesviruses rule!. *Journal of Virology, 83*(3), 1168–1172.

Lee, H.-H., et al. (2008). Essential role of PKCδ in histone deacetylase inhibitor-induced Epstein–Barr virus reactivation in nasopharyngeal carcinoma cells. *Journal of General Virology, 89*(4), 878–883.

Lee, J.-S., et al. (2009b). FLIP-mediated autophagy regulation in cell death control. *Nature Cell Biology, 11*(11), 1355–1362.

Lee, S. Y., et al. (2019). Endoplasmic reticulum stress enhances the antigen-specific T cell immune responses and therapeutic antitumor effects generated by therapeutic HPV vaccines. *Journal of Biomedical Science, 26*(1), 41.

Lee, T., et al. (1984). Human T-cell leukemia virus-associated membrane antigens: Identity of the major antigens recognized after virus infection. *Proceedings of the National Academy of Sciences, 81*(12), 3856–3860.

Lei, Y., et al. (2021). HBx induces hepatocellular carcinogenesis through ARRB1-mediated autophagy to drive the G(1)/S cycle. *Autophagy, 17*(12), 4423–4441.

Leidal, A. M., et al. (2012). Subversion of autophagy by Kaposi's sarcoma-associated herpesvirus impairs oncogene-induced senescence. *Cell Host & Microbe, 11*(2), 167–180.

Leung, H. J., et al. (2012). Activation of the unfolded protein response by 2-deoxy-D-glucose inhibits Kaposi's sarcoma-associated herpesvirus replication and gene expression. *Antimicrobial Agents and Chemotherapy, 56*(11), 5794–5803.

Levrero, M., Testoni, B., & Zoulim, F. (2016). HBV cure: Why, how, when? *Current Opinion in Virology, 18*, 135–143.

Li, B., et al. (2007). Hepatitis B virus X protein (HBx) activates ATF6 and IRE1-XBP1 pathways of unfolded protein response. *Virus Research, 124*(1-2), 44–49.

Li, J., et al. (2022). Interferon alpha induces cellular autophagy and modulates hepatitis B virus replication. *Frontiers in Cellular and Infection Microbiology, 12*, 804011.

Li, J., et al. (2011). Subversion of cellular autophagy machinery by hepatitis B virus for viral envelopment. *Journal of Virology, 85*(13), 6319–6333.

Li, N., et al. (2019b). Autophagy-related 5 gene rs510432 polymorphism is associated with hepatocellular carcinoma in patients with chronic hepatitis B virus infection. *Immunological Investigations, 48*(4), 378–391.

Li, S., et al. (2016). Fine-tuning of the Kaposi's sarcoma-associated herpesvirus life cycle in neighboring cells through the RTA-JAG1-Notch pathway. *PLoS Pathogens, 12*(10), e1005900.

Li, X., et al. (2015). Autophagy knocked down by high-risk HPV infection and uterine cervical carcinogenesis. *International Journal of Clinical and Experimental Medicine, 8*(7), 10304–10314.

Li, Y., et al. (2019a). Hepatitis B surface antigen activates unfolded protein response in forming ground glass hepatocytes of chronic hepatitis B. *Viruses, 11*(4).

Liang, Q., et al. (2013). Kaposi's sarcoma-associated herpesvirus K7 modulates Rubicon-mediated inhibition of autophagosome maturation. *Journal of Virology, 87*(22), 12499–12503.

Lin, J. H., et al. (2007). IRE1 signaling affects cell fate during the unfolded protein response. *Science (New York, N. Y.), 318*(5852), 944–949.

Lin, W. L., Hung, J. H., & Huang, W. (2020a). Association of the hepatitis B virus large surface protein with viral infectivity and endoplasmic reticulum stress-mediated liver carcinogenesis. *Cells, 9*(9).

Lin, Y., et al. (2020c). Glucosamine promotes hepatitis B virus replication through its dual effects in suppressing autophagic degradation and inhibiting MTORC1 signaling. *Autophagy, 16*(3), 548–561.

Lin, Y., et al. (2019b). Hepatitis B virus is degraded by autophagosome-lysosome fusion mediated by Rab7 and related components. *Protein Cell, 10*(1), 60–66.

Lin, Y., et al. (2020b). Interplay between cellular autophagy and hepatitis B virus replication: A systematic review. *Cells, 9*(9).

Lin, Y., et al. (2019a). Synaptosomal-associated protein 29 is required for the autophagic degradation of hepatitis B virus. *The FASEB Journal, 33*(5), 6023–6034.

Liu, D. X., et al. (2019). Exosomes derived from HBV-associated liver cancer promote chemoresistance by upregulating chaperone-mediated autophagy. *Oncology Letters, 17*(1), 323–331.

Liu, Q., et al. (2022). Hepatocyte steatosis inhibits hepatitis B virus secretion via induction of endoplasmic reticulum stress. *Molecular and Cellular Biochemistry, 477*(11), 2481–2491.

Liu, Y., et al. (2021). Apolipoprotein H drives hepatitis B surface antigen retention and endoplasmic reticulum stress during hepatitis B virus infection. *The International Journal of Biochemistry & Cell Biology, 131*, 105906.

Liu, Z., et al. (2015). Protein kinase R-like ER kinase and its role in endoplasmic reticulum stress-decided cell fate. *Cell Death & Disease, 6*(7), e1822.
Llovet, J. M., et al. (2021). Hepatocellular carcinoma. *Nature Reviews Disease Primers, 7*(1), 6.
Locarnini, S., et al. (2015). Strategies to control hepatitis B: Public policy, epidemiology, vaccine and drugs. *Journal of Hepatology, 62*(1 Suppl), S76–S86.
Lorin, S., et al. (2013). *Autophagy regulation and its role in cancer. Seminars in cancer biology.* Elsevier.
Ma, Y., & Hendershot, L. M. (2004). The role of the unfolded protein response in tumour development: Friend or foe? *Nature Reviews. Cancer, 4*(12), 966–977.
Mao, J., et al. (2019). Autophagy and viral infection. *Advances in Experimental Medicine and Biology, 1209*, 55–78.
Marcellin, P., et al. (2019). Ten-year efficacy and safety of tenofovir disoproxil fumarate treatment for chronic hepatitis B virus infection. *Liver International: Official Journal of the International Association for the Study of the Liver, 39*(10), 1868–1875.
Martinez, M. G., et al. (2021). Covalently closed circular DNA: The ultimate therapeutic target for curing HBV infections. *Journal of Hepatology, 75*(3), 706–717.
Matar, C. G., et al. (2014). Murine gammaherpesvirus 68 reactivation from B cells requires IRF4 but not XBP-1. *Journal of Virology, 88*(19), 11600–11610.
McKnight, N. C., & Yue, Z. (2013). Beclin 1, an essential component and master regulator of PI3K-III in health and disease. *Current Pathobiology Reports, 1*(4), 231–238.
Mehrbod, P., et al. (2019). The roles of apoptosis, autophagy and unfolded protein response in arbovirus, influenza virus, and HIV infections. *Virulence, 10*(1), 376–413.
Mesnard, J.-M., Barbeau, B. T., & Devaux, C. (2006). HBZ, a new important player in the mystery of adult T-cell leukemia. *Blood, 108*(13), 3979–3982.
Mesnard, J.-M., et al. (2015). Roles of HTLV-1 basic zip factor (HBZ) in viral chronicity and leukemic transformation. Potential new therapeutic approaches to prevent and treat HTLV-1-related diseases. *Viruses, 7*(12), 6490–6505.
Miyakawa, K., et al. (2022). Galectin-9 restricts hepatitis B virus replication via p62/SQSTM1-mediated selective autophagy of viral core proteins. *Nature Communications, 13*(1), 531.
Mizushima, N., & Komatsu, M. (2011). Autophagy: Renovation of cells and tissues. *Cell, 147*(4), 728–741.
Mohanty, S., & Harhaj, E. W. (2020). Mechanisms of oncogenesis by HTLV-1 Tax. *Pathogens, 9*(7), 543.
Moore-Carrasco, R., et al. (2009). Both AP-1 and NF-kappaB seem to be involved in tumour growth in an experimental rat hepatoma. *Anticancer Research, 29*(4), 1315–1317.
Moore-Carrasco, R., et al. (2007). The AP-1/NF-kappaB double inhibitor SP100030 can revert muscle wasting during experimental cancer cachexia. *International Journal of Oncology, 30*(5), 1239–1245.
Mosier, D. E., et al. (1988). Transfer of a functional human immune system to mice with severe combined immunodeficiency. *Nature, 335*(6187), 256–259.
Mowers, E. E., Sharifi, M. N., & Macleod, K. F. (2018). Functions of autophagy in the tumor microenvironment and cancer metastasis. *The FEBS Journal, 285*(10), 1751–1766.
Mui, U. N., Haley, C. T., & Tyring, S. K. (2017a). Viral oncology: Molecular biology and pathogenesis. *Journal of Clinical Medicine, 6*(12), 111.
Mui, U. N., Haley, C. T., & Tyring, S. K. (2017b). Viral oncology: Molecular biology and pathogenesis. *Journal of Clinical Medicine, 6*(12).
Mukai, R., & Ohshima, T. (2014). HTLV-1 HBZ positively regulates the mTOR signaling pathway via inhibition of GADD34 activity in the cytoplasm. *Oncogene, 33*(18), 2317–2328.
Nagelkerke, A., et al. (2014). The unfolded protein response as a target for cancer therapy. *Biochimica et Biophysica Acta, 1846*(2), 277–284.

Nair, A., et al. (2003). NF-kappaB is constitutively activated in high-grade squamous intraepithelial lesions and squamous cell carcinomas of the human uterine cervix. *Oncogene, 22*(1), 50–58.

Nazio, F., et al. (2019). Autophagy and cancer stem cells: Molecular mechanisms and therapeutic applications. *Cell Death & Differentiation, 26*(4), 690–702.

Nerenberg, M., et al. (1987). The tat gene of human T-lymphotropic virus type 1 induces mesenchymal tumors in transgenic mice. *Science (New York, N. Y.), 237*(4820), 1324–1329.

Nicot, C., et al. (2000). Tax oncoprotein trans-represses endogenous B-myb promoter activity in human T cells. *AIDS Research and Human Retroviruses, 16*(16), 1629–1632.

Nicot, C. (2015). HTLV-I Tax-mediated inactivation of cell cycle checkpoints and DNA repair pathways contribute to cellular transformation: "a random mutagenesis model". *Journal of Cancer Sciences, 2*(2).

NIH-NCI. (2017). *HIV infection and cancer risk*. National Cancer Institute.

Okada, S., Goto, H., & Yotsumoto, M. (2014). Current status of treatment for primary effusion lymphoma. *Intractable & Rare Diseases Research, 3*(3), 65–74.

Okochi, K., Sato, H., & Hinuma, Y. (1984). A retrospective study on transmission of adult T cell leukemia virus by blood transfusion: Seroconversion in recipients. *Vox Sanguinis, 46*(5), 245–253.

Oliveira, P. D., Farre, L., & Bittencourt, A. L. (2016). Adult T-cell leukemia/lymphoma. *Revista da Associacao Medica Brasileira (1992), 62*(7), 691–700.

Osame, M., et al. (1986). HTLV-I associated myelopathy, a new clinical entity. *The Lancet, 327*(8488), 1031–1032.

Pang, X., Zhang, Y., & Zhang, S. (2017). High-mobility group box 1 is overexpressed in cervical carcinoma and promotes cell invasion and migration in vitro. *Oncology Reports, 37*(2), 831–840.

Pankiv, S., et al. (2007). p62/SQSTM1 binds directly to Atg8/LC3 to facilitate degradation of ubiquitinated protein aggregates by autophagy. *Journal of Biological Chemistry, 282*(33), 24131–24145.

Park, S. W., & Ozcan, U. (2013). Potential for therapeutic manipulation of the UPR in disease. *Seminars in Immunopathology, 35*(3), 351–373.

Pattingre, S., et al. (2005). Bcl-2 antiapoptotic proteins inhibit Beclin 1-dependent autophagy. *Cell, 122*(6), 927–939.

Peterman, T. A., Jaffe, H. W., & Beral, V. (1993). Epidemiologic clues to the etiology of Kaposi's sarcoma. *AIDS (London, England), 7*(5), 605–612.

Poillet-Perez, L., & White, E. (2019). Role of tumor and host autophagy in cancer metabolism. *Genes & Development, 33*(11-12), 610–619.

Prasad, V., & Greber, U. F. (2021). The endoplasmic reticulum unfolded protein response—homeostasis, cell death and evolution in virus infections. *FEMS Microbiology Reviews, 45*(5).

Pujari, R., et al. (2013). A20-mediated negative regulation of canonical NF-κB signaling pathway. *Immunologic Research, 57*, 166–171.

Division of STD Prevention, National Center for HIV, Viral Hepatitis, STD, and TB Prevention, Centers for Disease Control and Prevention, April 12, 2022—*Fact sheet.* Accessed at ⟨www.cdc.gov/std/HPV/STDFact-HPV.htm⟩.

Quest, C. (2016). *Origins of HPV transmission from Neanderthals?* Emory Winship Cancer Institute.

Raimondo, G., et al. (2019). Update of the statements on biology and clinical impact of occult hepatitis B virus infection. *Journal of Hepatology, 71*(2), 397–408.

Ramayanti, O., et al. (2018). Curcuminoids as EBV lytic activators for adjuvant treatment in EBV-positive carcinomas. *Cancers (Basel), 10*(4).

Rao, H.-L., et al. (2012). Increased intratumoral neutrophil in colorectal carcinomas correlates closely with malignant phenotype and predicts patients' adverse prognosis. *PLoS One, 7*(1), e30806.

Rauch, D., et al. (2009). Imaging spontaneous tumorigenesis: Inflammation precedes development of peripheral NK tumors. *Blood, The Journal of the American Society of Hematology, 113*(7), 1493–1500.

Ren, T., et al. (2015). HTLV-1 Tax deregulates autophagy by recruiting autophagic molecules into lipid raft microdomains. *Oncogene, 34*(3), 334–345.

Revill, P. A., et al. (2019). A global scientific strategy to cure hepatitis B. *The Lancet Gastroenterology and Hepatology, 4*(7), 545–558.

Rivas, A., Vidal, R. L., & Hetz, C. (2015). Targeting the unfolded protein response for disease intervention. *Expert Opinion on Therapeutic Targets, 19*(9), 1203–1218.

Robek, M. D., & Ratner, L. (1999). Immortalization of CD4(+) and CD8(+) T lymphocytes by human T-cell leukemia virus type 1 Tax mutants expressed in a functional molecular clone. *Journal of Virology, 73*(6), 4856–4865.

Roca Suarez, A. A., et al. (2018). Viral manipulation of STAT3: Evade, exploit, and injure. *PLoS Pathogens, 14*(3), e1006839.

Ron, D., & Walter, P. (2007). Signal integration in the endoplasmic reticulum unfolded protein response. *Nature Reviews. Molecular Cell Biology, 8*(7), 519–529.

Rutkowski, D. T., & Hegde, R. S. (2010). Regulation of basal cellular physiology by the homeostatic unfolded protein response. *The Journal of Cell Biology, 189*(5), 783–794.

Saito, M., et al. (2009). In vivo expression of the HBZ gene of HTLV-1 correlates with proviral load, inflammatory markers and disease severity in HTLV-1 associated myelopathy/tropical spastic paraparesis (HAM/TSP). *Retrovirology, 6*(1), 1–11.

Saji, C., et al. (2011). Proteasome inhibitors induce apoptosis and reduce viral replication in primary effusion lymphoma cells. *Biochemical and Biophysical Research Communications, 415*(4), 573–578.

Sano, R., & Reed, J. C. (2013). ER stress-induced cell death mechanisms. *Biochimica et Biophysica Acta (BBA)-Molecular Cell Research, 1833*(12), 3460–3470.

Sant, M., et al. (2010). Incidence of hematologic malignancies in Europe by morphologic subtype: Results of the HAEMACARE project. *Blood, 116*(19), 3724–3734.

Santarelli, R., et al. (2019). STAT3 phosphorylation affects p53/p21 axis and KSHV lytic cycle activation. *Virology, 528*, 137–143.

Satou, Y., et al. (2006). HTLV-I basic leucine zipper factor gene mRNA supports proliferation of adult T cell leukemia cells. *Proceedings of the National Academy of Sciences, 103*(3), 720–725.

Schneider, J. W., & Dittmer, D. P. (2017). Diagnosis and treatment of Kaposi Sarcoma. *American Journal of Clinical Dermatology, 18*(4), 529–539.

Schwob, A., et al. (2019). SQSTM-1/p62 potentiates HTLV-1 Tax-mediated NF-κB activation through its ubiquitin binding function. *Scientific reports, 9*(1), 16014.

Seeger, C., & Mason, W. S. (2015). Molecular biology of hepatitis B virus infection. *Virology, 479-480*, 672–686.

Senft, D., & Ronai, Z. A. (2015). UPR, autophagy, and mitochondria crosstalk underlies the ER stress response. *Trends in Biochemical Sciences, 40*(3), 141–148.

Sharma, A., et al. (2020). The autophagy gene ATG16L1 (T300A) variant is associated with the risk and progression of HBV infection. *Infection, Genetics and Evolution, 84*, 104404.

Shibata, Y., et al. (2012). p47 negatively regulates IKK activation by inducing the lysosomal degradation of polyubiquitinated NEMO. *Nature Communications, 3*(1), 1061.

Shiels, M. S., & Engels, E. A. (2017). Evolving epidemiology of HIV-associated malignancies. *Current Opinion in HIV and AIDS, 12*(1), 6.

Shiels, M. S., et al. (2009). A meta-analysis of the incidence of non-AIDS cancers in HIV-infected individuals. *Journal of Acquired Immune Deficiency Syndromes (1999), 52*(5), 611.

Shigemi, Z., et al. (2016). Effects of ER stress on unfolded protein responses, cell survival, and viral replication in primary effusion lymphoma. *Biochemical and Biophysical Research Communications, 469*(3), 565–572.

Silverman, L. R., et al. (2004). Human T-cell lymphotropic virus type 1 open reading frame II-encoded p30II is required for in vivo replication: Evidence of in vivo reversion. *Journal of Virology, 78*(8), 3837–3845.

Sir, D., et al. (2010). The early autophagic pathway is activated by hepatitis B virus and required for viral DNA replication. *Proceedings of the National Academy of Sciences of the United States of America, 107*(9), 4383–4388.

Smith, A., et al. (2011). Incidence of haematological malignancy by sub-type: A report from the Haematological Malignancy Research Network. *British Journal of Cancer, 105*(11), 1684–1692.

Society, A. C. (2017). *Cancer prevention & early detection facts & figures 2017–2018*. Atlanta, GA: American Cancer Society.

Son, J., et al. (2021). Hepatitis B virus X protein promotes liver cancer progression through autophagy induction in response to TLR4 stimulation. *Immune Network, 21*(5), e37.

Stănescu, L., et al. (2007). Kaposi's sarcoma associated with AIDS. *Romanian Journal of Morphology and Embryology = Revue Roumaine de Morphologie et Embryologie, 48*(2), 181–187.

Stelzle, D., et al. (2021). Estimates of the global burden of cervical cancer associated with HIV. *The Lancet Global Health, 9*(2), e161–e169.

Su, H.-L., Liao, C.-L., & Lin, Y.-L. (2002). Japanese encephalitis virus infection initiates endoplasmic reticulum stress and an unfolded protein response. *Journal of Virology, 76*(9), 4162–4171.

Su, I. J., et al. (2008). Ground glass hepatocytes contain pre-S mutants and represent pre-neoplastic lesions in chronic hepatitis B virus infection. *Journal of Gastroenterology and Hepatology, 23*(8 Pt 1), 1169–1174.

Sudarshan, S. R., Schlegel, R., & Liu, X. (2010). The HPV-16 E5 protein represses expression of stress pathway genes XBP-1 and COX-2 in genital keratinocytes. *Biochemical and Biophysical Research Communications, 399*(4), 617–622.

Sun, S.-C., & Yamaoka, S. (2005). Activation of NF-κB by HTLV-I and implications for cell transformation. *Oncogene, 24*(39), 5952–5964.

Sunami, Y., et al. (2016). Canonical NF-kappaB signaling in hepatocytes acts as a tumor-suppressor in hepatitis B virus surface antigen-driven hepatocellular carcinoma by controlling the unfolded protein response. *Hepatology (Baltimore, Md.), 63*(5), 1592–1607.

Surviladze, Z., et al. (2013). Cellular entry of human papillomavirus type 16 involves activation of the phosphatidylinositol 3-kinase/Akt/mTOR pathway and inhibition of autophagy. *Journal of Virology, 87*(5), 2508–2517.

Suthaus, J., et al. (2012). HHV-8-encoded viral IL-6 collaborates with mouse IL-6 in the development of multicentric Castleman disease in mice. *Blood, The Journal of the American Society of Hematology, 119*(22), 5173–5181.

Suwanmanee, Y., Wada, M., & Ueda, K. (2021). Functional roles of GRP78 in hepatitis B virus infectivity and antigen secretion. *Microbiology and Immunology, 65*(5), 189–203.

Suzuki, F., et al. (2019). Long-term outcome of entecavir treatment of nucleos(t)ide analogue-naive chronic hepatitis B patients in Japan. *Journal of Gastroenterology, 54*(2), 182–193.

Svenning, S., & Johansen, T. (2013). Selective autophagy. *Essays in Biochemistry, 55*, 79–92.

Sychev, Z. E., et al. (2017). Integrated systems biology analysis of KSHV latent infection reveals viral induction and reliance on peroxisome mediated lipid metabolism. *PLoS Pathogens, 13*(3), e1006256.

Takeda, S., et al. (2004). Genetic and epigenetic inactivation of tax gene in adult T-cell leukemia cells. *International Journal of Cancer, 109*(4), 559–567.

Tanaka, A., et al. (1990). Oncogenic transformation by the tax gene of human T-cell leukemia virus type I in vitro. *Proceedings of the National Academy of Sciences, 87*(3), 1071–1075.

Tang, H., et al. (2009). Hepatitis B virus X protein sensitizes cells to starvation-induced autophagy via up-regulation of beclin 1 expression. *Hepatology (Baltimore, Md.), 49*(1), 60–71.

Tang, S. W., et al. (2012). Impact of cellular autophagy on viruses: Insights from hepatitis B virus and human retroviruses. *Journal of Biomedical Science, 19*(1), 92.

Tang, S.-W., et al. (2013). The cellular autophagy pathway modulates human T-cell leukemia virus type 1 replication. *Journal of Virology, 87*(3), 1699–1707.

Tardif, K. D., et al. (2004). Hepatitis C virus suppresses the IRE1-XBP1 pathway of the unfolded protein response. *Journal of Biological Chemistry, 279*(17), 17158–17164.

Terrault, N. A., et al. (2018). Update on prevention, diagnosis, and treatment of chronic hepatitis B: AASLD 2018 hepatitis B guidance. *Hepatology (Baltimore, Md.), 67*(4), 1560–1599.

Thakker, S., & Verma, S. C. (2016). Co-infections and pathogenesis of KSHV-associated malignancies. *Frontiers in Microbiology, 7*, 151.

Tian, Y., et al. (2011). Autophagy required for hepatitis B virus replication in transgenic mice. *Journal of Virology, 85*(24), 13453–13456.

Tian, Y., Wang, L.-Y., & Ou, J.-H. J. (2015). *Autophagy and Hepatitis B Virus*, 169–176.

Tian, Z., et al. (2019). Expression of autophagy-modulating genes in peripheral blood mononuclear cells from familial clustering patients with chronic hepatitis B virus infection. *Archives of Virology, 164*(8), 2005–2013.

Towers, C. G., & Thorburn, A. (2016). Therapeutic targeting of autophagy. *EBioMedicine, 14*, 15–23.

Usui, T., et al. (2008). Characteristic expression of HTLV-1 basic zipper factor (HBZ) transcripts in HTLV-1 provirus-positive cells. *Retrovirology, 5*, 1–11.

Vescovo, T., et al. (2020a). Regulation of autophagy in cells infected with oncogenic human viruses and its impact on cancer development. *Frontiers in Cell and Developmental Biology, 8*, 47.

Vescovo, T., et al. (2020b). Regulation of autophagy in cells infected with oncogenic human viruses and its impact on cancer development. *Frontiers in Cell and Developmental Biology, 8*, 47.

Vo, M. T., & Choi, Y. B. (2021). Herpesvirus regulation of selective autophagy. *Viruses, 13*(5), 820.

Vo, M. T., et al. (2019). Activation of NIX-mediated mitophagy by an interferon regulatory factor homologue of human herpesvirus. *Nature Communications, 10*(1), 3203.

Wakeman, B. S., Izumiya, Y., & Speck, S. H. (2017). Identification of novel Kaposi's sarcoma-associated herpesvirus Orf50 transcripts: Discovery of new RTA isoforms with variable transactivation potential. *Journal of Virology, 91*(1) p. 10.1128/jvi. 01434-16.

Wali, V. B., Bachawal, S. V., & Sylvester, P. W. (2009). Endoplasmic reticulum stress mediates γ-tocotrienol-induced apoptosis in mammary tumor cells. *Apoptosis: An International Journal on Programmed Cell Death, 14*, 1366–1377.

Walter, P., & Ron, D. (2011). The unfolded protein response: From stress pathway to homeostatic regulation. *Science (New York, N. Y.), 334*(6059), 1081–1086.

Wang, D., et al. (2018). XBP1 activation enhances MANF expression via binding to endoplasmic reticulum stress response elements within MANF promoter region in hepatitis B. *The International Journal of Biochemistry & Cell Biology, 99*, 140–146.

Wang, J., et al. (2019b). Hepatitis B virus induces autophagy to promote its replication by the Axis of miR-192-3p-XIAP through NF kappa B signaling. *Hepatology (Baltimore, Md.), 69*(3), 974–992.

Wang, J., et al. (1990). Hepatitis B virus integration in a cyclin A gene in a hepatocellular carcinoma. *Nature, 343*(6258), 555–557.

Wang, J., Shi, Y., & Yang, H. (2010). Infection with hepatitis B virus enhances basal autophagy. *Wei Sheng Wu Xue Bao = Acta Microbiologica Sinica, 50*(12), 1651–1656.

Wang, L., Ye, X., & Zhao, T. (2019a). The physiological roles of autophagy in the mammalian life cycle. *Biological Reviews, 94*(2), 503–516.

Wang, S. Y., et al. (2011). Core signaling pathways of survival/death in autophagy-related cancer networks. *The International Journal of Biochemistry & Cell Biology, 43*(9), 1263–1266.

Wang, X., et al. (2020b). AMPK and Akt/mTOR signalling pathways participate in glucose-mediated regulation of hepatitis B virus replication and cellular autophagy. *Cellular Microbiology, 22*(2), e13131.

Wang, X., et al. (2020a). O-GlcNAcylation modulates HBV replication through regulating cellular autophagy at multiple levels. *The FASEB Journal, 34*(11), 14473–14489.

Watanabe, T. (2017). Adult T-cell leukemia: Molecular basis for clonal expansion and transformation of HTLV-1–infected T cells. *Blood, The Journal of the American Society of Hematology, 129*(9), 1071–1081.

Wen, H.-J., et al. (2010). Enhancement of autophagy during lytic replication by the Kaposi's sarcoma-associated herpesvirus replication and transcription activator. *Journal of Virology, 84*(15), 7448–7458.

White, E. (2016). Autophagy and p53. *Cold Spring Harbor Perspectives in Medicine, 6*(4).

Wilson, S. J., et al. (2007). X box binding protein XBP-1s transactivates the Kaposi's sarcoma-associated herpesvirus (KSHV) ORF50 promoter, linking plasma cell differentiation to KSHV reactivation from latency. *Journal of Virology, 81*(24), 13578–13586.

Wójcik, C., Yano, M., & DeMartino, G. N. (2004). RNA interference of valosin-containing protein (VCP/p97) reveals multiple cellular roles linked to ubiquitin/proteasome-dependent proteolysis. *Journal of Cell Science, 117*(2), 281–292.

Wooddell, C. I., et al. (2017). RNAi-based treatment of chronically infected patients and chimpanzees reveals that integrated hepatitis B virus DNA is a source of HBsAg. *Science Translational Medicine, 9*(409).

Wu, S. X., et al. (2022). Hepatitis B virus small envelope protein promotes hepatocellular carcinoma angiogenesis via endoplasmic reticulum stress signaling to upregulate the expression of vascular endothelial growth factor A. *Journal of Virology, 96*(4), e0197521.

Wu, S. Y., Lan, S. H., & Liu, H. S. (2019). Degradative autophagy selectively regulates CCND1 (cyclin D1) and MIR224, two oncogenic factors involved in hepatocellular carcinoma tumorigenesis. *Autophagy, 15*(4), 729–730.

Yang, J. D., & Roberts, L. R. (2010). Hepatocellular carcinoma: A global view. *Nature Reviews Gastroenterology & Hepatology, 7*(8), 448–458.

Yang, J. D., et al. (2019). A global view of hepatocellular carcinoma: Trends, risk, prevention and management. *Nature Reviews Gastroenterology & Hepatology, 16*(10), 589–604.

Yiu, S. P. T., et al. (2019). Autophagy-dependent reactivation of epstein-barr virus lytic cycle and combinatorial effects of autophagy-dependent and independent lytic inducers in nasopharyngeal carcinoma. *Cancers, 11*(12), 1871.

Yiu, S. P. T., et al. (2018). Intracellular iron chelation by a novel compound, C7, reactivates epstein–barr virus (EBV) lytic cycle via the ERK-autophagy Axis in EBV-positive epithelial cancers. *Cancers, 10*(12), 505.

Young, L., et al. (1987). New type B isolates of Epstein—Barr virus from Burkitt's lymphoma and from normal individuals in endemic areas. *Journal of General Virology, 68*(11), 2853–2862.

Yu, C.-Y., et al. (2006). Flavivirus infection activates the XBP1 pathway of the unfolded protein response to cope with endoplasmic reticulum stress. *Journal of Virology, 80*(23), 11868–11880.

Yu, F., et al. (2007). B cell terminal differentiation factor XBP-1 induces reactivation of Kaposi's sarcoma-associated herpesvirus. *FEBS Letters, 581*(18), 3485–3488.

Yuen, M. F., et al. (2018). Hepatitis B virus infection. *Nature Reviews Disease Primers, 4*, 18035.

Yun, C. W., & Lee, S. H. (2018). The roles of autophagy in cancer. *International Journal of Molecular Sciences, 19*(11), 3466.

Zhang, K., et al. (2006). Endoplasmic reticulum stress activates cleavage of CREBH to induce a systemic inflammatory response. *Cell, 124*(3), 587–599.

Zhang, L. (2020). Autophagy in hepatitis B or C virus infection: An incubator and a potential therapeutic target. *Life Sciences, 242*, 117206.

Zhang, L.-l, et al. (2017). Human T-cell lymphotropic virus type 1 and its oncogenesis. *Acta Pharmacologica Sinica, 38*(8), 1093–1103.

Zhang, S., et al. (2021). Difference between acyclovir and ganciclovir in the treatment of children with Epstein-Barr virus-associated infectious mononucleosis. *Evidence-Based Complementary and Alternative Medicine: eCAM*, 8996934.

Zhang, S., et al. (2022). HBXIP is a novel regulator of the unfolded protein response that sustains tamoxifen resistance in ER+ breast cancer. *The Journal of Biological Chemistry, 298*(3), 101644.

Zhang, T., et al. (2015). G-protein-coupled receptors regulate autophagy by ZBTB16-mediated ubiquitination and proteasomal degradation of Atg14L. *elife, 4*, e06734.

Zhang, X., et al. (2019). Nitric oxide inhibits autophagy and promotes apoptosis in hepatocellular carcinoma. *Cancer Science, 110*(3), 1054–1063.

Zhang, Y., et al. (2020). HBx-associated long non-coding RNA activated by TGF-beta promotes cell invasion and migration by inducing autophagy in primary liver cancer. *International Journal of Oncology, 56*(1), 337–347.

Zhao, T. (2016). The role of HBZ in HTLV-1-induced oncogenesis. *Viruses, 8*(2), 34.

Zhi, H., et al. (2011). NF-κB hyper-activation by HTLV-1 tax induces cellular senescence, but can be alleviated by the viral anti-sense protein HBZ. *PLoS Pathogens, 7*(4), e1002025.

Zhu, X., et al. (2017). NF-κB pathway link with ER stress-induced autophagy and apoptosis in cervical tumor cells. *Cell Death Discovery, 3*(1), 17059.

Zinszner, H., et al. (1998). CHOP is implicated in programmed cell death in response to impaired function of the endoplasmic reticulum. *Genes & Development, 12*(7), 982–995.

Zur Hausen, H. (2009). The search for infectious causes of human cancers: Where and why. *Virology, 392*(1), 1–10.

# CHAPTER FOUR

# The crosstalk between miRNAs and signaling pathways in human cancers: Potential therapeutic implications

Ritu Shekhar[a,*,1], Sujata Kumari[b,1], Satyam Vergish[c], and Prajna Tripathi[d]

[a]Department of Molecular Genetics and Microbiology, University of Florida, Gainesville, FL, USA
[b]Department of Zoology, Magadh Mahila College, Patna University, Patna, India
[c]Department of Plant Pathology, University of Florida, Gainesville, FL, USA
[d]Department of Microbiology and Immunology, Weill Cornell Medical College, New York, USA
*Corresponding author. e-mail address: ritushekhar@ufl.edu

## Contents

| | |
|---|---|
| 1. Introduction | 134 |
| 1.1 Biogenesis of miRNAs | 135 |
| 1.2 Regulation of gene expression by miRNAs | 136 |
| 2. Overall action of miRNAs in the regulation of physiological processes | 137 |
| 3. MicroRNAs as potential oncogenes and tumor suppressors | 138 |
| 4. MicroRNAs as modulators of various signaling pathways in human cancers | 140 |
| 4.1 TGF-β signaling pathway | 140 |
| 4.2 MAPK signaling pathway | 143 |
| 4.3 PI3K/Akt signaling | 149 |
| 5. Discussion | 151 |
| 5.1 Diagnostic applications of miRNAs in cancer | 153 |
| 5.2 Therapeutic applications of miRNAs in cancer | 153 |
| 5.3 Limitations | 154 |
| 6. Conclusion | 155 |
| References | 155 |

## Abstract

MicroRNAs (miRNAs) are increasingly recognized as central players in the regulation of eukaryotic physiological processes. These small double stranded RNA molecules have emerged as pivotal regulators in the intricate network of cellular signaling pathways, playing significant roles in the development and progression of human cancers. The central theme in miRNA-mediated regulation of signaling pathways involves their ability to target and modulate the expression of pathway components. Aberrant expression of miRNAs can either promote or suppress key signaling events,

[1] Contributed equally.

influencing critical cellular processes such as proliferation, apoptosis, angiogenesis, and metastasis. For example, oncogenic miRNAs often promote cancer progression by targeting tumor suppressors or negative regulators of signaling pathways, thereby enhancing pathway activity. Conversely, tumor-suppressive miRNAs frequently inhibit oncogenic signaling by targeting key components within these pathways. This complex regulatory crosstalk underscores the significance of miRNAs as central players in shaping the signaling landscape of cancer cells.

Furthermore, the therapeutic implications of targeting miRNAs in cancer are substantial. miRNAs can be manipulated to restore normal signaling pathway activity, offering a potential avenue for precision medicine. The development of miRNA-based therapeutics, including synthetic miRNA mimics and miRNA inhibitors, has shown promise in preclinical and clinical studies. These strategies aim to either enhance the activity of tumor-suppressive miRNAs or inhibit the function of oncogenic miRNAs, thereby restoring balanced signaling and impeding cancer progression.

In conclusion, the crosstalk between miRNAs and signaling pathways in human cancers is a dynamic and influential aspect of cancer biology. Understanding this interplay provides valuable insights into cancer development and progression. Harnessing the therapeutic potential of miRNAs as regulators of signaling pathways opens up exciting opportunities for the development of innovative cancer treatments with the potential to improve patient outcomes. In this chapter, we provide an overview of the crosstalk between miRNAs and signaling pathways in the context of cancer and highlight the potential therapeutic implications of targeting this regulatory interplay.

## 1. Introduction

Cancer stands as a leading global cause of fatalities. The principal hallmark of cancer is the rapid and uncontrolled division of cells due to genetic mutations. A harmonic balance between mitogenic and anti-proliferative signals is essential for maintaining tight control of cell division in mammalian cells. Aberrations in signaling pathways perturbing cell division can frequently trigger oncogenesis. With advancements in molecular and clinical research, researchers now have the tools to systematically investigate and identify the genetic alterations responsible for driving oncogenic transformations by modulating the cell signaling pathways. Traditionally, the classical approach for comprehending signaling pathways has been the identification of protein-coding genes. However, microRNAs (miRNAs), a class of small non-coding RNAs, have garnered significant attention for their role in regulating gene expression. The significance of miRNAs in diverse physiological processes including proliferation, development and apoptosis, and their implications in pathogenesis of various diseases including cancers has now been widely recognized. miRNAs are small ~22 nucleotide(nt) single-stranded RNA molecules that are known to regulate the expression of target genes by primarily binding to the 3'UTR of messenger RNA (mRNA). Although miRNAs are expressed from only less than one percent of the human genome, they exert their regulatory influence over nearly all of our developmental and pathological processes (Ha & Kim, 2014). Remarkably, more than 60% of human protein-coding genes contain at least one conserved and various non-conserved miRNA-binding sites underscoring the extensive regulatory potential wielded by miRNAs (Ebert & Sharp, 2012). This chapter aims to elucidate the intricate interplay between miRNAs and signaling pathways in human cancers, as well as to explore their potential therapeutic implications. In light of this, it becomes essential to initially delve into the fundamentals of miRNA generation and provide a concise overview of how miRNAs regulate gene expression. The upcoming sections will therefore establish a foundational understanding of the biological mechanism of functioning and importance of miRNAs in the context of cancer.

## 1.1 Biogenesis of miRNAs

miRNA precursors are transcribed as monocistronic or polycistronic units from both the intergenic and the intragenic regions of human genome (Fazi & Nervi, 2008). Transcription of miRNA genes is mediated by RNA polymerase II, however, the possibility that a few miRNAs might be transcribed by other RNA polymerases cannot be excluded (Borchert, Lanier, & Davidson, 2006; Lee et al., 2004). The primary transcripts (pri-miRNA) containing miRNA sequences embedded in local stem-loop structure undergo a stepwise process for genesis of a functionally mature miRNA (Yoontae Lee, Jeon, Lee, & Narry Kim, 2002). To begin with, primary miRNAs (pri-miRNAs) undergo cleavage into approximately 70 nucleotide precursor miRNAs (pre-miRNAs) through the action of the Drosha-DGCR8 microprocessor complex (Denli, Tops, Plasterk, Ketting, & Hannon, 2004; Han et al., 2004). Following the initial cleavage by Drosha, pre-miRNAs are exported to the cytoplasm by Exportin 5 (Yi, Qin, Macara, & Cullen, 2003) and then cleaved into ~22 nt mature miRNA duplex by RNase III, Dicer (Yoontae Lee et al., 2002). The mature miRNA with a relatively unstable 5' end (guide strand) is incorporated into the miRNA-induced silencing complex (miRISC) while the complementary strand (passenger strand) is quickly degraded

(Gregory, Chendrimada, Cooch, & Shiekhattar, 2005). In some cases, both strands of miRNA duplex have the potential of being incorporated into the RISC complex and are referred to as miR-5p or miR-3p, based on their proximity to the 5′ or 3′ end of pre-miRNA, respectively.

## 1.2 Regulation of gene expression by miRNAs

The recognition of a target mRNA by miRNA relies heavily on base pairing between the seed sequence (residues 2–8 at the 5′end) of miRNA guide and miRNA recognition elements (MREs), that lie usually within the 3′UTR of the target mRNAs. The mature 22-nt miRNA recognizes the MREs in the target mRNAs and guides the activated RISC complex to suppress gene expression by inhibiting translation, promoting mRNA decay or both (Bartel, 2009). The gene silencing mechanism is contingent on the strength and nature of the complementarity between a miRNA and its target site (Brennecke, Stark, Russell, & Cohen, 2005). An extensive base-pairing between the miRNA and target generally leads to Ago2 mediated endonucleolytic cleavage of the target (Gunter, & Thomas, 2004), whereas the presence of multiple complementary sites with only moderate (limited) base-pairing per site commonly results in translation inhibition. Presence of a G:U wobble also affects the specificity and activity of miRNA (Doench & Sharp, 2004; Macfarlane & Murphy, 2010).

### 1.2.1 Identification of the target genes of miRNAs

In order to understand the functional role of miRNAs, the foremost requirement is to determine their direct target genes. Each miRNA may regulate multiple genes in response different developmental or environmental cues. Target prediction of miRNAs is classically performed by computational algorithms based on seed pairing i.e. complementarity between miRNA seed region and 3′ UTR of target mRNA (Bartel, 2009; John et al., 2004) (Lewis, Burge, & Bartel, 2005). Evidence of miRNA-mRNA interactions lacking absolute seed pairing (Lal et al., 2009; Grimson et al., 2007) and the presence of MREs in the transcript regions other than 3′ UTR including the coding sequence, 5′ UTR as well as promoter sequences (Hausser, Syed, Bilen, & Zavolan, 2013; Kim et al., 2011; Xu et al., 2014; Xu Lucas et al., 2014) have founded the development of more efficient strategies for miRNA target prediction. Updated algorithms for in silico miRNA target prediction focus variably on multiple features of miRNA–mRNA interaction, including seed pairing, conservation of target site, free energy of miRNA–mRNA heteroduplex, G:U wobble and MRE sites (Thomas, Lieberman, & Lal, 2010). Functional classification of predicted candidate targets by using gene ontology and interactome analysis can help to define miRNA function and pinpoint biologically relevant target genes from hundreds of identified candidates (Lal et al., 2009). Strategies for in vitro identification of miRNA-mRNA binding based on immunoprecipitation of miRISC associated proteins, followed by subsequent analysis of mRNAs precipitated with RISC proteins, have also emerged as strong tools for deciphering miRNA targets (Hendrickson et al., 2009; Karginov et al., 2007; Easow, Teleman, & Cohen, 2007). High-throughput sequencing of RNA isolated by crosslinking immunoprecipitation and photo-activatable-ribonucleoside-enhanced crosslinking and immunoprecipitation have further advanced the identification of miRNA-mRNA interactions in-vitro (Chi, Zang, Mele, & Darnell, 2009; Hafner et al., 2010). In addition to prediction of a miRNA target gene using the target prediction algorithms or a high-throughput genome-wide approach, individual experiments analyzing the effect of miRNA perturbation on target gene expression are considered as the most significant validation for determining a gene as the target of a particular miRNA (Thomson, Bracken, & Goodall, 2011).

## 2. Overall action of miRNAs in the regulation of physiological processes

The role of miRNAs was primarily recognized in regulation of developmental transitions in *Caenorhabditis elegans*, by suppressing the residual transcripts of the previous stage (Lee, Feinbaum, & Ambrost, 1993). Thereafter, miRNAs have been implicated in almost all aspects of animal and plant cell physiology, including cell differentiation, proliferation, development, apoptosis and a wide repertoire of other signaling pathways (He & Hannon, 2004). Besides acting as binary-off-switches, miRNAs are also recognized for fine-tuning the expression of target genes during altered intracellular conditions. Therefore, the loss of miRNAs can result either in catastrophic defects or imprecise, variable phenotypes depending on their extent of gene regulation.

While most of the miRNA studies are focused primarily on identifying targets of individual miRNAs, a few recent studies have recognized miRNAs as the key components of vast regulatory networks, highlighting the crosstalk between functionally associated genes and their respective targeting miRNAs (Peter, 2010). Furthermore, the potential of miRNAs to target multiple genes allows for simultaneous modulation of widespread pathways or/and enhanced effect on single pathway through suppression of different target genes. Evidence of polycistronic (clustered) miRNAs and the discovery of miRNA families with miRNAs comprising the same seed sequence, have indicated that multiple miRNAs can work in combination to accomplish their function across different cellular processes (Wu et al., 2010; Hashimoto, Akiyama, & Yuasa, 2013; Pekarsky & Croce, 2015; Truscott, Islam, & Frolov, 2016). Cumulative reports suggest that the degree of repression of a gene by miRNAs varies proportionally with the number of MREs present in the gene transcript for the cell/tissue specific combinations of expressed miRNAs. In conclusion, to understand the biological function of miRNAs, it is important to analyze both the simultaneous effect of a miRNA on multiple target genes and the synergic effect of co-expressed miRNAs regulating common targets. A plethora of studies have reported deregulated miRNA expression in pathogenesis of various diseases including diabetes, immunological and neurological disorders and nearly all categories of cancers (Garofalo, Condorelli, & Croce, 2008; Ma & Weinberg, 2008).

## 3. MicroRNAs as potential oncogenes and tumor suppressors

The initial evidence for miRNAs' involvement in cancers came from a molecular study that characterized deletion of 13q14 locus, residing miR-15a and miR16-1, in human chronic lymphocytic leukemia (CLL) (Adrian Calin et al., 2002). Later, it was identified that both miR-15a and miR-16-1 target Bcl-2, an anti-apoptotic gene, resulting in reduced cell death, which is a characteristic of CLL. Multiple miRNAs have since been revealed as modulators of oncogenic and/or tumor suppressor genes (Zhang, Pan, Cobb, & Anderson, 2007; Tong & Nemunaitis, 2008; Ma & Weinberg, 2008). Genome-wide studies have demonstrated that miRNAs are frequently located in the cancer associated or fragile sites of the genome (Adrian Calin et al., 2004). Gene expression profiling of cancer vs. normal cells have further added to the evidence for dysregulation of miRNAs in the majority of cancers (Maire et al., 2011; Baumhoer et al., 2012). Based on their target genes and the affected pathways, cancer associated miRNAs are classified as

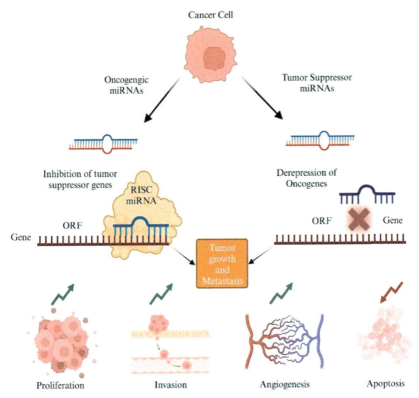

Fig. 1 **miRNAs classified as tumor suppressors and oncogenes:** The amplification or overexpression of an oncogenic miRNA eliminates the expression of its target tumor suppressor gene. The overall outcome might involve increased proliferation, invasiveness, angiogenesis or decreased levels of apoptosis, ultimately leading to tumor formation. In contrast, a reduction in the levels of tumor suppressor miRNA leads to the augmented expression of the target oncoprotein, also resulting in tumor formation or metastasis.

oncomirs and tumor suppressors (Fig. 1) (Garzon, Calin, & Croce, 2009; Di Leva, Garofalo, & Croce, 2014). A microRNA is termed as tumor suppressor when its loss-of-function initiates or contributes to the malignant transformation of a normal cell (Zhang et al., 2007). For instance, loss-of-function of miR-16 and miR-34 family miRNAs in osteosarcoma has been revealed to cause aggressive cancer phenotypes (Shekhar et al., 2019). In contrast to this, if amplification or gain-of-function of a miRNA promotes tumor development, it is classified as an oncomir (Esquela-Kerscher & Slack, 2006). It is important to note that the dysregulated miRNAs may not actually represent the initiating or driving factors of cancer and can rather be the consequence of tumorigenesis.

## 4. MicroRNAs as modulators of various signaling pathways in human cancers

Most physiological aspects of multicellular organisms are regulated through a cascade of molecules that transfer signals from outside or within the cell. Human growth factors, cytokines and other internal and external molecules have been shown to act through various signaling cascades for modulating gene expression. Multiple studies have demonstrated gene regulation by miRNAs in conserved signaling cascades including the transforming growth factor (TGF)-β, Notch, Hedgehog, and mitogen-activated protein kinase (MAPK) pathways (Suzuki, 2018). Dysregulation of these pathways is frequently associated with malignant phenotypes. In the following sections we provide an overview of the key signaling pathways involved in cancer and their multidimension crosstalk with miRNAs.

### 4.1 TGF-β signaling pathway

The TGF-β family comprises of more than 35 pleiotropic cytokines including TGFβ1, TGFβ2, TGFβ3, bone morphogenetic proteins, growth differentiation factors, activins, nodal and inhibins (Derynck & Budi, 2019). TGF-βs are synthesized as precursors and undergo proteolytic cleavage or other structural modifications to form active homodimer or heterodimer complexes that can bind to the receptors (Tzavlaki & Moustakas, 2020; Ten Dijke & Arthur, 2007). TGF-β receptors are classified into three groups known as type I, type II and type III receptors and transduce signal through Suppressor of Mothers against Decapentaplegic (Smad) and non-Smad pathways. The intracellular signaling is initiated by binding of a TGF-β ligand to TGFβR2 in the heterotetrameric cell surface receptor complex. Upon ligand binding, TGFβR2 phosphorylates a glycine-serine rich region (GS box) in the cytoplasmic domain of TGFβR1 and activates the downstream signaling. Canonically, activation of TGFβR1 leads to phosphorylation of SMAD2/3 proteins that further interact with Smad4 and translocate into the nucleus to regulate the transcription of TGF-β-target genes. Proteins Smad6/7 inhibit TGF-β signaling by obstructing receptor activation, Smad2/3-Smad4 complex formation and binding to the promoter of target genes (Rahimi & Leof, 2007; Bierie & Moses, 2006). In the non-canonical pathway, TGF-β activates ubiquitin ligase TRAF6 that ubiquitinylates TGFβR1 followed by its proteolytic cleavage. The relieved intercellular domain of TGFβR1 is then translocated to the nucleus where it functions as transcription factor for various genes and signaling molecule for other pathways.

### 4.1.1 TGF-β signaling in cancer
Many studies have shown that alterations in TGF-β pathway due to mutations in the individual signaling components correlate to pathogenesis of various diseases including tumorigenesis. Dysregulation of TGF-β signaling has been shown to associate with neoplasms in breast, prostate, endometrium, colorectum, thyroid, parathyroid and pancreas. Due to the diverse range of cell-type-dependent and context-dependent effects of TGF-β signaling, this pathway is found to influence all the critical steps of cancer including proliferation, migration, and invasion. TGF-β signaling is demonstrated to suppress proliferation by inhibiting cell cycle progression through activation of cyclin dependent kinase inhibitors and promoting G1 arrest, whereas cell-autonomous TGF-β signaling has been shown to enhance invasion and metastasis by promoting epithelial-mesenchymal transitions (EMTs) in later stages of cancer. Overall, it is the stimulation and tissue specific effect of TGF-β signaling that determines its consequence on tumor progression.

### 4.1.2 miRNAs targeting TGF-β signaling
Numerous miRNAs have been identified as regulators of oncogenic and tumor suppressive effects of TGF-β signaling. The effect of these miRNAs is mediated by their targeting of one or more components of this pathway, subject to distinct cancer types. As demonstrated by previous research, TGFβR1, TGFβR2, SMAD4 and SMAD2 are the most frequently mutated genes of the TGF-β pathway and therefore the miRNAs targeting these genes have most pronounced effects on tumorigenesis (Fig. 2).

Targeting of TGFβR2 by miRNAs is the most frequently studied case of miRNA mediated TGF-β receptor gene regulation. Multiple miRNAs including miR-145, miR-373–3p, miR-590–5p, miR-21 and miR-211 target TGFβR2 in different cancer types and impose a tissue-specific impact on cancer progression. Upregulated miR-211 and reduced expression of its direct target, TGFβR2, are associated with poor prognosis of head and neck squamous cell carcinoma (Chu et al., 2013). Further, expression of TGFβR2 is regulated by miR-590–5p to control proliferation and invasion in hepatocellular carcinoma and by miR-21 to induce stemness in colon cancer cells (Jiang et al., 2012; Yu et al., 2012). Ectopic expression of miR-145, reduces migration, invasion, and EMT of bladder

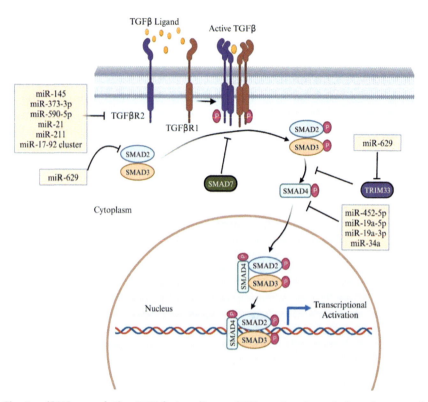

**Fig. 2 miRNAs regulating TGF-β signaling:** miRNA mediated regulation of canonical TGF-β signaling pathway proteins identified in different cancer types.

cancer cells by targeting TGFβR2 and SMAD3. This miR-145-TGFβR2 axis of TGF-β pathway is also determined as one of the action mechanisms of celecoxib, a drug used as chemo-preventive against several cancers (Liu et al., 2019). Similarly, targeting of TGFβR2/SMAD3 by miR-373–3p is determined as potential therapeutic to suppress prostate cancer metastasis (Qiu et al., 2015).

Another set of miRNAs regulate TGF-β signaling by targeting the effector genes of TGF-β signaling. Elevated levels of miR-452-5p significantly correlate with reduced expression of SMAD4 and poor prognosis of renal cancer patients (Zhai et al., 2018). The targeting of miR-452-5p by Sunitinib, a targeted RTK inhibitor has also been identified as therapeutic approach for inhibiting migration and invasion of renal cancer. Occasionally, mature miRNAs derived from two arms of same precursor are observed to inflict contrary impacts in different cancers through TGF-β signaling. For instance, upregulated miR-19a-5p targets PTEN and

SMAD4 genes to promote cell proliferation and leads to poor prognosis in renal carcinomas (Ma et al., 2016). On the contrary, reduced expression of miR-19a-3p, leads to bone metastasis of prostate cancer by constitutive activation of TGF-β signaling (Wa et al., 2018). It is reported that the restoration of miR-19a-3p inhibits the invasion and migration abilities of prostate cancer cells via targeted reduction of SMAD2 and SMAD4 expression that impedes TGF-β signaling. Indirect targeting of SMAD4 by miRNAs has also been observed frequently. For instance, upregulated miR-629 in renal cancer targets TRIM33, which inhibits the TGF-β-Smad4 binding and/or affects the expression or cellular localization of Smad4. Therefore, targeting of TRIM33 by miR-629 results in enhanced EMT in renal cancers through excessive TGFβ1 signaling (Jingushi et al., 2015). Members of the miR-200 family can target ZEB1 and ZEB2, which are transcription factors involved in the EMT induced by TGF-β signaling (Burk et al., 2008; Park, Gaur, Lengyel, & Peter, 2008; Korpal, Lee, Hu, & Kang, 2008). Suppression of ZEB1 and ZEB2 by miR-200 family members helps maintain an epithelial phenotype and inhibits tumor progression.

Many miRNAs previously classified either as oncogenic or tumor suppressor have later been found to be involved in regulating the TGF-β signaling components. For instance, miR-17–92 cluster that is often activated in cancer cells has been identified to target key components of TGF-β pathway and its downstream effectors in aggressive neuroblastoma tumors for evading cytostasis (Mestdagh et al., 2010). Inhibition of this oncogenic miRNA cluster in neuroblastomas has been found to suppress cell proliferation through de-repression of TGFβR2 and relieved expression of other tumor suppressor target genes of this miRNA cluster. Similarly, the well-known tumor suppressor miRNA, miR-34a, is found to regulate SMAD4 and TGFβR2 in order to control TGF-β signaling mediated EMT in various cancers (Qiao et al., 2015).

In the context-dependent TGF-β signaling, various miRNAs target key components of this pathway to either promote or inhibit cancer progression. The therapeutic implications of miRNA-mediated regulations of proteins in the TGF-β signaling pathway are significant and hold promise for cancer treatment and other diseases associated with aberrant TGF-β signaling.

## 4.2 MAPK signaling pathway

MAPK signaling pathway is another highly conserved signaling cascade that regulates various cellular processes such as differentiation, proliferation, and response to external stimuli (Morrison, 2012). MAPK signaling pathway consists of sequential kinase-based reactions through which MAPK is activated by upstream kinases. The activation starts at cell membrane where small GTPases and/or protein kinases downstream from cell surface receptors phosphorylates MAPK kinase Kinase or Raf (Rapidly Accelerated Fibrosarcoma) which in turn phosphorylates and activates MAPK Kinase (MAPKK) or MEK, which activates MAPK or ERK (extracellular signal regulated kinase). Subsequently, MAPKs activate multiple substrates to modulate transcription factors that act as effectors of cellular responses to MAPK pathway activation. MAPKs are categorized in seven distinct groups in mammals viz Jun N-terminal kinase (JNK)1/2/3, (ERK)1/2, ERK5, p38 isoforms $\alpha/\beta/\gamma/\delta$, ERK3/4, ERK7, and Nemo like Kinases (Thatcher, 2010; Chen & Thorner, 2007).

### 4.2.1 MAPK signaling in cancer

Aberrant MAPK signaling pathway is often observed as a common feature in various human cancers. A genome sequence study by Musalula Sinkala et al., analyzed 101 diverse cancer types obtained from 40,848 cancer patients and found mutations in the genes involved in MAPK signaling pathway in nearly 58% of the tumor samples. It was evident therefore that aberrant MAPK pathway promotes tumor development (Sinkala, Nkhoma, Mulder, & Martin, 2021). ERK protein of MAPK pathway is found upregulated in different human cancer such as breast, ovary, colon, and lung cancer (Rao & Herr, 2017; Stutvoet et al., 2019; Cole, Dahl, & Cowden Dahl, 2021). Additionally, other key components of MAPK siganaling pathway, including BRAF (V-raf murine sarcoma viral oncogene homolog B1) and NRAS (neuroblastoma RAS viral (v-ras) oncogene homolog) are reported as frequently mutated genes in tumor tissues (Edlundh-Rose et al., 2006).

### 4.2.2 Regulation of MAPK signaling by miRNAs

In the context of MAPK signaling, several miRNAs have been identified as molecular switches that dampen or enhance this pathway by targeting expression of its components. Dysregulation of this crosstalk between miRNA and MAPK pathway is associated with various hallmarks of cancer including uncontrolled cell proliferation, resistance to apoptosis, and enhanced cell migration and invasion. The regulation of MAPK pathway components by miRNAs has been extensively studied in almost all prevalent forms of cancer. In this chapter, we have comprehended this regulation in some of the major human cancer types (Table 1).

Table 1 Dysregulated expression of miRNAs that regulate MAPK signaling pathway components in different human cancers.

| Cancer type | mi-RNA | Target | Expression | References |
|---|---|---|---|---|
| Colorectal | miR-335 | RASA1 | High | Lu et al. (2016) |
| | miR-31 | RASA1 | High | Sun et al. (2013) |
| | miR-650 | ING4 | High | You et al. (2018) |
| | miR-525-3p | MAP2K6 | High | Rasmussen et al. (2016) |
| | miR-487b | KRAS | Low | Hata et al. (2017) |
| | miR-543 | KRAS, MAP1, HMGA2 | Low | Fan et al. (2016) |
| | miR-337 | KRAS | Low | Liu, Wang, and Zhao (2017) |
| | miR-143 | KRAS | Low | Akao et al. (2010) |
| | miR-145 | PAK4 | Low | Wang et al. (2012) |
| | miR-1246 | SPRED2 | Low | Peng et al. (2019) |
| Ovarian | miR-141, miR-200 | p38a | High | Mateescu et al. (2011) |
| | miR-497 | VEGFA | Low | Wang et al. (2014) |
| | miR-133b | ERK1/2, AKT | Low | Liu and Li, (2015) |
| | miR 320 | MAPK1 | Low | Xu et al. (2017) |
| | miR-508 | MAPK1 | Low | Hong et al. (2018) |

(continued)

**Table 1** Dysregulated expression of miRNAs that regulate MAPK signaling pathway components in different human cancers. (cont'd)

| Cancer type | mi-RNA | Target | Expression | References |
|---|---|---|---|---|
| Lung | miR-16 | MEK, pERK1/2 | Low | Chen et al. (2019) |
| | miR-7 | FAK | Low | Wu, Liu, Liang, Zhou, and Liu (2017) |
| | miR-34c-3p | PAC1 | Low | Zhou, Xu, & Qiao, (2015a) |
| | miR-148b | pJNK | Low | Lu et al. (2019) |
| | miR-338-3p | CHL1 | Low | Tian et al. (2021) |
| Glioma | miR-130b | MEK1/2, ERK1/2, MAPK, JNK1/2/3 | High | Li et al. (2017) |
| | miR-338-5p | FOXD1, MEK2, ERK1, | Low | Ma et al. (2018) |
| Breast | miR-433 | RAP1A | Low | Zhang et al. (2018) |
| | miR-564 | AKT2, GNA12, GYS1, SRF | Low | Mutlu et al. (2016) |
| | MiR-143-3p | MAPK7 | Low | Xia et al. (2018) |
| | MicroRNA-188-5p | RAP2C | Low | Zhu et al. (2020) |
| | MicroRNA 603 | eEF2K | Low | Bayraktar et al. (2017) |
| Melanoma | miR-340 | ERK1/2 | Low | Poenitzsch Strong, Setaluri, and Spiegelman (2014) |

Several studies have reported miRNA mediated regulation of MAPK signaling in ovarian cancers. Mateescu et al., have demonstrated that miR-141 and miR-200a target p38a MAPK protein, which functions as a sensor of oxidative stress and suppressor of tumor development. Upregulated miR-200a is predicted to target p38a expression in ovarian adenocarcinomas. In addition, the fibroblast cells, overexpressing miR-141 and miR-200a, when xenografted in nude mice lead to increased tumor size (Mateescu et al., 2011). In contrast, miR-497 exhibits tumor suppressive effects in ovarian cancer cells by targeting vascular endothelial growth factor A (VEGFA) expression and resulting in the attenuation of vascular endothelial growth factor receptor-2 mediated PI3K/AKT and MAPK/ERK pathways that are responsible for angiogenesis (Wang et al., 2014, 2019). Further, miR-133b which regulates MAPK signaling through targeting of epidermal growth factor receptor (EGFR) is also reported to be downregulated in ovarian cancer cells. Over expression of miR-133b in cancer cell lines reduces the expression of EGFR by inhibition of Erk 1/2 and Akt phosphorylation (Liu & Li, 2015). The miRNAs predicted to target MAPK1, such as miR-320 and miR-508, are also found frequently downregulated in ovarian cancers (Xu, Hu, Zhang, & Liu, 2017). Downregulation of these miRNAs activate MAPK signaling through enhanced the expression of MAPK1 and ERK. Furthermore, treatment of ovarian cancer cells with MAPK1/ERK signaling pathway inhibitor results in the reversal of miR-508 mediated effects in ovarian cancer such as reduced cell proliferation, migration, and invasion while enhancing the expression of E-cadherin (Hong, Wang, Chen, & Yang, 2018).

Aberrant activation of MAPK signaling pathway in breast cancer is also frequently observed. A study reports that miR-564 that target a set of genes namely, AKT2, GNA12, GYS1 and SRF to regulate PI3K/MAPK signaling is downregulated and acts as tumor suppressor in breast cancer (Mutlu et al., 2016). The expression of miR-564 inhibits breast cancer cell proliferation, invasion and epithelial–mesenchymal transition. Interestingly, siRNA mediated inhibition of miR-564 target genes was also found to result in G1 arrest of breast cancer cells, therefore mimicking the growth inhibition effect of miR-564. In another study, a MAPK7 targeting miRNA, miR-143–3p is also reported as downregulated in breast cancer and its overexpression is reported to reduce the progression of breast cancer (Xia, Yang, Kong, Kong, & Shan, 2018). Moreover, Rap2c and Rap1a targeting miRNAs, miR-188 and miR-433 respectively, were independently found as tumor suppressor miRNAs in breast cancer (Zhang, Zhang, Lv, Wang, & Zhang, 2018; Zhang, Lakshmanan, et al., 2018; Zhu et al., 2020). These miRNAs negatively regulate their target genes involved in activation of MAPK signaling and induction of cancer associated phenotypes. Therefore, ectopic expression of miR-188 and miR-433 in breast cancer cell lines is found to inhibit cell proliferation and induced apoptosis.

Evidences of miRNA mediated RAS/MAPK signaling dysregulation have also been reported in colorectal, lung and gastric cancers. Various tumor suppressor miRNAs have been identified to reduce hyperactivation of RAS/MAPK signaling in colorectal cancer. For instance, miR-487b inhibits colorectal cancer cell proliferation and metastasis by reduction in KRAS/MAPK signaling pathways (Hata et al., 2017). Similarly, miR-126 shows its antitumor effects by targeting CXC chemokine receptor 4 and suppressing ERK1/2 and AKT signaling (Liu et al., 2014). Another example of miRNA mediated regulation of MAPK signaling in colorectal cancer is regulation of AKT and ERK phosphorylation by miR-1. Reduced miR-1 expression has been observed in colorectal cancers and re-introduction of exogenous miR-1 in colorectal cancer cell lines has been shown to effectively impede tumor cell growth and prevent EMT (Xu et al., 2014). MEK1 and p-ERK1/2, the regulatory molecules of MAPK signaling pathway, have been shown to be inhibited by miR-16 in lung cancer (Chen et al., 2019). Additionally, miR-34c-3p that targets PAC1/MAPK pathway and miR-148b that suppresses phosphorylation of JNK in MAPK signaling are both recognized as tumor suppressor miRNAs in non-small cell lung cancer (NSCLC) (Zhou et al., 2015; Lu et al., 2019).

Contrary to the tumor suppressor phenotype, some miRNAs may also exhibit oncogenic effect in cancer progression through their regulation of MAPK pathway. In gastric cancer miR-592 serves as an oncogene, where its upregulation induces gastric cancer cell proliferation, migration, and promotes EMT. miR-592 inhibits Spry2 involved in MAPK/ERK signaling pathway (He et al., 2018). Similarly, miR-181a-5p exerts its tumor promoting activity by inhibiting Ras association domain family member 6 in gastric cancer patients and miR-203 by inhibition of PIK3/c-Jun and p38 MAPK in liver cancers (Mi et al., 2017; Zhang et al., 2018).

In summary, the miRNAs play a crucial role in modulating MAPK signaling by either enhancing or suppressing the pathway's activity. The interplay between miRNAs and the MAPK signaling pathway is a complex regulatory network that influences gene expression and cellular responses.

Dysregulation of this miRNA-MAPK crosstalk is associated with cancer hallmarks such as uncontrolled proliferation, apoptosis resistance and enhanced cell migration. Understanding the crosstalk between miRNAs and the MAPK pathway provides insights into the underlying mechanisms of various cancers and may offer new therapeutic strategies to target these pathways.

## 4.3 PI3K/Akt signaling

Phosphatidylinositol 3-kinases (PI3Ks) are family of kinases which catalyze the phosphorylation of 3'-OH group of the inositol ring of phosphatidylinositides (PtdIns) (Fresno Vara et al., 2004). They are categorized into three classes: (a) class I kinases, which are further subdivided into two subclasses IA and IB, (b) class II that comprise of three isoforms PI3K-C2α, PI3K-C2β, and PI3K-C2γ, and (c) class III includes PI3K-C3. Class I kinases are most studied and exist as heterodimers comprising of a catalytic subunit (p110) and regulatory subunit (p85). They catalyze the phosphorolysis of phosphatidylinositol-(4,5)-bisphosphate (PIP2) to phosphatidylinositol-(3,4,5)-trisphosphate (PIP3). The PIP3 acts as second messenger and serves as docking site for pleckstrin homology domains of proteins including AKT recruiting them to the membrane. AKT is activated by phosphorylation at amino acid residues threonine 308 and serine 473 by PDK1 and mTORC2, respectively, (Martini et al., 2014) and can modulate several downstream targets upon activation. PI3K signaling is inhibited by PTEN (Phosphatase and tensin homolog deleted on chromosome 10) which dephosphorylates PIP3 to PIP2.

### 4.3.1 PI3K/Akt signaling in cancer

The PI3K/AKT pathway plays crucial role in various cell processes such as cell survival, proliferation, apoptosis, and growth, thus making it an important regulatory axis in the context of cancer. Dysregulation of PI3K can occur through several mechanisms, including mutations in the PIK3CA gene, amplification of PI3K pathway components, or altered upstream signaling such as receptor tyrosine kinases (RTKs) (Yuan & Cantley, 2008). Activation of PI3K leads to the production of phosphatidylinositol 3,4,5-trisphosphate (PIP3), which in turn activates downstream effectors like Akt and mTOR, promoting cell survival and proliferation. Numerous studies have underscored the significance of PI3K dysregulation in cancers, with therapeutic interventions targeting this pathway showing promise (Fruman & Rommel, 2014).

### 4.3.2 Role of miRNAs in regulation PI3K/Akt signaling in cancer

PI3K/Akt signaling is also found to be modulated by miRNAs thus associated its dysregulation in different cancers including breast, hepatocellular, prostate, lung, and other cancer types. Studies have shown that miR-99a is downregulated in breast cancer, while the restoration of its expression resulted in the suppression of tumor growth (Yang, Han, Cheng, Zhang, & Wang, 2014). Bioinformatic analyses have predicted that it is the mTOR molecule that acts as a target of miR-99a. Similarly, the expression of miR-122 was found to be downregulated in breast cancer. It is of interest that miR-122 inhibits tumor growth by regulating PI3K signaling which induces G1 cell cycle arrest (Wang et al., 2012; Wang Wang et al., 2012). Further, Zhang Li et al. (2018), has reported the reduced expression of miR-147 that regulates Akt/mTOR signaling pathway in breast cancer cells. Ectopic expression of miR-147 in the invasive cancer cell line was found to inhibit cancer cell proliferation, invasion, and migration (Zhang, Zhang, & Liu, 2016). Consistent with this, miR-204 was is also downregulated in breast cancer cell lines and has been identified to induce apoptosis and G2/M cell cycle arrest through targeting PI3K/Akt signaling (Zhou, Fan, Liu, Xiong, & Li, 2019).

A study in liver cancer has established miR-7 as a tumor suppressor through its role in targeting PI3K catalytic subunit delta. miR-7 downregulates several molecules involved in PI3K/Akt pathway including Akt, mTOR and p70S6K (Wu et al., 2017). Further, in colon cancer, miR-218 was found to inhibit the PI3K/Akt signaling. Reduced expression of miR-218 and hyperactivity of PI3K/Akt pathway were observed associated to colon cancer tissue samples (Zhang et al., 2015). Individual studies have identified miR-34b and miR-133a-3p as tumor suppressor miRNAs with their target genes in PI3K signaling found downregulated in prostate cancer (Majid et al., 2013; Shao et al., 2019). On the contrary, expression of miR-146b and miR-410 are upregulated in prostate cancer and reported to promote tumor growth and progression by inhibiting autophagy through PTEN/Akt/mTOR signaling pathway (Gao et al., 2019; Zhang et al., 2018).

Furthermore, abnormal expression of miRNA related with PI3K/Akt pathway plays a crucial role in the initiation and progression of lung cancer. The upregulated miR-23a in NSCLC cell line contributes to erlotinib resistance of tumor cells. Suppression of miR-23a enhances the antitumor effect of erlotinib via PTEN/PI3K/Akt pathway (Han et al., 2017). Also, downregulation of miR-181, which regulates cisplatin resistance through PI3K/Akt pathway, promotes cell growth and metastasis in lung cancer (Liu, Xing, & Rong, 2018). Other miRNAs which modulates PI3K/Akt pathway to alleviate lung cancer pathogenesis are miR-153 and miR-29c (Sun et al., 2018; Yuan et al., 2015).

miRNA mediated regulation of PI3K pathway is found in other cancer types as well. In gastric cancer, overexpression of miR-589 and miR-107 is associated with promotion of tumor via Akt activation, whereas the expression of miR-340 and miR-567 results in tumor suppression by inhibiting PI3K/Akt pathway activation (Wang, Li, Wang, Shi, & Yang, 2019; Yu et al., 2017; Zhang et al., 2018, 2019). A study has found miR-195 downregulation in ovarian cancer and established its association with inhibition of phosphorylated Akt (Chen, 2018). In osteosarcoma Akt is targeted by miR-564 resulting in reduction in tumor cell proliferation (Ru et al., 2018). Additionally, in bladder cancer, overexpression of miR-608 induces tumor cells G1-phase arrest via AKT/FOXO3a signaling (Liang et al., 2017) and in pancreatic cancer miR-30a is found to regulate response to chemotherapy via SNAI1/IRS/Akt signaling (Wang et al., 2019). Studies have also identified miRNAs that target the expression of PTEN, which is an inhibitor of PI3K signaling. In glioma patients, miR-155 is upregulated and promotes cell growth and proliferation through suppression of PTEN expression (Wu and Wang, 2020). On the contrary, the expression of miR-451 is low in glioma patients and it exerts antitumor effect by inhibiting PI3K/Akt pathway (Tian et al., 2012).

Similar to other signaling pathways, the complexity of miRNA mediated regulation of PI3K signaling, is diverse with different miRNAs identified to target same PI3K components in different types of cancer types (Fig. 3). All together, these individual findings of miRNA crosstalk with signaling pathways have helped us to gain broader insights towards understanding the role of miRNAs in regulatory networks and their classification as oncogenic or tumor suppressor molecules in tissue and context specific manner.

## 5. Discussion

Taken together, the intricate crosstalk between miRNAs and signaling pathways in human cancers is a multifaceted phenomenon with significant therapeutic implications. miRNAs have emerged as versatile

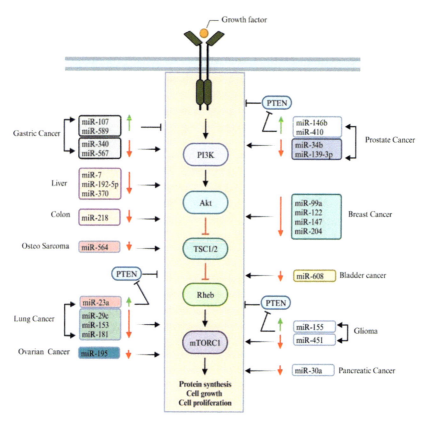

**Fig. 3 Tissue specific regulation of PI3K pathway by different miRNAs:** The tumor suppressor miRNAs downregulated in different cancers (shown in red arrows) are observed to enhance PI3K signaling leading to tumorigenesis while a few miRNAs (shown in green arrows) that are upregulated in cancers also identified to enhance PI3K signaling through their targeting of PTEN, which is a recognized inhibitor of PI3K pathway.

molecules with immense potential in treatment of cancer. Of particular interest are the miRNAs identified as tumor suppressors or oncogenes through their roles in regulation of cancer associated signaling pathways. Furthermore, miRNAs show promise as diagnostic and prognostic markers for tissue specific cancer. Specific miRNA profiles associated with different cancer types offer opportunities for early detection and more accurate prognostication. The following sections aim to delve into a comprehensive discussion of both the applications and constraints associated with miRNAs within the context of cancer.

## 5.1 Diagnostic applications of miRNAs in cancer

### 5.1.1 Circulating miRNAs as biomarkers

The small non-coding RNA molecules can be detected in various body fluids, including blood, urine, and saliva. Circulating miRNAs have been recognized as potential biomarkers to detect the presence and progression of cancer (Mitchell et al., 2008; Sohel, 2020). For instance, elevated levels of miR-21 have been observed in the blood of patients with colorectal, breast, and lung cancers, making it a potential diagnostic biomarker (Bautista-Sánchez et al., 2020). Similarly, miR-155 and miR-210 have shown promise in identifying lymphoma and glioblastoma, respectively (Due et al., 2016; Lai et al., 2015). These circulating miRNAs offer a minimally invasive method for early cancer detection and monitoring treatment response.

### 5.1.2 Tissue-specific miRNA signatures

Tumor tissues often exhibit unique miRNA expression patterns compared to healthy tissues. These signatures can aid in differentiating cancer types and subtypes, guiding clinicians in making more accurate diagnoses. For example, miR-15a and miR-16-1 downregulation is commonly associated with CLL, while miR-143 and miR-145 are downregulated in colorectal cancer (Pekarsky & Croce, 2015; Li et al., 2019). These distinct miRNA profiles provide valuable diagnostic insights.

## 5.2 Therapeutic applications of miRNAs in cancer

### 5.2.1 miRNA replacement therapy

One of the most promising therapeutic strategies for miRNA-based cancer treatments is miRNA replacement therapy (Rupaimoole & Slack, 2017). In cancers where specific tumor-suppressive miRNAs are downregulated, synthetic miRNA mimics can be introduced into cancer cells to restore their function. MiR-34, a p53-regulated miRNA, has shown significant anti-tumor effects in preclinical studies (Bader, 2012; Bouchie, 2013). Clinical trials involving miR-34 mimics have demonstrated safety and efficacy, indicating the potential of miRNA replacement therapy in cancer treatment (Van Zandwijk et al., 2017).

### 5.2.2 Inhibition of oncogenic miRNAs

In contrast to replacement of tumor suppressor miRNAs, therapeutic inhibition of oncogenic miRNAs is another approach being pursued for cancer treatments. Locked nucleic acids, antagomirs, and other antisense oligonucleotides have been developed to specifically block the function of oncogenic miRNAs. Currently, the most promising candidate for this approach is miR-21, which is found to be frequently overexpressed in various cancers (Nedaeinia et al., 2017). Inhibition of miR-21 with antagomirs has shown promise in suppressing tumor growth and metastasis with high efficacy.

### 5.2.3 Combination therapy

Clinical outcomes of miRNA therapy combined with conventional cancer treatments are also being investigated. Combining miRNA-based therapies with existing treatment modalities is predicted to enhance therapeutic outcomes. In particular, combining the modulation of miRNA with chemotherapy or targeted therapy can overcome resistance mechanisms and improve treatment response. For example, miR-205 increases the sensitivity of prostate cancer cells to radiation therapy by targeting the DNA repair protein RAD52 and miR-122, sensitizes hepatic cancer cells to chemotherapy and sorafenib, when restored by replacement along with treatment (Bai et al., 2009; El Bezawy et al., 2019).

### 5.2.4 Precision medicine

The development of miRNA-based biomarkers can also aid in patient stratification and personalized treatment plans. Profiling miRNA expression in tumor tissues or liquid biopsies can provide valuable diagnostic and prognostic information, helping clinicians tailor therapies to individual patients (Kim & Croce, 2023; Menon, Abd-Aziz, Khalid, Poh, & Naidu, 2022; Sempere, Azmi, & Moore, 2021).

## 5.3 Limitations

Despite the promising potential of miRNA-based therapies, researchers have been facing many challenges to establish these molecules in cancer treatment. One of the major challenges that we face in utilizing miRNAs for therapeutics is their efficient delivery to the target tissues. Advances in nanotechnology and targeted delivery systems are being explored to improve miRNA stability and specificity (Zhang, Cheng, Wang, & Han, 2021). Nanoparticles, liposomes, and viral vectors are among the promising platforms for enhancing miRNA delivery (Dasgupta & Chatterjee, 2021; Lee et al., 2019). The second main concern for using miRNAs in cancer therapeutics is the off-target effects (Segal & Slack, 2020). It is prerequisite to scan for all putative targets of therapeutic candidate miRNAs on a genome-wide scale for confirming least off-target effects and long-term safety in treated individuals. Hence, understanding the complex regulatory networks involving multiple miRNAs and their targets is critical for the development of effective therapies.

## 6. Conclusion

In summary, the interplay between miRNAs and signaling pathways in human cancers represents a complex regulatory network. Understanding the precise roles of miRNAs in these pathways not only deepens our comprehension of cancer biology but also opens new avenues for targeted therapeutic interventions and development of diagnostic and prognostic tools. The unique roles of miRNAs in gene regulation, combined with their accessibility in various bodily fluids, makes them valuable biomarkers for early cancer detection and monitoring. Moreover, miRNA-based therapies including miRNA replacement and inhibition strategies, to target specific nodes within intricate signaling networks, hold immense potential to revolutionize cancer treatment, offering new avenues for personalized medicine and improved patient outcomes in the fight against cancer. Continued research on both molecular and translational fronts are essential to harness the full therapeutic potential of miRNA-based approaches in cancer treatment. As research continues to unveil the roles of miRNAs in cancer biology and therapeutic strategies evolve, the integration of miRNA- based therapies into clinical practice holds the promise of improving patient outcomes and advancing precision medicine in oncology.

## References

Adrian Calin, G., Dan Dumitru, C., Shimizu, M., Bichi, R., Zupo, S., Noch, E., et al. (2002). Frequent deletions and down-regulation of micro-RNA genes miR15 and miR16 at 13q14 in chronic lymphocytic leukemia. *Proceedings of the National Academy of Sciences of the United States of America, 99*, 15524–15529.

Adrian Calin, G., Sevignani, C., Dan Dumitru, C., Hyslop, T., Noch, E., Yendamuri, S., et al. (2004). Human microRNA genes are frequently located at fragile sites and genomic regions involved in cancers. *Proceedings of the National Academy of Sciences of the United States of America, 101*, 2999–3004.

Akao, Y., Nakagawa, Y., Hirata, I., Iio, A., Itoh, T., Kojima, K., et al. (2010). Role of anti-oncomirs miR-143 and-145 in human colorectal tumors. *Cancer Gene Therapy, 17*, 398–408. https://doi.org/10.1038/cgt.2009.88.

Bader, A. G. (2012). MiR-34 – A microRNA replacement therapy is headed to the clinic. *Frontiers in Genetics, 3*, 120. https://doi.org/10.3389/fgene.2012.00120.

Bai, S., Nasser, M. W., Wang, B., Hsu, S. H., Datta, J., Kutay, H., et al. (2009). MicroRNA-122 inhibits tumorigenic properties of hepatocellular carcinoma cells and sensitizes these cells to sorafenib. *Journal of Biological Chemistry, 284,* 32015–32027. https://doi.org/10.1074/jbc.M109.016774.

Bartel, D. P. (2009). MicroRNAs: Target recognition and regulatory functions. *Cell, 136,* 215–233. https://doi.org/10.1016/j.cell.2009.01.002.

Baumhoer, D., Zillmer, S., Unger, K., Rosemann, M., Atkinson, M. J., Irmler, M., et al. (2012). MicroRNA profiling with correlation to gene expression revealed the oncogenic miR-17-92 cluster to be up-regulated in osteosarcoma. *Cancer Genetics, 205,* 212–219. https://doi.org/10.1016/j.cancergen.2012.03.001.

Bautista-Sánchez, D., Arriaga-Canon, C., Pedroza-Torres, A., De La Rosa-Velázquez, I. A., González-Barrios, R., Contreras-Espinosa, L., et al. (2020). The promising role of miR-21 as a cancer biomarker and its importance in RNA-based therapeutics. *Molecular Therapy Nucleic Acids, 20,* 409–420. https://doi.org/10.1016/j.omtn.2020.03.003.

Bayraktar, R., Pichler, M., Kanlikilicer, P., Ivan, C., Bayraktar, E., Kahraman, N., et al. (2017). MicroRNA 603 acts as a tumor suppressor and inhibits triple-negative breast cancer tumorigenesis by targeting elongation factor 2 kinase. *Oncotarget, 8,* 11641–11658.

El Bezawy, R., Tinelli, S., Tortoreto, M., Doldi, V., Zuco, V., Folini, M., et al. (2019). MiR-205 enhances radiation sensitivity of prostate cancer cells by impairing DNA damage repair through PKCε and ZEB1 inhibition. *Journal of Experimental and Clinical Cancer Research, 38,* 51. https://doi.org/10.1186/s13046-019-1060-z.

Bierie, B., & Moses, H. L. (2006). Tumour microenvironment – TGFB: The molecular Jekyll and Hyde of cancer. *Nature Reviews. Cancer, 6,* 506–520. https://doi.org/10.1038/nrc1926.

Borchert, G. M., Lanier, W., & Davidson, B. L. (2006). RNA polymerase III transcribes human microRNAs. *Nature Structural & Molecular Biology, 13,* 1097–1101. https://doi.org/10.1038/nsmb1167.

Bouchie, A. (2013). First microRNA mimic enters clinic. *Nature Biotechnology, 31,* 577. https://doi.org/10.1038/nbt0713-577.

Brennecke, J., Stark, A., Russell, R. B., & Cohen, S. M. (2005). Principles of microRNA-target recognition. *PLoS Biology, 3,* e85. https://doi.org/10.1371/journal.pbio.0030085.

Burk, U., Schubert, J., Wellner, U., Schmalhofer, O., Vincan, E., Spaderna, S., et al. (2008). A reciprocal repression between ZEB1 and members of the miR-200 family promotes EMT and invasion in cancer cells. *EMBO Reports, 9,* 582–589. https://doi.org/10.1038/embor.2008.74.

Chen, J. (2018). miRNA-195 suppresses cell proliferation of ovarian cancer cell by regulating VEGFR2 and AKT signaling pathways. *Molecular Medicine Reports, 18,* 1666–1673. https://doi.org/10.3892/mmr.2018.9098.

Chen, R. E., & Thorner, J. (2007). Function and regulation in MAPK signaling pathways: Lessons learned from the yeast *Saccharomyces cerevisiae*. *Biochimica et Biophysica Acta (BBA) - Molecular Cell Research, 1773,* 1311–1340. https://doi.org/10.1016/j.bbamcr.2007.05.003.

Chen, T. M., Xiao, Q., Wang, X. J., Wang, Z. Q., Hu, J. W., Zhang, Z., et al. (2019). miR-16 regulates proliferation and invasion of lung cancer cells via the ERK/MAPK signaling pathway by targeted inhibition of MAPK kinase 1 (MEK1). *Journal of International Medical Research, 47,* 5194–5204. https://doi.org/10.1177/0300060519856505.

Chi, S. W., Zang, J. B., Mele, A., & Darnell, R. B. (2009). Argonaute HITS-CLIP decodes microRNA-mRNA interaction maps. *Nature, 460,* 479–486. https://doi.org/10.1038/nature08170.

Chu, T. H., Yang, C. C., Liu, C. J., Lui, M. T., Lin, S. C., & Chang, K. W. (2013). MiR-211 promotes the progression of head and neck carcinomas by targeting TGFβRII. *Cancer Letters, 337,* 115–124. https://doi.org/10.1016/j.canlet.2013.05.032.

Cole, J. M., Dahl, R., & Cowden Dahl, K. D. (2021). MAPK signaling is required for generation of tunneling nanotube-like structures in ovarian cancer cells. *Cancers (Basel), 13*, 1–16. https://doi.org/10.3390/cancers13020274.

Dasgupta, I., & Chatterjee, A. (2021). Recent advances in miRNA delivery systems. *Methods and Protocols, 4*, 1–18. https://doi.org/10.3390/mps4010010.

Denli, A. M., Tops, B. B. J., Plasterk, R. H. A., Ketting, R. F., & Hannon, G. J. (2004). Processing of primary microRNAs by the microprocessor complex. *Nature, 432*, 231–235.

Derynck, R., & Budi, E. H. (2019). Specificity, versatility, and control of TGF-β family signaling. *Science Signaling, 12*, eaav5183.

Ten Dijke, P., & Arthur, H. M. (2007). Extracellular control of TGFβ signaling in vascular development and disease. *Nature Reviews Molecular Cell Biology, 8*, 857–869. https://doi.org/10.1038/nrm2262.

Doench, J. G., & Sharp, P. A. (2004). Specificity of microRNA target selection in translational repression. *Genes & Development, 18*, 504–511. https://doi.org/10.1101/gad.1184404.

Due, H., Svendsen, P., Bødker, J. S., Schmitz, A., Bøgsted, M., Johnsen, H. E., et al. (2016). MiR-155 as a biomarker in B-cell malignancies. *BioMed Research International, 2016*, 9513037. https://doi.org/10.1155/2016/9513037.

Easow, G., Teleman, A. A., & Cohen, S. M. (2007). Isolation of microRNA targets by miRNP immunopurification. *RNA (New York, N. Y.), 13*, 1198–1204. https://doi.org/10.1261/rna.563707.

Ebert, M. S., & Sharp, P. A. (2012). Roles for MicroRNAs in conferring robustness to biological processes. *Cell, 149*, 515–524. https://doi.org/10.1016/j.cell.2012.04.005.

Edlundh-Rose, E., Egyházi, B. S., Omholt, K., Månsson-Brahme, E., Platz, A., Hansson, J., et al. (2006). NRAS and BRAF mutations in melanoma tumours in relation to clinical characteristics: a study based on mutation screening by pyrosequencing. *Melanoma Research, 16*, 471–478.

Esquela-Kerscher, A., & Slack, F. J. (2006). Oncomirs – MicroRNAs with a role in cancer. *Nature Reviews Cancer, 6*, 259–266. https://doi.org/10.1038/nrc1840.

Fan, C., Lin, Y., Mao, Y., Huang, Z., Liu, A. Y., Ma, H., et al. (2016). MicroRNA-543 suppresses colorectal cancer growth and metastasis by targeting KRAS, MTA1 and HMGA2. *Oncotarget, 7*, 21825–21839.

Fazi, F., & Nervi, C. (2008). MicroRNA: Basic mechanisms and transcriptional regulatory networks for cell fate determination. *Cardiovascular Research, 79*, 553–561. https://doi.org/10.1093/cvr/cvn151.

Fresno Vara, J.Á., Casado, E., de Castro, J., Cejas, P., Belda-Iniesta, C., & González-Barón, M. (2004). PI3K/Akt signaling pathway and cancer. *Cancer Treatment Reviews, 30*, 193–204. https://doi.org/10.1016/j.ctrv.2003.07.007.

Fruman, D. A., & Rommel, C. (2014). PI3K and cancer: Lessons, challenges and opportunities. *Nature Reviews Drug Discovery, 13*, 140–156. https://doi.org/10.1038/nrd4204.

Gao, S., Zhao, Z., Wu, R., Wu, L., Tian, X., & Zhang, Z. (2019). Correction: MiR-146b inhibits autophagy in prostate cancer through affecting PTEN/AKT/mTOR signaling pathway. *Aging (Albany NY), 11*, 284 https://doi.org/10.18632/aging.101534.

Garofalo, M., Condorelli, G., & Croce, C. M. (2008). MicroRNAs in diseases and drug response. *Current Opinion in Pharmacology, 8*, 661–667. https://doi.org/10.1016/j.coph.2008.06.005.

Garzon, R., Calin, G. A., & Croce, C. M. (2009). MicroRNAs in cancer. *Annual Review of Medicine, 60*, 167–179. https://doi.org/10.1146/annurev.med.59.053006.104707.

Gregory, R. I., Chendrimada, T. P., Cooch, N., & Shiekhattar, R. (2005). Human RISC couples microRNA biogenesis and posttranscriptional gene silencing. *Cell, 123*, 631–640. https://doi.org/10.1016/j.cell.2005.10.022.

Grimson, A., Farh, K. K. H., Johnston, W. K., Garrett-Engele, P., Lim, L. P., & Bartel, D. P. (2007). MicroRNA targeting specificity in mammals: Determinants beyond seed pairing. *Molecular Cell, 27*, 91–105. https://doi.org/10.1016/j.molcel.2007.06.017.

Gunter, M., & Thomas, T. (2004). Mechanisms of gene silencing by double-stranded RNA. *Nature, 431*, 343–349. https://doi.org/10.1038/nature02873.

Ha, M., & Kim, V. N. (2014). Regulation of microRNA biogenesis. *Nature Reviews Molecular Cell Biology, 15*, 509–524. https://doi.org/10.1038/nrm3838.

Hafner, M., Landthaler, M., Burger, L., Khorshid, M., Hausser, J., Berninger, P., et al. (2010). Transcriptome-wide identification of RNA-binding protein and MicroRNA target sites by PAR-CLIP. *Cell, 141*, 129–141. https://doi.org/10.1016/j.cell.2010.03.009.

Han, J., Lee, Y., Yeom, K. H., Kim, Y. K., Jin, H., & Kim, V. N. (2004). The Drosha-DGCR8 complex in primary microRNA processing. *Genes & Development, 18*, 3016–3027. https://doi.org/10.1101/gad.1262504.

Han, Z., Zhou, X., Li, S., Qin, Y., Chen, Y., & Liu, H. (2017). Inhibition of miR-23a increases the sensitivity of lung cancer stem cells to erlotinib through PTEN/PI3K/Akt pathway. *Oncology Reports, 38*, 3064–3070. https://doi.org/10.3892/or.2017.5938.

Hashimoto, Y., Akiyama, Y., & Yuasa, Y. (2013). Multiple-to-multiple relationships between MicroRNAs and target genes in gastric cancer. *PLoS One, 8*, e62589. https://doi.org/10.1371/journal.pone.0062589.

Hata, T., Mokutani, Y., Takahashi, H., Inoue, A., Munakata, K., Nagata, K., et al. (2017). Identification of microRNA-487b as a negative regulator of liver metastasis by regulation of KRAS in colorectal cancer. *International Journal of Oncology, 50*, 487–496. https://doi.org/10.3892/ijo.2016.3813.

Hausser, J., Syed, A. P., Bilen, B., & Zavolan, M. (2013). Analysis of CDS-located miRNA target sites suggests that they can effectively inhibit translation. *Genome Research, 23*, 604–615. https://doi.org/10.1101/gr.139758.112.

He, L., & Hannon, G. J. (2004). MicroRNAs: Small RNAs with a big role in gene regulation. *Nature Reviews. Genetics, 5*, 522–531. https://doi.org/10.1038/nrg1379.

He, Y., Ge, Y., Jiang, M., Zhou, J., Luo, D., Fan, H., et al. (2018). MiR-592 promotes gastric cancer proliferation, migration, and invasion through the PI3K/AKT and MAPK/ERK signaling pathways by targeting Spry2. *Cellular Physiology and Biochemistry, 47*, 1465–1481. https://doi.org/10.1159/000490839.

Hendrickson, D. G., Hogan, D. J., McCullough, H. L., Myers, J. W., Herschlag, D., Ferrell, J. E., et al. (2009). Concordant regulation of translation and mRNA abundance for hundreds of targets of a human microRNA. *PLoS Biology, 7*, e1000238. https://doi.org/10.1371/journal.pbio.1000238.

Hong, L., Wang, Y., Chen, W., & Yang, S. (2018). MicroRNA-508 suppresses epithelial-mesenchymal transition, migration, and invasion of ovarian cancer cells through the MAPK1/ERK signaling pathway. *Journal of Cellular Biochemistry, 119*, 7431–7440. https://doi.org/10.1002/jcb.27052.

Jiang, X., Xiang, G., Wang, Y., Zhang, L., Yang, X., Cao, L., et al. (2012). MicroRNA-590-5p regulates proliferation and invasion in human hepatocellular carcinoma cells by targeting TGF-β RII. *Molecules and Cells, 33*, 545–551. https://doi.org/10.1007/s10059-012-2267-4.

Jingushi, K., Ueda, Y., Kitae, K., Hase, H., Egawa, H., Ohshio, I., et al. (2015). MiR-629 targets TRIM33 to promote TGFβ/smad signaling and metastatic phenotypes in ccRCC. *Molecular Cancer Research, 13*, 565–574. https://doi.org/10.1158/1541-7786.MCR-14-0300.

John, B., Enright, A. J., Aravin, A., Tuschl, T., Sander, C., & Marks, D. S. (2004). Human microRNA targets. *PLoS Biology, 2*, e363. https://doi.org/10.1371/journal.pbio.0020363.

Karginov, F. V., Conaco, C., Xuan, Z., Schmidt, B. H., Parker, J. S., Mandel, G., et al. (2007). A biochemical approach to identifying microRNA targets. *Proceedings of the National Academy of Sciences of the United States of America, 104*, 19291–19296.

Kim, D. H., Park, S. E., Kim, M., Ji, Y. I., Kang, M. Y., Jung, E. H., et al. (2011). A functional single nucleotide polymorphism at the promoter region of cyclin A2 is associated with increased risk of colon, liver, and lung cancers. *Cancer, 117*, 4080–4091. https://doi.org/10.1002/cncr.25930.

Kim, T., & Croce, C. M. (2023). MicroRNA: Trends in clinical trials of cancer diagnosis and therapy strategies. *Experimental & Molecular Medicine, 55*, 1314–1321. https://doi.org/10.1038/s12276-023-01050-9.

Korpal, M., Lee, E. S., Hu, G., & Kang, Y. (2008). The miR-200 family inhibits epithelial-mesenchymal transition and cancer cell migration by direct targeting of E-cadherin transcriptional repressors ZEB1 and ZEB2. *Journal of Biological Chemistry, 283*, 14910–14914. https://doi.org/10.1074/jbc.C800074200.

Lai, N. S., Wu, D. G., Fang, X. G., Lin, Y. C., Chen, S. S., Li, Z. B., et al. (2015). Serum microRNA-210 as a potential noninvasive biomarker for the diagnosis and prognosis of glioma. *British Journal of Cancer, 112*, 1241–1246. https://doi.org/10.1038/bjc.2015.91.

Lal, A., Navarro, F., Maher, C. A., Maliszewski, L. E., Yan, N., O'Day, E., et al. (2009). miR-24 inhibits cell proliferation by targeting E2F2, MYC, and other cell-cycle genes via binding to "seedless" 3′ UTR MicroRNA recognition elements. *Molecular Cell, 35*, 610–625. https://doi.org/10.1016/j.molcel.2009.08.020.

Lee, R. C., Feinbaum, R. L., & Ambrost, V. (1993). The C. elegans heterochronic gene lin-4 encodes small RNAs with antisense complementarity to lin-14. *Cell, 75*, 843–854.

Lee, S. W. L., Paoletti, C., Campisi, M., Osaki, T., Adriani, G., Kamm, R. D., et al. (2019). MicroRNA delivery through nanoparticles. *Journal of Controlled Release, 313*, 80–95. https://doi.org/10.1016/j.jconrel.2019.10.007.

Lee, Y., Kim, M., Han, J., Yeom, K. H., Lee, S., Baek, S. H., et al. (2004). MicroRNA genes are transcribed by RNA polymerase II. *EMBO Journal, 23*, 4051–4060. https://doi.org/10.1038/sj.emboj.7600385.

Di Leva, G., Garofalo, M., & Croce, C. M. (2014). MicroRNAs in cancer. *Annual Review of Pathology: Mechanisms of Disease, 9*, 287–314. https://doi.org/10.1146/annurev-pathol-012513-104715.

Lewis, B. P., Burge, C. B., & Bartel, D. P. (2005). Conserved seed pairing, often flanked by adenosines, indicates that thousands of human genes are microRNA targets. *Cell, 120*, 15–20. https://doi.org/10.1016/j.cell.2004.12.035.

Li, B., Liu, Y. H., Sun, A. G., Huan, L. C., Li, H. D., & Liu, D. M. (2017). MiR-130b regulated ERK/MAPK pathway. *European Review for Medical and Pharmacological Sciences, 21*, 2840–2846.

Li, C., Yan, G., Yin, L., Liu, T., Li, C., & Wang, L. (2019). Prognostic roles of microRNA 143 and microRNA 145 in colorectal cancer: A meta-analysis. *International Journal of Biological Markers, 34*, 6–14. https://doi.org/10.1177/1724600818807492.

Liang, Z., Wang, X., Xu, X., Xie, B., Ji, A., Meng, S., et al. (2017). MicroRNA-608 inhibits proliferation of bladder cancer via AKT/FOXO3a signaling pathway. *Molecular Cancer, 16*, 96. https://doi.org/10.1186/s12943-017-0664-1.

Liu, J., Xing, Y., & Rong, L. (2018). MiR-181 regulates cisplatin-resistant non-small cell lung cancer via downregulation of autophagy through the PTEN/PI3K/AKT pathway. *Oncology Reports, 39*, 1631–1639. https://doi.org/10.3892/or.2018.6268.

Liu, X., & Li, G. (2015). MicroRNA-133b inhibits proliferation and invasion of ovarian cancer cells through Akt and Erk1/2 inactivation by targeting epidermal growth factor receptor. *International Journal of Clinical and Experimental Pathology, 8*, 10605–10614.

Liu, X., Wang, Y., & Zhao, J. (2017). MicroRNA-337 inhibits colorectal cancer progression by directly targeting KRAS and suppressing the AKT and ERK pathways. *Oncology Reports, 38*, 3187–3196. https://doi.org/10.3892/or.2017.5997.

Liu, X., Wu, Y., Zhou, Z., Huang, M., Deng, W., Wang, Y., et al. (2019). Celecoxib inhibits the epithelial-to-mesenchymal transition in bladder cancer via the miRNA-145/TGFBR2/Smad3 axis. *International Journal of Molecular Medicine, 44*, 683–693. https://doi.org/10.3892/ijmm.2019.4241.

Liu, Y., Zhou, Y., Feng, X., An, P., Quan, X., Wang, H., et al. (2014). MicroRNA-126 functions as a tumor suppressor in colorectal cancer cells by targeting CXCR4 via the AKT and ERK1/2 signaling pathways. *International Journal of Oncology, 44*, 203–210. https://doi.org/10.3892/ijo.2013.2168.

Lu, L., Liu, Q., Wang, P., Wu, Y., Liu, X., Weng, C., et al. (2019). MicroRNA-148b regulates tumor growth of non-small cell lung cancer through targeting MAPK/JNK pathway. *BMC Cancer, 19*, 209. https://doi.org/10.1186/s12885-019-5400-3.

Lu, Y., Yang, H., Yuan, L., Liu, G., Zhang, C., Hong, M., et al. (2016). Overexpression of miR-335 confers cell proliferation and tumour growth to colorectal carcinoma cells. *Molecular and Cellular Biochemistry, 412*, 235–245. https://doi.org/10.1007/s11010-015-2630-9.

Ma, L., & Weinberg, R. A. (2008). Micromanagers of malignancy: Role of microRNAs in regulating metastasis. *Trends in Genetics, 24*, 448–456. https://doi.org/10.1016/j.tig.2008.06.004.

Ma, Q., Peng, Z., Wang, L., Li, Y., Wang, K., Zheng, J., et al. (2016). MiR-19a correlates with poor prognosis of clear cell renal cell carcinoma patients via promoting cell proliferation and suppressing PTEN/SMAD4 expression. *International Journal of Oncology, 49*, 2589–2599. https://doi.org/10.3892/ijo.2016.3746.

Ma, X. L., Shang, F., Ni, W., Zhu, J., Luo, B., & Zhang, Y. Q. (2018). MicroRNA-338-5p plays a tumor suppressor role in glioma through inhibition of the MAPK-signaling pathway by binding to FOXD1. *Journal of Cancer Research and Clinical Oncology, 144*, 2351–2366. https://doi.org/10.1007/s00432-018-2745-y.

Macfarlane, L.-A., & Murphy, P. R. (2010). MicroRNA: Biogenesis, function and role in cancer. *Current Genomics, 11*, 537–561.

Maire, G., Martin, J. W., Yoshimoto, M., Chilton-MacNeill, S., Zielenska, M., & Squire, J. A. (2011). Analysis of miRNA-gene expression-genomic profiles reveals complex mechanisms of microRNA deregulation in osteosarcoma. *Cancer Genet, 204*, 138–146. https://doi.org/10.1016/j.cancergen.2010.12.012.

Majid, S., Dar, A. A., Saini, S., Shahryari, V., Arora, S., Zaman, M. S., et al. (2013). miRNA-34b inhibits prostate cancer through demethylation, active chromatin modifications, and AKT pathways. *Clinical Cancer Research, 19*, 73–84. https://doi.org/10.1158/1078-0432.CCR-12-2952.

Martini, M., De Santis, M. C., Braccini, L., Gulluni, F., & Hirsch, E. (2014). PI3K/AKT signaling pathway and cancer: an updated review. *Annals of Medicine, 46*, 372–383. https://doi.org/10.3109/07853890.2014.912836.

Mateescu, B., Batista, L., Cardon, M., Gruosso, T., De Feraudy, Y., Mariani, O., et al. (2011). MiR-141 and miR-200a act on ovarian tumorigenesis by controlling oxidative stress response. *Nature Medicine, 17*, 1627–1635. https://doi.org/10.1038/nm.2512.

Menon, A., Abd-Aziz, N., Khalid, K., Poh, C. L., & Naidu, R. (2022). miRNA: A promising therapeutic target in cancer. *International Journal of Molecular Sciences, 23*, 11502. https://doi.org/10.3390/ijms231911502.

Mestdagh, P., Boström, A. K., Impens, F., Fredlund, E., Van Peer, G., De Antonellis, P., et al. (2010). The miR-17-92 MicroRNA cluster regulates multiple components of the TGF-β pathway in neuroblastoma. *Molecular Cell, 40*, 762–773. https://doi.org/10.1016/j.molcel.2010.11.038.

Mi, Y., Zhang, D., Jiang, W., Weng, J., Zhou, C., Huang, K., et al. (2017). miR-181a-5p promotes the progression of gastric cancer via RASSF6-mediated MAPK signaling activation. *Cancer Letters, 389*, 11–22. https://doi.org/10.1016/j.canlet.2016.12.033.

Mitchell, P. S., Parkin, R. K., Kroh, E. M., Fritz, B. R., Wyman, S. K., Pogosova-Agadjanyan, E. L., et al. (2008). Circulating microRNAs as stable blood-based markers for cancer detection. *Proceedings of the National Academy of Sciences of the United States of America, 105*, 10513–10518.

Morrison, D. K. (2012). MAP kinase pathways. *Cold Spring Harbor Perspectives in Biology, 4*, a011254. https://doi.org/10.1101/cshperspect.a011254.

Mutlu, M., Saatci, Ö., Ansari, S. A., Yurdusev, E., Shehwana, H., Konu, Ö., et al. (2016). MiR-564 acts as a dual inhibitor of PI3K and MAPK signaling networks and inhibits proliferation and invasion in breast cancer. *Scientific Reports, 6*, 32541. https://doi.org/10.1038/srep32541.

Nedaeinia, R., Sharifi, M., Avan, A., Kazemi, M., Nabinejad, A., Ferns, G. A., et al. (2017). Inhibition of microRNA-21 via locked nucleic acid-anti-miR suppressed metastatic features of colorectal cancer cells through modulation of programmed cell death 4. *Tumor Biology, 39*. https://doi.org/10.1177/1010428317692261.

Park, S. M., Gaur, A. B., Lengyel, E., & Peter, M. E. (2008). The miR-200 family determines the epithelial phenotype of cancer cells by targeting the E-cadherin repressors ZEB1 and ZEB2. *Genes & Development, 22*, 894–907. https://doi.org/10.1101/gad.1640608.

Pekarsky, Y., & Croce, C. M. (2015). Role of miR-15/16 in CLL. *Cell Death and Differentiation, 22*, 6–11. https://doi.org/10.1038/cdd.2014.87.

Peng, W., Li, J., Chen, R., Gu, Q., Yang, P., Qian, W., et al. (2019). Upregulated METTL3 promotes metastasis of colorectal Cancer via miR-1246/SPRED2/MAPK signaling pathway. *Journal of Experimental and Clinical Cancer Research, 38*, 393. https://doi.org/10.1186/s13046-019-1408-4.

Peter, M. E. (2010). Targeting of mRNAs by multiple miRNAs: The next step. *Oncogene, 29*, 2161–2164. https://doi.org/10.1038/onc.2010.59.

Poenitzsch Strong, A. M., Setaluri, V., & Spiegelman, V. S. (2014). MicroRNA-340 as a modulator of RAS-RAF-MAPK signaling in melanoma. *Archives of Biochemistry and Biophysics, 563*, 118–124. https://doi.org/10.1016/j.abb.2014.07.012.

Qiao, P., Li, G., Bi, W., Yang, L., Yao, L., & Wu, D. (2015). microRNA-34a inhibits epithelial mesenchymal transition in human cholangiocarcinoma by targeting Smad4 through transforming growth factor-beta/Smad pathway. *BMC Cancer, 15*, 469. https://doi.org/10.1186/s12885-015-1359-x.

Qiu, X., Zhu, J., Sun, Y., Fan, K., Yang, D.-R., Li, G., et al. (2015). TR4 nuclear receptor increases prostate cancer invasion via decreasing the miR-373-3p expression to alter TGFβR2/p-Smad3 signals. *Oncotarget, 6*, 15397–15409.

Rahimi, R. A., & Leof, E. B. (2007). TGF-β signaling: A tale of two responses. *Journal of Cellular Biochemistry, 102*, 593–608. https://doi.org/10.1002/jcb.21501.

Rao, A., & Herr, D. R. (2017). G protein-coupled receptor GPR19 regulates E-cadherin expression and invasion of breast cancer cells. *Biochimica et Biophysica Acta - Molecular Cell Research, 1864*, 1318–1327. https://doi.org/10.1016/j.bbamcr.2017.05.001.

Rasmussen, M. H., Lyskjær, I., Jersie-Christensen, R. R., Tarpgaard, L. S., Primdal-Bengtson, B., Nielsen, M. M., et al. (2016). MiR-625-3p regulates oxaliplatin resistance by targeting MAP2K6 p38 signaling in human colorectal adenocarcinoma cells. *Nature Communications, 7*, 12436. https://doi.org/10.1038/ncomms12436.

Ru, N., Zhang, F., Liang, J., Du, Y., Wu, W., Wang, F., et al. (2018). MiR-564 is down-regulated in osteosarcoma and inhibits the proliferation of osteosarcoma cells via targeting Akt. *Gene, 645*, 163–169. https://doi.org/10.1016/j.gene.2017.12.028.

Rupaimoole, R., & Slack, F. J. (2017). MicroRNA therapeutics: Towards a new era for the management of cancer and other diseases. *Nature Reviews. Drug Discovery, 16*, 203–222. https://doi.org/10.1038/nrd.2016.246.

Segal, M., & Slack, F. J. (2020). Challenges identifying efficacious miRNA therapeutics for cancer. *Expert Opinion on Drug Discovery, 16*, 203–222. https://doi.org/10.1080/17460441.2020.1765770.

Sempere, L. F., Azmi, A. S., & Moore, A. (2021). microRNA-based diagnostic and therapeutic applications in cancer medicine. *Wiley Interdisciplinary Reviews. RNA, 12*, e1662. https://doi.org/10.1002/wrna.1662.

Shao, Y., Chong, L., Lin, P., Li, H., Zhu, L., Wu, Q., et al. (2019). MicroRNA-133a alleviates airway remodeling in asthtama through PI3K/AKT/mTOR signaling pathway by targeting IGF1R. *Journal of Cellular Physiology, 234*, 4068–4080. https://doi.org/10.1002/jcp.27201.

Shekhar, R., Priyanka, P., Kumar, P., Ghosh, T., Khan, M., Nagarajan, P., et al. (2019). The microRNAs miR-449a and miR-424 suppress osteosarcoma by targeting cyclin A2 expression. *Journal of Biological Chemistry, 294*, 4381–4400. https://doi.org/10.1074/jbc.RA118.005778.

Sinkala, M., Nkhoma, P., Mulder, N., & Martin, D. P. (2021). Integrated molecular characterisation of the MAPK pathways in human cancers reveals pharmacologically vulnerable mutations and gene dependencies. *Communications Biology, 4*, 9. https://doi.org/10.1038/s42003-020-01552-6.

Sohel, M. M. H. (2020). Circulating microRNAs as biomarkers in cancer diagnosis. *Life Sciences, 248*, 117473. https://doi.org/10.1016/j.lfs.2020.117473.

Stutvoet, T. S., Kol, A., de Vries, E. G. E., de Bruyn, M., Fehrmann, R. S. N., Terwisscha van Scheltinga, A. G. T., et al. (2019). MAPK pathway activity plays a key role in PD-L1 expression of lung adenocarcinoma cells. *Journal of Pathology, 249*, 52–64. https://doi.org/10.1002/path.5280.

Sun, D., Yu, F., Ma, Y., Zhao, R., Chen, X., Zhu, J., et al. (2013). MicroRNA-31 activates the RAS pathway and functions as an oncogenic MicroRNA in human colorectal cancer by repressing RAS p21 GTPase activating protein 1 (RASA1). *Journal of Biological Chemistry, 288*, 9508–9518. https://doi.org/10.1074/jbc.M112.367763.

Sun, D. M., Tang, B. F., Li, Z. X., Guo, H. B., Cheng, J. L., Song, P. P., et al. (2018). MiR-29c reduces the cisplatin resistance of non-small cell lung cancer cells by negatively regulating the PI3K/Akt pathway. *Scientific Reports, 8*, 8007. https://doi.org/10.1038/s41598-018-26381-w.

Suzuki, H. I. (2018). MicroRNA control of TGF-β signaling. *International Journal of Molecular Sciences, 19*, 1901. https://doi.org/10.3390/ijms19071901.

Thatcher, J. D. (2010). The Ras-MAPK signal transduction pathway. *Science Signaling, 3*, tr1. https://doi.org/10.1126/scisignal.3119tr1.

Thomas, M., Lieberman, J., & Lal, A. (2010). Desperately seeking microRNA targets. *Nature Structural & Molecular Biology, 17*, 1169–1174. https://doi.org/10.1038/nsmb.1921.

Thomson, D. W., Bracken, C. P., & Goodall, G. J. (2011). Experimental strategies for microRNA target identification. *Nucleic Acids Research, 39*, 6845–6853. https://doi.org/10.1093/nar/gkr330.

Tian, W., Yang, X., Yang, H., Lv, M., Sun, X., & Zhou, B. (2021). Exosomal miR-338-3p suppresses non-small-cell lung cancer cells metastasis by inhibiting CHL1 through the MAPK signaling pathway. *Cell Death & Disease, 12*, 1030. https://doi.org/10.1038/s41419-021-04314-2.

Tian, Y., Nan, Y., Han, L., Zhang, A., Wang, G., Jia, Z., et al. (2012). MicroRNA miR-451 downregulates the PI3K/AKT pathway through CAB39 in human glioma. *International Journal of Oncology, 40*, 1105–1112. https://doi.org/10.3892/ijo.2011.1306.

Tong, A. W., & Nemunaitis, J. (2008). Modulation of miRNA activity in human cancer: A new paradigm for cancer gene therapy? *Cancer Gene Therapy, 15*, 341–355. https://doi.org/10.1038/cgt.2008.8.

Truscott, M., Islam, A. B. M. M. K., & Frolov, M. V. (2016). Novel regulation and functional interaction of polycistronic miRNAs. *RNA (New York, N. Y.), 22*, 129–138. https://doi.org/10.1261/rna.053264.115.

Tzavlaki, K., & Moustakas, A. (2020). TGF-B signaling. *Biomolecules, 10*, 487. https://doi.org/10.3390/biom10030487.

Wa, Q., Li, L., Lin, H., Peng, X., Ren, D., Huang, Y., et al. (2018). Downregulation of MIR-19a-3p promotes invasion, migration and bone metastasis via activating TGF-β signaling in prostate cancer. *Oncology Reports, 39*, 81–90. https://doi.org/10.3892/or.2017.6096.

Wang, B., Wang, H., & Yang, Z. (2012). MiR-122 inhibits cell proliferation and tumorigenesis of breast cancer by targeting IGF1R. *PLoS One, 7*, e47053. https://doi.org/10.1371/journal.pone.0047053.

Wang, L., Li, K., Wang, C., Shi, X., & Yang, H. (2019). miR-107 regulates growth and metastasis of gastric cancer cells via activation of the PI3K-AKT signaling pathway by downregulating FAT4. *Cancer Medicine, 8*, 5264–5273. https://doi.org/10.1002/cam4.2396.

Wang, T., Chen, G., Ma, X., Yang, Y., Chen, Y., Peng, Y., et al. (2019). MiR-30a regulates cancer cell response to chemotherapy through SNAI1/IRS1/AKT pathway. *Cell Death &Disease, 10*, 153. https://doi.org/10.1038/s41419-019-1326-6.

Wang, W., Ren, F., Wu, Q., Jiang, D., Li, H., & Shi, H. (2014). MicroRNA-497 suppresses angiogenesis by targeting vascular endothelial growth factor A through the PI3K/AKT and MAPK/ERK pathways in ovarian cancer. *Oncology Reports, 32*, 2127–2133. https://doi.org/10.3892/or.2014.3439.

Wang, Z., Zhang, X., Yang, Z., Du, H., Wu, Z., Gong, J., et al. (2012). MiR-145 regulates PAK4 via the MAPK pathway and exhibits an antitumor effect in human colon cells. *Biochemical and Biophysical Research Communications, 427*, 444–449. https://doi.org/10.1016/j.bbrc.2012.06.123.

Wu, D., & Wang, C. (2020). miR-155 regulates the proliferation of glioma cells through PI3K/AKT signaling. *Frontiers in Neurology, 11*, 297. https://doi.org/10.3389/fneur.2020.00297.

Wu, S., Huang, S., Ding, J., Zhao, Y., Liang, L., Liu, T., et al. (2010). Multiple microRNAs modulate p21Cip1/Waf1 expression by directly targeting its 3′ untranslated region. *Oncogene, 29*, 2302–2308. https://doi.org/10.1038/onc.2010.34.

Wu, W., Liu, S., Liang, Y., Zhou, Z., & Liu, X. (2017). MiR-7 inhibits progression of hepatocarcinoma by targeting KLF-4 and promises a novel diagnostic biomarker. *Cancer Cell International, 17*, 31. https://doi.org/10.1186/s12935-017-0386-x.

Xia, C., Yang, Y., Kong, F., Kong, Q., & Shan, C. (2018). MiR-143-3p inhibits the proliferation, cell migration and invasion of human breast cancer cells by modulating the expression of MAPK7. *Biochimie, 147*, 98–104. https://doi.org/10.1016/j.biochi.2018.01.003.

Xu, L., Zhang, Y., Wang, H., Zhang, G., Ding, Y., & Zhao, L. (2014). Tumor suppressor miR-1 restrains epithelial-mesenchymal transition and metastasis of colorectal carcinoma via the MAPK and PI3K/AKT pathway. *Journal of Translational Medicine, 12*, 244. https://doi.org/10.1186/s12967-014-0244-8.

Xu, W., Lucas, A. S., Wang, Z., & Liu, Y. (2014). Identifying microRNA targets in different gene regions. *BMC Bioinformatics, 15*, S4. https://doi.org/10.1186/1471-2105-15-S7-S4.

Xu, Y., Hu, J., Zhang, C., & Liu, Y. (2017). MicroRNA-320 targets mitogen-activated protein kinase 1 to inhibit cell proliferation and invasion in epithelial ovarian cancer. *Molecular Medicine Reports, 16*, 8530–8536. https://doi.org/10.3892/mmr.2017.7664.

Yang, Z., Han, Y., Cheng, K., Zhang, G., & Wang, X. (2014). miR-99a directly targets the mTOR signaling pathway in breast cancer side population cells. *Cell Proliferation, 47*, 587–595. https://doi.org/10.1111/cpr.12146.

Yi, R., Qin, Y., Macara, I. G., & Cullen, B. R. (2003). Exportin-5 mediates the nuclear export of pre-microRNAs and short hairpin RNAs. *Genes & Development, 17,* 3011–3016. https://doi.org/10.1101/gad.1158803.

Yoontae Lee, K., Jeon, J.-T., Lee, S. K., & Narry Kim, V. (2002). MicroRNA maturation: stepwise processing and subcellular localization. *The EMBO Journal, 21,* 4663–4670.

You, Q., Li, H., Liu, Y., Xu, Y., Miao, S., ... Yao, G., et al. (2018). MicroRNA-650 targets inhibitor of growth 4 to promote colorectal cancer progression via mitogen activated protein kinase signaling. *Oncology Letters, 16,* 2326–2334. https://doi.org/10.3892/ol.2018.8910.

Yu, J., Wang, R., Chen, J., Wu, J., Dang, Z., Zhang, Q., et al. (2017). miR-340 inhibits proliferation and induces apoptosis in gastric cancer cell line SGC-7901, possibly via the AKT pathway. *Medical Science Monitor, 23,* 71–77. https://doi.org/10.12659/MSM.898449.

Yu, Y., Kanwar, S. S., Patel, B. B., Oh, P. S., Nautiyal, J., Sarkar, F. H., et al. (2012). MicroRNA-21 induces stemness by downregulating transforming growth factor beta receptor 2 (TGFβr2) in colon cancer cells. *Carcinogenesis, 33,* 68–76. https://doi.org/10.1093/carcin/bgr246.

Yuan, T. L., & Cantley, L. C. (2008). PI3K pathway alterations in cancer: Variations on a theme. *Oncogene, 27,* 5497–5510. https://doi.org/10.1038/onc.2008.245.

Yuan, Y., Du, W., Wang, Y., Xu, C., Wang, J., Zhang, Y., et al. (2015). Suppression of AKT expression by miR-153 produced anti-tumor activity in lung cancer. *International Journal of, 136,* 1333–1340. https://doi.org/10.1002/ijc.29103.

Van Zandwijk, N., Pavlakis, N., Kao, S. C., Linton, A., Boyer, M. J., Clarke, S., et al. (2017). Safety and activity of microRNA-loaded minicells in patients with recurrent malignant pleural mesothelioma: a first-in-man, phase 1, open-label, dose-escalation study. *The Lancet Oncology, 18,* 1386–1396. https://doi.org/10.1016/S1470-2045(17)30621-6.

Zhai, W., Li, S., Zhang, J., Chen, Y., Ma, J., ... Kong, W., et al. (2018). Sunitinib-suppressed miR-452-5p facilitates renal cancer cell invasion and metastasis through modulating SMAD4/SMAD7 signals. *Molecular Cancer, 17,* 157. https://doi.org/10.1186/s12943-018-0906-x.

Zhang, A., Lakshmanan, J., Motameni, A., & Harbrecht, B. G. (2018). MicroRNA-203 suppresses proliferation in liver cancer associated with PIK3CA, p38 MAPK, c-Jun, and GSK3 signaling. *Molecular and Cellular Biochemistry, 441,* 89–98. https://doi.org/10.1007/s11010-017-3176-9.

Zhang, B., Pan, X., Cobb, G. P., & Anderson, T. A. (2007). microRNAs as oncogenes and tumor suppressors. *Developmental Biology, 302,* 1–12. https://doi.org/10.1016/j.ydbio.2006.08.028.

Zhang, F., Li, K., Pan, M., Li, W., Wu, J., ... Li, M., et al. (2018). MiR-589 promotes gastric cancer aggressiveness by a LIFR-PI3K/AKT-c-Jun regulatory feedback loop. *Journal of Experimental and Clinical Cancer Research, 37,* 152. https://doi.org/10.1186/s13046-018-0821-4.

Zhang, F., Li, K., Yao, X., Wang, H., Li, W., Wu, J., et al. (2019). A miR-567-PIK3AP1-PI3K/AKT-c-Myc feedback loop regulates tumour growth and chemoresistance in gastric cancer. *EBioMedicine, 44,* 311–321. https://doi.org/10.1016/j.ebiom.2019.05.003.

Zhang, S., Cheng, Z., Wang, Y., & Han, T. (2021). The risks of mirna therapeutics: In a drug target perspective. *Drug Design, Development and Therapy, 15,* 721–733. https://doi.org/10.2147/DDDT.S288859.

Zhang, T., Jiang, K., Zhu, X., Zhao, G., Wu, H., Deng, G., et al. (2018). miR-433 inhibits breast cancer cell growth via the MAPK signaling pathway by targeting Rap1a. *International Journal of Biological Sciences, 14,* 622–632. https://doi.org/10.7150/ijbs.24223.

Zhang, X., Shi, H., Tang, H., Fang, Z., Wang, J., & Cui, S. (2015). MiR-218 inhibits the invasion and migration of colon cancer cells by targeting the PI3K/Akt/mTOR signaling pathway. *International Journal of Molecular Medicine, 35*, 1301–1308. https://doi.org/10.3892/ijmm.2015.2126.

Zhang, Y., Zhang, D., Lv, J., Wang, S., & Zhang, Q. (2018). miR-410-3p promotes prostate cancer progression via regulating PTEN/AKT/mTOR signaling pathway. *Biochemical and Biophysical Research Communications, 503*, 2459–2465. https://doi.org/10.1016/j.bbrc.2018.06.176.

Zhang, Y., Zhang, H., & Liu, Z. (2016). MicroRNA-147 suppresses proliferation, invasion and migration through the AKT/mTOR signaling pathway in breast cancer. *Oncology Letters, 11*, 405–410. https://doi.org/10.3892/ol.2015.3842.

Zhou, W., Fan, X., Liu, G., Xiong, Q., & Li, Z. (2019). MicroRNA-204 inhibits the proliferation and metastasis of breast cancer cells by targeting PI3K/AKT pathway. *Jbuon, 24*, 1054–1059.

Zhou, Y.-L., Xu, Y.-J., & Qiao, C.-W. (2015). MiR-34c-3p suppresses the proliferation and invasion of non-small cell lung cancer (NSCLC) by inhibiting PAC1/MAPK pathway. *International Journal of Clinical and Experimental Pathology, 8*, 6312–6322.

Zhu, X., Qiu, J., Zhang, T., Yang, Y., Guo, S., Li, T., et al. (2020). MicroRNA-188-5p promotes apoptosis and inhibits cell proliferation of breast cancer cells via the MAPK signaling pathway by targeting Rap2c. *Journal of Cellular Physiology, 235*, 2389–2402. https://doi.org/10.1002/jcp.29144.

CHAPTER FIVE

# Targeting KRAS and SHP2 signaling pathways for immunomodulation and improving treatment outcomes in solid tumors

**Priyanka Sahu[a], Ankita Mitra[a], and Anirban Ganguly[b],\***

[a]Laura and Isaac Perlmutter Cancer Center, New York University Langone Medical Center, New York, NY, United States
[b]Department of Biochemistry, All India Institute of Medical Sciences, Deoghar, Jharkhand, India
*Corresponding author. e-mail address: anirban.biochemistry@aiimsdeoghar.edu.in

## Contents

| | |
|---|---|
| 1. RAS signaling pathway and its brief biological activity and importance in solid tumors | 168 |
| 2. KRAS structure and function | 169 |
| 3. KRAS signaling in health and disease | 170 |
|    3.1 RAF-MEK-ERK pathway | 171 |
|    3.2 PI3K pathway | 173 |
|    3.3 Post translational modification | 173 |
| 4. KRAS mutant solid tumors | 174 |
|    4.1 Non-small cell lung cancer | 174 |
|    4.2 Colorectal carcinoma | 175 |
|    4.3 Pancreatic ductal adenocarcinoma | 175 |
| 5. KRAS inhibitors | 176 |
| 6. KRAS targeted resistance | 178 |
|    6.1 Intrinsic resistance | 178 |
|    6.2 Acquired resistance | 178 |
| 7. KRAS and SHP2 pathways linked with immunomodulation and TME | 179 |
|    7.1 Regulation of cytokines and chemokine by Ras signaling | 180 |
|    7.2 Modulation of immune cells in the TME by Ras signaling | 188 |
|    7.3 Alternate targets for Ras mutant cancers in the immune-community | 193 |
| 8. SHP-2 tyrosine phosphatase | 196 |
|    8.1 Structure and biological function of SHP2 | 196 |
|    8.2 SHP-2 at the interface of KRAS and immune signaling | 197 |
|    8.3 SHP2 inhibitors emerge as potential therapy for KRAS-mutant cancers | 199 |
|    8.4 SHP2 inhibitors used as monotherapy | 201 |

9. Combination of SHP2 inhibitors with other drugs in modulating KRAS driven solid tumours    202
  9.1 SHP2 in combination with MEK (mitogen activated ERK kinase) inhibitors    202
  9.2 SHP2 in combination with KRAS inhibitors    203
  9.3 SHP2 in combination with anti-CXCR1/2    205
  9.4 SHP2 in combination with other drugs    206
10. Future directions    207
References    207

## Abstract

Historically, KRAS has been considered 'undruggable' inspite of being one of the most frequently altered oncogenic proteins in solid tumors, primarily due to the paucity of pharmacologically 'druggable' pockets within the mutant isoforms. However, pioneering developments in drug design capable of targeting the mutant KRAS isoforms especially KRAS$^{G12C}$-mutant cancers, have opened the doors for emergence of combination therapies comprising of a plethora of inhibitors targeting different signaling pathways. SHP2 signaling pathway, primarily known for activation of intracellular signaling pathways such as KRAS has come up as a potential target for such combination therapies as it emerged to be the signaling protein connecting KRAS and the immune signaling pathways and providing the link for understanding the overlapping regions of RAS/ERK/MAPK signaling cascade. Thus, SHP2 inhibitors having potent tumoricidal activity as well as role in immunomodulation have generated keen interest in researchers to explore its potential as combination therapy in KRAS mutant solid tumors. However, the excitement with these combination therapies need to overcome challenges thrown up by drug resistance and enhanced toxicity. In this review, we will discuss KRAS and SHP2 signaling pathways and their roles in immunomodulation and regulation of tumor microenvironment and also analyze the positive effects and drawbacks of the different combination therapies targeted at these signaling pathways along with their present and future potential to treat solid tumors.

## 1. RAS signaling pathway and its brief biological activity and importance in solid tumors

Ras is a small guanosine triphosphate (GTPase), ubiquitously expressed in all cells, and a member of the G protein (guanine nucleotide binding protein) family (Chen, Zhang, Qian, & Wang, 2021). GTPases are prime regulators that adds or removes a phosphate group to transmit signals within the cell by switching from an active (GTP) to inactive (GDP) state and vice versa (Chen et al., 2021). This activity of Ras is mainly regulated by three crucial factors: Guanosine nucleotide dissociation inhibitors (GDIs), guanine nucleotide exchange factors (GEFs) and GTPase activator proteins (GAPs)

(Buday & Downward, 2008; Liu, Yan, & Chan, 2017; Vigil, Cherfils, Rossman, & Der, 2010). GAPs function to hydrolyze GTP to GDP whereas GEFs works by stimulating the exchange of GDP for GTP while GDIs prevent the exchange. (Qu et al., 2019). Ras signaling is crucial for the intracellular signaling networks in normal health for cellular proliferation and survival. However, the Ras- cascade when mutated has a dark side and becomes a pivotal factor in cells for becoming cancer-inducing state (Laskovs, Partridge, & Slack, 2022). The mutational status in RAS and its upstream/downstream network has direct influence on the clinicopathological traits of patients such as poor differentiation, response to therapies, etc (Chen et al., 2021).

The three members of RAS superfamily are H-Ras, K-Ras, and N-Ras and consists of almost 150 proteins which can be further subclassified as Ras, Rab, Rho, Ran, and Arf families (Qu et al., 2019). Out of all human cancer patients, almost 25% carries a Ras mutation with a worse prognosis and shorter overall survival (OS) (Cox, Fesik, Kimmelman, Luo, & Der, 2014). The active GTP bound Ras initiates signaling through RAS–RAF–MEK–ERK and RAS–PI3K-AKT–mTORC pathways (Chen et al., 2021; Krygowska & Castellano, 2018). Mutations noted in the isoforms RASG12, RASG13, and RASQ61 abrogates the GAPs, increases the GEF-mediated exchange (Chen et al., 2021), thus remain in an active GTP-bound state promoting oncogenesis.

RAS, a proto-oncogene, is a driver of tumorigenesis in various lethal solid tumors with Kirsten rat sarcoma viral oncogene homologue (KRAS) being the most frequently identified mutated isoform (Indini, Rijavec, Ghidini, Cortellini, & Grossi, 2021). The lack of small inhibitory molecule binding pocket on the surface of Kras proteins made the researchers to term it as "undruggable" (Huang, Guo, Wang, & Fu, 2021). However, recent introduction of novel small inhibitory molecules has shown promising results, suggesting improved potential of the current therapeutic strategy (Indini et al., 2021).

## 2. KRAS structure and function

KRAS is the most frequently mutated Ras isoform in humans and is fatally associated with solid tumors such as nonsmall-cell lung cancer (NSCLC), colorectal cancer (CRC) and pancreatic ductal adenocarcinoma (PDAC) (Yang, Zhang, Huang, & Chu, 2023). KRAS homologue was first identified in human lung cancer cells and was located on the short arm of chromosome 12

(12p11.1–12p12.1) (McBride et al., 1983). The KRAS gene generates two similar protein isoforms, KRAS-4B and KRAS-4A, due to the splice site variation in the 4th exon, consisting of 188 and 189 amino acids respectively, which is (Tsai et al., 2015). According to TCGA, KRAS mostly has 21 missense mutations including G12D (29.19%), G12V (22.97%), and G12C (13.43%). These being the most common mutations, abrogates GTP hydrolysis through the steric hindrance of GAPs or promotes the GTP bound KRAS active state by GEFs, such as son of sevenless isoform 1 (SOS1) protein (Yang et al., 2023). KRAS protein activates the cellular processes by bridging the growth factor receptors to the intracellular signaling pathways. The crystal structure of KRAS indicates few separate domains: The N terminus with 85 amino acids is identical in the three forms of RAS. The second domain has low identical sequence between the three RAS and jointly forms a catalytic G domain which binds to GTP and a membrane-targeting hypervariable region (HVR) domain at the C-terminus (the third domain) (Vetter & Wittinghofer, 2001). The G domain consisting of switch I, switch II, and P loop region, binds to GDP/GTP thus inducing the signal transduction. The HVR has a CAAX motif and is responsible for membrane localization (Bourne, Sanders, & McCormick, 1991). Once the KRAS gets activated by binding to GTP, it activates and initiates almost 80 downstream signaling pathways, such as the MAPK–MAPK kinase (MEK), phosphoinositide 3-kinase (PI3K)–AKT–mechanistic target of rapamycin (mTOR). It also triggers rapidly accelerated fibrosarcoma (RAF)–MEK–extracellular signal-regulated kinase (ERK) along with triggering of transcription factors, such as ELK, JUN, and MYC etc. (Indini et al., 2021). In normal health, KRAS binds to GDP which when receives an internal stimulus such as growth factors like EGF, is replaced by GTP through GEFs and thus acquire an activation, mediating the cascade of signaling pathways. GAPS will promote the binding of KRAS to back to GDP, thus maintaining an inactive state. Missense mutations in KRAS impedes the activity of GTPase, and prevents GAPs from converting GTP to GDP thus permanently binds GTP to KRAS, activating downstream signaling cascade and nuclear transcription factors, leading to excessive cell proliferation and survival (Buscail, Bournet, & Cordelier, 2020).

## 3. KRAS signaling in health and disease

KRAS protein with a molecular mass of 23.2 kDa, performs as the first trigger for the intracellular signaling pathways. In human cells,

KRAS4B is profoundly expressed compared to KRAS4A, which has weaker expression (Jančík, Drábek, Radzioch, & Hajdúch, 2010). Protein tyrosine kinase receptors in healthy condition, triggers and activates the KRAS signaling pathways, by binding to the ligands and inducing oligomerization of the receptor (Lemmon & Schlessinger, 2010). PTPN11 encoding SHP2, is a tyrosine phosphatase which has its activities downstream of RTK and is a prime regulator of KRAS activation. SHP2 has a catalytic protein-tyrosine phosphatase (PTP) domain and a C-terminal domain with two tyrosine residues. This C-terminal domain binds to growth factor receptor-bound protein 2 (Grb2), when phosphorylated and recruits GEFs to interact with K-ras protein to instigate the exchange of GDP for GTP and activates downstream canonical signaling (Fedele et al., 2020). Once activated, K-ras can activate almost 80 downstream signaling pathways including MAPK–MEK, PI3K–AKT–mTOR, and RAF–MEK–ERK (Buscail et al., 2020). The mutated form of KRAS prefers to attach to GTP over GDP, thus activating the downstream pathways without the presence of a trigger such as the growth factors (Molina & Adjei, 2006) (Fig. 1).

## 3.1 RAF-MEK-ERK pathway

The Ras/Raf/MAPK is one of the best characterized and also the first Ras effector pathway that was identified. The Ras/Raf/MAPK pathway transduces signals from the extracellular sphere to the nucleus, regulating various cellular functions including cell growth, division and differentiation, tissue repair, integrin signaling, angiogenesis etc. which are essential in tumorigenesis (Molina & Adjei, 2006). The pathway is activated as soon as the ligand binds to the best-known growth factor receptors such epidermal growth factor receptor (EGFR) and platelet-derived growth factor receptor (Cantrell, 2003). This binding leads to the oligomerization of the receptor eventually resulting in contiguity of the cytoplasmic and catalytic domains allowing activation of the kinase activity and transphosphorylation (Schlessinger, 2000). Sequence homology 2 (SH2) domains such as Shc are now being able to be identified by adaptor proteins such as Grb2, thus recruiting GEFs and exchange of GDP for GTP (Molina & Adjei, 2006). Thus, Ras is activated and followed by this Raf is transported to the plasma membrane through binding to the switch I domain of Ras (Marais, Light, Paterson, & Marshall, 1995). Raf is a known member of the serine/tyrosine kinase family and is sub-categorized as Raf-1, A-Raf, and B-Raf (Molina & Adjei, 2006). The activated Ras interacts with Raf-1 through guanylyl-imidodiphosphate (GMP-PNP), activating Raf and phosphorylating by protein kinase (Moodie, Willumsen, Weber, & Wolfman, 1993). The activated

## KRAS Signalling in Health

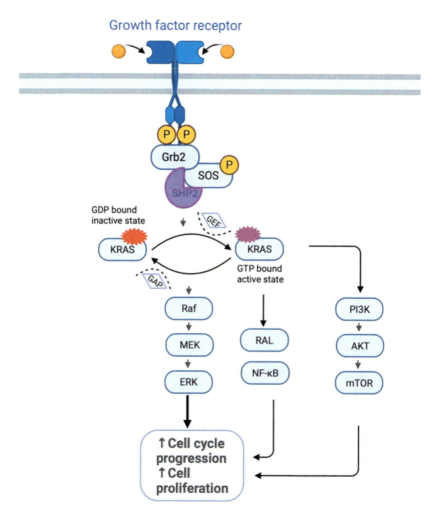

Fig. 1 KRAS signaling pathway in health and disease: KRAS cycles from 'inactive' state bound to GDP and switches to 'active' state bound to GTP in response to extracellular receptors. This switch is catalyzed by the guanine nucleotide exchange factor (GEF). The GTPase-activating protein (GAP) stops the Kras signaling by switching Kras into an inactive GDP-bound signaling state.

and phosphorylated Raf triggers a signaling cascade by phosphorylating MAPK, which now phosphorylates and activates the downstream proteins ERK1 and ERK2, finally inducing cellular response (Molina & Adjei, 2006). Other than

Raf and MAPK, PI3K cell survival pathway, the small GTP-binding proteins Rac and Rho, the stress-activated protein kinase (JNK) pathway are also the downstream effectors of K-ras (Molina & Adjei, 2006).

## 3.2 PI3K pathway

PI3Ks is second most known Ras effector family and is responsible for K-Ras mediated cell survival and proliferation. PI3K pathway activation is triggered by binding of RTK to a ligand, leading to dimerization and phosphorylation which further leads to linking with PI3K via Src homology 2 (SH2) domains (Schlessinger, 2002).

GRB2, adaptor protein of PI3K, binds to phosphor-YXN motifs of the RTK and activates SOS which leads to Ras activation (Pawson, 2004).

PI3K when activated converts phosphatidylinositol (4,5)-bisphosphate ($PIP_2$) into phosphatidylinositol (3,4,5)-trisphosphate ($PIP_3$) which in turn binds to pleckstrin homology (PH) domain of Akt/PKB (Kim, Lee, Jeong, & Jang, 2021). This stimulates kinase activity of PIP3 resulting in the phosphorylation of a host of other proteins that affect cell growth, cell cycle entry, and cell survival (Cantley, 2002). The PI3K signaling and its effect in the Ras signaling pathway is prime regulator in cellular oncogenic signaling.

## 3.3 Post translational modification

Ras proteins attaches with cellular membranes in order to transduce signaling activity. Membrane association constrains Ras thereby initiates its interaction with GEFs and GAPs. Ras GEFs are essentially regulated through their translocation to membranes via membrane binding protein domains (e.g. SH2) or membrane phospholipids (e.g. PH). Activated Ras, would thus, recruit its effectors to membranes(Ahearn, Haigis, Bar-Sagi, & Philips, 2011).

Post-translational modifications (PTMs) of Ras proteins leads them to various cellular membranes and thus help modulate GTP–GDP exchange. Essential Ras PTMs include the proteolysis and methylation, palmitoylation, phosphorylation, peptidyl-proly isomerisation, mono- and di-ubiquitination, nitrosylation, ADP ribosylation and glucosylation (Ahearn et al., 2011).

The prime function of the PTMs of Ras is to deliver the molecule to the right place within the cell. PTMs influence Ras trafficking to the plasma membrane and its partitioning into membrane microdomains which in turn causes downstream signaling activation. These PTMs are thus the targets for the development of small molecule inhibitors that might limit Ras activity.

## 4. KRAS mutant solid tumors

Wild type KRAS is a tumor suppressor gene and is frequently found to mutated during tumor progression in many cancers (Jančík et al., 2010). KRAS mutations are predominantly found in NSCLC, CRC, PDAC, Low-Grade Serous Ovarian Carcinoma, and Endometrial Cancer (Indini et al., 2021).

### 4.1 Non-small cell lung cancer

KRAS mutations are most commonly identified in NSCLC patients in the western countries. Approximately 25–30% of adenocarcinoma patients, specially with a smoking history, have mutations in KRAS gene, and have been associated with poor prognosis and lack of survival benefit (Riely et al., 2008). Almost 95% of KRAS mutations in NSCLC are noted in codons 12 in which either a pyrimidine is substituted for a purine or purine for a pyrimidine (transversion) indicating a molecular signature for the carcinogenic effects of cigarette smoke (Ahrendt et al., 2001). KRAS G12C (glycine 12 to cysteine) is profoundly found in patients accounting for 50% of all KRAS mutations, along with KRAS G12V and KRAS G12D mutations (Campbell et al., 2016). In spite of KRAS mutation by itself having a negative prognostic feature, tumor protein p53 gene (TP53), serine/threonine kinase 11 gene (STK11), kelch like ECH associated protein 1 gene (KEAP1), MNNG HOS transforming gene (MET) amplifications, erb-b2 receptor tyrosine kinase 2 gene amplifications (exclusively in G12C) are common co-occurring mutations in KRAS mutated NSCLC (Scheffler et al., 2019). Although KRAS NSCLC patients has improved OS when given immunotherapy compared to chemotherapy, patients show an inflammatory phenotype with adaptive immune resistance, with increased CD8+ tumor-infiltrating lymphocytes (TILs) and high tumor mutational burden (TMB) (Liu et al., 2020a; Liu et al., 2020b).

In the past, studies involving the inhibition of KRAS downstream signaling pathways, such as RAF/MEK/ERK and PI3K/AKT/mTOR pathways, failed due to the feedback mechanisms, and did not improve progression-free survival (PFS) and response rates (RR) (Jänne et al., 2013). Several potent and selective inhibitors targeting KRAS G12C, such as Adagrasib (MRTX849) and Sotorasib (AMG 510) have been investigated in clinical trials. Both demonstrated a manageable safety profile, the common adverse events are diarrhea, fatigue, nausea, vomiting, and elevations of aminotransferase levels (Indini et al., 2021).

## 4.2 Colorectal carcinoma

KRAS mutations constitute of almost 86% of all the RAS mutated CRC with a majority of mutation in the Codon 12 of exon (G12C, G12V, G12D) and also identified to have a negative prognostic factor (Indini et al., 2021). Different therapeutic strategies such as targeting mutant KRAS, inhibition of downstream effectors, and targeting KRAS-membrane association have been elucidated. Like NSCLC, Sotorasib produced irreversible KRAS G12C inhibitor in a phase I trial (Fakih et al., 2020). Both Sotorasib and Adagrasib showed promising tumor regression and safety in preclinical and clinical studies. However, some patients with $KRAS^{G12C}$ mutation either do not respond or has less benefit to monotherapy, likely due to treatment induced adaptive resistance. Thus combination therapy such as anti-EGFR along anti- $KRAS^{G12C}$ is underway (Ji, Wang, & Fakih, 2022). Targeting the KRAS-membrane structure has been investigated in preclinical studies. EMICORON is a synthetic compound that binds to G4 structures (G-quadruplex (G4) ligands are small molecules that bind to and stabilize G4 structures at the telomeric ends of chromosomes) and downregulates the KRAS mRNA in CRC cell lines and tumor regression in patient derived xenograft models (Porru et al., 2015).

## 4.3 Pancreatic ductal adenocarcinoma

KRAS oncogene with a mutation on codon 12 of exon 2 is identified as an initiating agent in 75–90% PDAC cases (Buscail et al., 2020). G12D and G12V accounts for almost 80% of all the KRAS mutation and has been identified in preneoplastic pancreatic lesions, as pancreatic intraepithelial neoplasia (PanIN) and intraductal papillary mucinous neoplasm (IPMN), indicating its role in initiating carcinogenesis (Delpu et al., 2011). KRAS protein interplays along with immune cells and tumor associated fibroblasts in tumor microenvironment (TME) through paracrine secretion of chemokines and is associated with poor prognosis in patients (Jonckheere, Vasseur, & Van Seuningen, 2017).

KRAS G12C inhibitors has achieved limited success in treating PDAC as majority is represented by G12D and G12V (Indini et al., 2021). However, delivering small interfering RNAs with a vector such as Local Drug Eluter, has shown to suppress KRAS mutated cell in preclinical studies (Yuan et al., 2014). This strategy combined with chemotherapy FOLFIRINOX and PROTACT, has established improved OS in a small phase I–IIa study (Indini et al., 2021). The other strategy has been to target the RAS-binding pockets, anti-RAS vaccination, and the disruption of RAS membrane localization, however has produced limited success (Buscail et al., 2020). Targeting downstream effectors such as usage of MEK inhibitors resulted in low efficacy and tumor relapse, however dual inhibition with MEK and EGFR inhibitors, have shown positive results (Fountzilas et al., 2008).

## 5. KRAS inhibitors

The high incidence of occurrence and lack of drug binding pocket makes KRAS one of the most studied and challenging therapeutic targets. Typically, patients with KRAS-mutant solid tumors were treated with conventional chemotherapy, where median PFS of CRC patients was 11.6 months, and OS for NSCLC patients was approximately 2 years (Ceddia, Landi, & Cappuzzo, 2022; Zocche, Ramirez, Fontao, Costa, & Redal, 2015).

Immunotherapy and targeted therapy are combinedly under research by scientists to effectively inhibit KRAS and its downstream pathways. Several studies have established that KRAS mutations has a role in reshaping the TME. A study was conducted on lung adenocarcinoma patients and it was reported that TILs were abundant in almost 60% of the cases, and PD-L1 positivity was in 25% of cases (Pirlog et al., 2022). KRAS mutations are involved in secretion of neutrophil chemokines, the downregulation of MHC I, the induction of regulatory T (Treg) cells, and the upregulation of PD-L1 (Dias Carvalho, Machado, Martins, Seruca, & Velho, 2019) indicating the benefits that KRAS mutation harboring patients can have with immunotherapy. In a recent study with KRAS mutant NSCLC patients were treated with immune checkpoint inhibitors which resulted in a median PFS of 16.2 months and a median OS of 31.3 months (Uehara, Watanabe, Hakozaki, Yomota, & Hosomi, 2022). Another study showed that patients with KRAS G12C mutations and high PD-L1 expression, have a significantly longer PFS when treated with anti-PD-1 (Cefalì et al., 2022). Researchers have recently conducted a study where patients KRAS mutant NSCLC had a significantly longer survival when treated with chemoimmunotherapy together with immunotherapy when compared to monotherapy (median PFS 13.9 vs 5.2 months, $p = 0.049$) (Chen et al., 2022a). A clinical trial with 150 KRAS mutant NSCLC patients established atezolizumab plus bevacizumab and chemotherapy to be a successful first-line therapy for them with STK11, KEAP1, or TP53 co-mutations

(West et al., 2022). In summary, immunotherapy combined with chemotherapy have improved clinical efficacy in patients with KRAS mutations compared to immune monotherapy.

Recently after studying the structure of KRAS, direct inhibitors of KRAS are introduced which showed promising results. A study with 100 phase 2 single-arm trial was conducted for Sotorasib (AMG510), a tyrosine kinase inhibitor (TKI) that targets KRASG12C mutation in patients (Zheng et al., 2022). It is reported that Sotorasib alone had an ORR of 37% and a PFS of 6.7 months (Skoulidis et al., 2021). Adagrasib (MRTX849) on the other hand also showed anti-tumor efficacy in KRASG12C patients in phase 1 and phase 2 trials (Jänne et al., 2022). Another KRASG12C inhibitor is ASP2453, which has been tested in preclinical models showed improved success (Nakayama et al., 2022). G12C inhibitors are a breakthrough in KRAS targeting, for other subtypes research is ongoing and are to be expected. Indirect inhibitors that target upstream or downstream effector have been developed. A recent study established that AZD0424 (an SRC inhibitor), when combined with MEK inhibitors, shows antitumor efficacy compared to monotherapy (Dawson et al., 2022). Use of NCB-0846 (Traf2- and Nck-interacting protein kinase (TNIK) inhibitor) causes improved cell death by BCL-X(L) inhibitor ABT-263 in KRAST/BRAF-mutant CRC (Jung et al., 2021). Although direct and indirect inhibitors of KRAS has improved the life expectancy of KRAS mutant patients, complete responses are rare in clinical trials (Huang et al., 2021). The tumor stage and status, various intrinsic factors such as KRAS-mediated signaling and concurrent genetic alteration are few factors responsible for such sensitivity to drugs. Acquired drug resistance due to release of ERK-mediated feedback inhibition, activation of secondary effector, secondary KRAS mutations, after promising initial responses is one of the most common challenges in cancer monotherapy (Holohan, Van Schaeybroeck, Longley, & Johnston, 2013).

Combination therapy in KRAS mutant patients has shown to have better response compared to monotherapy. A phase I study with binimetinib (MEK inhibitor) and carboplatin and pemetrexed chemotherapy for NSCLC improved ORR for patients with KRAS/NRAS mutations (Fung et al., 2021). Another study with binimetinib plus pemetrexed and cisplatin reported an ORR of 33%, a median PFS of 5.7 months, and a median OS of 6.5 months (Froesch et al., 2021).

Despite extensive and significant research progress, clinicians often face challenges while treating patients with KRAS mutant Lung cancer patients. This primarily due acquired resistance in patients to drug target therapy. Therefore, advance research is needed to elucidate the underlying complex mechanisms responsible for such resistance to improve treatment outcomes in patients.

## 6. KRAS targeted resistance

Acquired resistance can occur for reasons such as de novo mutations, enhanced autophagy, feedback mechanisms of associated effectors. There are two key mechanisms responsible for resistance against KRAS inhibitors: intrinsic resistance and acquired resistance.

### 6.1 Intrinsic resistance

Resistance due to a low dependency of tumor development and progression on KRAS mutation is defined as primary resistance (Singh et al., 2009). A study conducted by researchers have found a subset of PDAC cell lines dependent on the PI3K-mediated MAPK pathway which can be activated by other signaling than that of KRAS (Muzumdar et al., 2017). Other studies established that transcriptional coactivator Yap1 and deubiquitin USP21 is able to drive KRAS-independent tumorigenesis (Hou et al., 2021; Kapoor et al., 2014). TME by recruiting tumor associated macrophages due to the effect of HDAC5 on chemokine, is also responsible for drug resistance (Yang et al., 2023).

There is also evidence that non-mutated wild-type (WT) RAS proteins play an important role in modulating downstream effector signaling thus driving therapeutic resistance in *RAS*-mutated cancers. This is a complex mechanism as different WT RAS family members have opposing functions. RTK-dependent activation of the WT RAS proteins from the two non-mutated WT *RAS* family members has tumor-promoting activity. Further, rebound activation of RTK–WT RAS signaling underlies therapeutic resistance to targeted therapeutics in *KRAS*-mutated cancers. WT RAS thus contributes significantly in proliferation and transformation of *KRAS*-mutated cancer cells and thus identifies upstream signaling molecules, including the phosphatase/adaptor SHP2 as potential therapeutic targets in *KRAS*-mutated cancers (Sheffels & Kortum, 2021).

### 6.2 Acquired resistance

Cellular plasticity, secondary genetic alterations in switch II binding pockets and codon 12, transactivating KRAS mutation, adaptive mechanism in downstream effector pathway, limits the efficacy of KRAS inhibitors (Reita et al., 2022). The transformation of one pathological form to another is commonly identified in lung cancer. This mechanism, termed as cellular plasticity, resulting in transformation of adenocarcinoma to squamous cell carcinoma leads to drug resistance in KRAS mutant solid tumors (Awad et al., 2021). Clinical studies have identified KRAS Y96D mutation affecting the Switch-II pocket and causing resistance to all KRAS G12C inhibitors (Tanaka et al., 2021). Negative feedback signaling from either upstream or downstream effectors is one of the key reasons to lower the efficacy of the KRAS inhibitors (Akhave, Biter, & Hong, 2021). KRAS inhibitors lead to rapid adaptive feedback reactivation triggered by RTK-mediated activation of wild-type RAS. This could not be abrogated by KRAS inhibitors alone thus is often combined with SHP2 inhibition to maintain antitumor efficacy (Ryan et al., 2020). Studies have also established the role of deregulated FAK-YAP signaling in lowering the antitumor efficacy of the inhibitors (Zhang et al., 2021).

Amplification of certain effectors such as MET is responsible to induce resistance to KRAS inhibitors through RAS-dependent and non-dependent pathways., which can be reversed via A combination therapy by using both MET and KRAS inhibitor causes to lower the effect of resistance to inhibitors (Suzuki et al., 2021). HER2 mediated drug resistance has also been noted by researchers and often SHP2 inhibition in combination with KRAS inhibitors is used to overcome the resistance (Ho et al., 2021).

Drug resistance is countered by using dual inhibition of another target along with KRAS inhibitors. SHP2 being one of the effectors in KRAS pathway is often studied and co-inhibited by researchers to overcome drug resistance.

## 7. KRAS and SHP2 pathways linked with immunomodulation and TME

The TME comprises of diverse range of cells- tumor cells, matrix-associated cells and immune cells. As our understanding of the TME continues to evolve, it becomes evident that comprehending the microenvironment extends beyond immune cells and tumor cells alone; it is crucial to also grasp their intercommunication. The intrinsic mechanisms of tumor cells can effectively orchestrate interactions between tumor cells and immune cells, thereby reshaping the tumor immune environment. This process facilitates immune evasion and promotes tumor progression. In this section, we will focus on the impact of different members of the Ras family, specifically KRAS, which have been implicated in the majority of immune modulation-related events.

A diverse range of innate and adaptive immune cell types play a crucial role in shaping the tumor immune microenvironment. This environment can be regulated through two primary mechanisms: by enhancing inflammation to facilitate the proliferation of tumor cells and the progression of cancer, while also promoting an immune-suppressive environment that hinders immune cell-mediated tumor clearance.

## 7.1 Regulation of cytokines and chemokine by Ras signaling

The regulation of tumor immune environment by oncogenic Ras involves a wide array of chemokines and cytokines. Cytokines and chemokines play a pivotal role in maintaining communication between tumor cells and immune cells, facilitated by a network of corresponding receptors expressed on the target cells. Among them, certain pro-inflammatory cytokines such as IL-6, IL-1β, TNF, IL-8, and IL-23 have been shown to promote tumor progression and angiogenesis. On the other hand, the cytokines and chemokines with immunosuppressive properties, including IL-10, TGF-β, CXCL1, CXCL2, and CXCL5, play a role in fostering the formation and infiltration of immunosuppressive cells, such as Myeloid-derived suppressor cells (MDSCs), Tregs, and tumor-associated macrophages (TAMs), within the TME. This collective support for the immunosuppressive cells aids in the sustenance of tumor growth and development (Deng, Clowers, Velasco, Ramos-Castaneda, & Moghaddam, 2019; Landskron, De La Fuente, Thuwajit, Thuwajit, & Hermoso, 2014; Ozga, Chow, & Luster, 2021).

### 7.1.1 Regulation of cytokines

Numerous studies have revealed that the activation of oncogenic KRAS in both murine and human tumors leads to the rapid release of various proinflammatory cytokines, including IL-6, IL-17a, IL-22, and IL-1β (Caetano et al., 2016; Chang et al., 2014; Ji et al., 2006). The significant association between Ras signaling and inflammation, which can exhibit either pro- or anti-tumor effects, represents an intriguing target area for potential interventions. Identifying and targeting specific cytokines and their functional roles can be instrumental in the effective design of therapeutics. Therefore, it becomes crucial to comprehensively understand the intricate interactions between each of these cytokines and Ras, as well as the regulatory mechanisms controlled by Ras, culminating in the final outcome. Such knowledge is pivotal for devising therapeutic strategies that can effectively modulate the TME and potentially impact tumor development and progression.

**(a) IL-6:** Interleukin-6 (IL-6) is a pleiotropic cytokine with inflammatory properties that plays a pivotal role in the progression of various types of human cancer. The diverse range of biological effects mediated by IL6 can be attributed to two distinct signaling pathways. The first pathway involves the membrane-bound IL-6 receptor (IL-6R$\alpha$), known as classical signaling, while the second pathway involves the soluble form of IL-6 receptor (sIL-6R$\alpha$), referred to as trans-signaling. Despite both pathways utilizing the signal transducing subunit glycoprotein 130 (gp130) for signaling, the differential expression of IL-6R$\alpha$ and the ubiquitous expression of gp130 contribute to the wide-ranging effects of IL6 in different tissues and cell types (Fisher, Appenheimer, & Evans, 2014). The IL-6/IL-6R$\alpha$ via gp130 initiates the signaling by phosphorylation and subsequent activation of Janus kinases (JAK) finally acting as docking sites for downstream signaling molecules like STAT3 (Johnson, O'Keefe, & Grandis, 2018; Tanaka, Narazaki, & Kishimoto, 2014). IL-6 is produced by various cell types within the TME, including MDSCs, Tumor associated macrophages (TAM), CD4 T cells, fibroblasts as well as tumor cells. The downstream signaling of IL-6 through the STAT3 pathway promotes angiogenesis by inducing factors such as VEGF and facilitates invasiveness or metastasis by activating matrix metalloproteases. STAT3 has been found to inhibit immune cell including neutrophils, including neutrophils (NK) cells, and Dendritic Cells (DC), while positively regulating Tregs and MDSCs. These contrasting effects of STAT3 on immune cells contribute to an overall immunosuppressive TME (8,9). An elevated serum IL-6 level have been correlated with poor OS, clinical outcome and recurrence in various cancer types (Chen et al., 2022b; Duffy et al., 2008; Feng et al., 2018; Holmer, Goumas, Waetzig, Rose-John, & Kalthoff, 2014; Keegan et al., 2020; Laino et al., 2020; Silva et al., 2017; Tartour et al., 1994; Tobin et al., 2019; Zhang et al., 2013).

It is intriguing to note elevated levels of IL-6 in cells with KRAS mutations. Ancrile et al. demonstrated that the oncogenic Ras$^{G12V}$ mutation can induce the expression of IL-6 gene and protein in various cell types, thereby promoting the tumorigenic effects of Ras (Ancrile, Lim, & Counter, 2007). Zhu et al. investigated the underlying mechanism and identified JAK/TBK1/IKKe as one of the mediators of KRAS-induced IL-6 activation, suggesting a potential mechanism of tumorigenesis involving proteins other than MAPK and PI3K (Zhu et al., 2014). Brooks et al. showed that the interaction between soluble IL-6 receptor (IL6R) and secreted IL-6 triggers STAT3 signaling, promoting KRAS-driven adenocarcinoma and upregulating antiapoptotic genes such as BCL-2/Bcl-xL, which enhance cancer cell survival (Brooks et al., 2016). Lesina et al. demonstrated that the activation of the STAT3/SOCS3 axis, dependent on IL6-IL6R trans-signaling, plays a crucial role in promoting KRAS$^{G12D}$-induced PanIN formation and the development of pancreatic cancer (Lesina et al., 2011). Fukuda et al. discovered that steady signaling of STAT3-MMP7, along with the expression of associated proteins such as IL-6, are critical mediators in the development and maintenance of ductal adenocarcinoma (PDA) driven by the KRAS$^{G12D}$ mutation (Fukuda et al., 2011). This was further validated by Corcoran et al., suggesting an upregulated IL6/STAT3 signaling axis in PDAC pathogenesis and indicating baseline phospho-STAT3 level as a biomarker for predicting response to JAK2 blockade (Corcoran et al., 2011). Consistent with aforementioned studies, Zhang et al. demonstrated that IL6 collaborates with KRAS to activate the MAPK/ERK signaling cascade, facilitating the progression of pancreatic cancer. These findings collectively establish the crucial involvement of a persistent IL-6/STAT3 signaling axis in the development of KRAS-driven PDA (Zhang et al., 2013).

Further investigation into the mechanisms by which the IL6/STAT3 axis regulates the TME has revealed contrasting roles of IL6 in the initiation and progression of KRAS-induced lung cancer. Mechanistic studies in murine models have shown that the IL6/STAT3 axis can prevent lung tumorigenesis by maintaining lung homeostasis through IL-10 production from macrophages and activating cytotoxic CD8 T cells. However, in contrast, the IL6/STAT3 axis appears to contribute to the growth and progression of KRASG12D-induced lung cancer (Corcoran et al., 2011). In this context, IL6 induces the expression of cell proliferation factors like Cyclin D in tumor cells, promoting uncontrolled growth and contributing to tumor development (Qu et al., 2015). Tan and colleagues investigated the role of IL-6 deficiency in KRAS-driven tumorigenesis and found that IL-6 deficiency increased the rate of tumorigenesis but delayed tumor progression, suggesting a complex role of IL-6 in cancer development and progression (Tan et al., 2013). Zhang et al. further elucidated the complex nature of IL-6 in shaping the TME. In an IL-6-deficient condition, there was a decrease in IL-1β or granulocyte macrophage colony-stimulating factor (GM-CSF)-like cytokines known as promoters of oncogenesis, accompanied by reduced recruitment of Tregs and MDSCs in the TME (Zhang et al., 2013). Further supported by corroborated by Caetano and colleagues, by demonstrating that pharmacological inhibition of IL6 in KRAS mutant lung cancer resulted in a reduction in tumor burden. Additionally, this inhibition led to a decrease in the frequency of tolerogenic macrophages, granulocytic MDSCs, and Tregs within the TME (Chang et al., 2014). These divergent functions of IL-6 in preserving lung homeostasis under stress conditions by inhibiting inducible nitric oxide synthase (iNOS) expression in macrophages and its involvement in compromising the anti-tumor immune response by promoting an immune-suppressive TME highlight the complex and context-dependent role of IL-6 in the context of KRAS-driven cancers. This dual role of IL6 in KRAS-induced lung cancer highlights the intricacies of the TME and the remodeling of the immune landscape by the various mediators of the tumor development and the need for further research to better understand these processes for potential therapeutic interventions.

**(b) IL-1β:** IL-1β is a pleiotropic pro-inflammatory cytokine, has shown elevated levels in both serum and tumors, with a consistent link to poor prognosis in lung cancer and various other cancer types (Apte et al., 2006; Barrera et al., 2015; Das, Shapiro, Vucic, Vogt, & Bar-Sagi, 2020; McLoed et al., 2016; Wu et al., 2016). As a proinflammatory cytokine, IL-1β can also induce the expression of other inflammatory cytokines, such as IL-6. Its overexpression in the TME contributes to carcinogenesis and tumor progression by promoting a local inflammatory response and enhancing tumor invasiveness (Apte et al., 2006). In the context of acute myeloid leukemia (AML), both Hras and Nras have been correlated with the expression of IL-1β, acting as autocrine growth factors for malignant cells (Castelli et al., 1994). A study by Beaupre and colleagues investigated the regulation of IL-1β by Ras genes in AML and myelogenic leukemia cell lines. Their findings revealed that targeted knockdown of KRAS specifically reduced the binding of Nuclear factor NF-IL-6/CREB to the IL-1β promoter, suggesting a mechanism of IL-1β regulation by KRAS (Beaupre et al., 1999).

Another study conducted by Yuan and colleagues observed elevated tumoral IL-1β levels in $KRAS^{G12D}$-driven lung cancer in mice. Antibody (Canakinumab) mediated blockade of IL-1β, not only reduced tumor burden but also demonstrated remodeling of the tumor immune environment. This was evident through increased infiltration of polyfunctional cytotoxic CD8 T cells and a subsequent reduction in the population of intratumoral neutrophils and MDSCs (Yuan et al., 2022).

Other pro-inflammatory cytokines: Aberrant Ras signaling has been observed to exhibit variable cytokine responses. Petanidis and colleagues demonstrated that in CRC patients with mutant KRAS, KRAS differentially regulates cytokines IL-17 and IL-23 (Petanidis, Anestakis, Argyraki, Hadzopoulou-Cladaras, & Salifoglou, 2013). Patients harboring mutant KRAS showed significantly higher levels of IL-17 and IL-23, both of which are associated with chronic inflammation. Notably, IL-23, in particular, is significantly linked to the progression and metastasis of colorectal and hepatocellular carcinoma (HCC) (Lan et al., 2011; Li et al., 2012). An intriguing mechanism for the regulation of IL-23 was identified by Kortelever and colleagues. In a $KRAS^{G12D}$-driven lung cancer model, Kortelever's team revealed that KRAS cooperates with Myc to elevate levels of IL-23 and CCL9, a chemokine that reshapes the TME. This leads to the blocking of T cell, B cell, and NK cell infiltration, resulting in an immunosuppressive TME (Kortlever et al., 2017).

(c) **TGFβ:** TGFβ, a pleiotropic cytokine, plays diverse roles in cell proliferation, survival, differentiation, and migration. In context of the immune system TGFβ exhibits multiple functions, including the regulation of T cell proliferation, differentiation into Th17 and Treg cells, B cell differentiation, survival, and IgA class switching. Additionally, it influences dendritic cell (DC) maturation and maintains homeostasis by suppressing the cytolytic and IFNγ-producing activities of NK cells (Li, Wan, Sanjabi, Robertson, & Flavell, 2006; Sanjabi, Oh, & Li, 2017). In the context of tumor progression, TGFβ consistently promotes tumor advancement and aggressive metastasis while simultaneously contributing to immune evasion. It achieves this by suppressing the activities of CD8 T cells and NK cells and promoting the formation of Treg cells. Notably, increased expression and secretion of TGFβ have been correlated with poor prognosis in various cancers, including HCC, NSCLC, gastric, colorectal, pancreatic, and breast cancer (Grusch et al., 2010; Padua & Massagué, 2009).

Studies by Zdanov and colleagues revealed that $KRAS^{G12V}$ induces the secretion of TGFβ from CRC cells through the regulation of the MEK/ERK/AP-1 pathway (Zdanov et al., 2016). Tsubaki and colleagues showed that inhibition of MAPK or PI3K downstream of Ras could suppress TGFβ, indicating both pathways being can act as mediators of Ras signaling to regulate TGFβ (Tsubaki et al., 2015) Additionally, Moo-Young and colleagues provided direct evidence of KRAS-induced TGFβ requirement in intratumoral Treg differentiation. They found that adoptive transfer of CD4+Foxp3- cells into mice implanted with KRAS-mutant, TGFβ-secreting tumor cells led to their differentiation into CD4+Foxp3+ T cells. This phenomenon was not observed in mice with wild-type KRAS, lacking TGFβ secretion (Moo-Young et al., 2009). While only a few studies have elucidated the regulation of TGFβ signaling by Ras signaling, evidence suggests that Ras and TGFβ signaling collaborate to regulate common downstream signaling proteins, such as Smad and MAPK signaling components (Liu, Iaria, Simpson, & Zhu, 2018; Suzuki, Wilkes, Garamszegi, Edens, & Leof, 2007). Liu and colleagues demonstrated that Ras enhances TGFβ signaling by blocking TGFβ receptors' negative regulator, subsequently enhancing Smad2/3 phosphorylation downstream of the TGFβ pathway (Liu et al., 2018). Thus TGFβ remains a highly complex factor within the tumor immune environment. The elevated levels of TGFβ and its association with poor prognosis in various Ras mutant cancers, along with its intricate interactions with other components of the immune system, emphasize its potential as an intriguing target for therapeutic intervention. Despite limited understanding of its direct regulation by Ras, it remains a relevant consideration.

**(d) IL-10:** IL-10 is an anti-inflammatory cytokine primarily associated with suppressive activities, predominantly secreted by Tregs, M2 macrophages, and Th2 cells. Additionally, B cells, mast cells, eosinophils, and DCs also contribute to its secretion (Cheng et al., 2019). Similar to IL6 signaling, IL-10 exerts its effects through STAT-mediated signaling. Specifically, it induces the activation of STAT3 in Tregs, leading to the suppression of immune responses, as mentioned in the previous section.

In a study conducted by Zdanov and colleagues, it was found that IL-10 and TGFβ are secreted and regulated by $KRAS^{G12D}$ via the MEK/ERK/AP-1 pathway (Zdanov et al., 2016). IL-10 plays a crucial role in immune regulation. It can directly act on Th1 and Th17 cells, suppressing immune responses. Moreover, it acts through the IL-10R and STAT3 in Tregs to limit Th17-mediated inflammation (Chaudhry et al., 2011; O'Garra, Vieira, Vieira, & Goldfeld, 2004; Stewart et al., 2013; Waugh and Wilson, 2008). While there is limited evidence regarding the direct link between IL-10 regulation by Ras, it is noteworthy that STAT serves as the common mediator connecting Ras and IL-10.

### 7.1.2 Regulation of chemokines

The IL-8 or CXCL8 chemokine, along with its receptors CXCR1 and CXCR2, establishes a signaling axis known to be upregulated in pro-inflammatory conditions, including tumor environments. This signaling axis has been linked to chemotherapeutic and immunotherapeutic resistance. In various cancers, CXCL8/CXCR1 and CXCL8/CXCR2 signaling have been associated with tumor cell proliferation and angiogenesis (Liu et al., 2016; Waugh & Wilson, 2008). CXCR1 also known as IL-8RA binds to CXCL8 with a higher affinity. Numerous studies have reported elevated levels of IL-8 in NSCLC, where it plays a pivotal role in angiogenesis and tumor progression (Chen et al., 2003; Masuya et al., 2001; Yuan et al., 2000; Zhu, Webster, Flower, & Woll, 2004). Notably, in lung cancer, increased CXCL8 serum levels were specifically linked to patients with mutant KRAS harboring tumors, indicating a direct relationship between KRAS mutation status and CXCL8 expression in human tumors (Tas et al., 2006).

Sparman and Bar-Sagi demonstrated that CXCL8 is induced by oncogenic HRAS$^{G12V}$ through concerted action of AP-1 and NF-κB. This effect led to robust infiltration of granulocytes and macrophages in the resulting tumor, thereby facilitating tumor growth and progression (Sparmann & Bar-Sagi, 2004). IL-8 overexpression was also observed in NSCLC tumor samples with KRAS mutations, and Sunaga et al. further showed that KRAS regulates IL-8 expression via the ERK-MAPK pathway. Additionally, NSCLC tumors with IL-8 overexpression exhibited a more aggressive phenotype and lower disease-free survival in KRAS mutant NSCLC patients with high IL-8 levels (Sunaga, Miura, Kasahara, & Sakurai, 2021). The regulation of IL-8 by KRAS via MAPK or pI3K signaling was further supported by studies in human cell lines of colon cancer (Mizukami et al., 2005) and ovarian cancer (Xu, Pathak, & Fukumura, 2004).

The signaling allies of CXCL8/IL-8, CXCR2 and CXCR1 are evidently regulated by the KRAS signaling axis. In KRAS$^{G12D}$ driven PDAC human cell lines and murine models, Purohit and colleagues demonstrated the upregulation of CXCR2 and its ligand CXCL8 by KRAS$^{G12D}$. As a reciprocal effect, elevated CXCR2 also regulated KRAS protein expression through activation of the ERK signaling pathways (Purohit et al., 2016). In human PDAC, the expression of CXCR2 and its ligands in the tumor stroma was associated with poor outcomes and accompanied by neutrophil infiltration in KRAS$^{G12D}$ driven PDAC (Steele et al., 2016). Awaji and colleagues demonstrated that oncogenic KRAS induces the upregulation of CXCR2 and its ligand. Consequently, the KRAS/CXCR2 axis has the ability to alter the functional properties of Cancer-Associated Fibroblasts (CAFs), leading them to secrete tumor-promoting cytokines such as IL-4, IL-10, and IL-13 (Awaji et al., 2020).

Various chemokines and their receptors, apart from the CXCL8-CXCR1/2 combination, play a significant role in the regulatory circuit of mutant Ras. CXCL8, classified as an ELR+ chemokine due to the presence of a Glu-Leu-Arg sequence preceding the Cys-X-Cys sequence, actively engages with CXC receptors (Cullis, Das, & Bar-Sagi, 2018). The ELR+ group of chemokines includes CXCL1 (GRO-α), CXCL2 (GRO-β), CXCL3 (GRO-γ), CXCL5 (ENA-78), CXCL6 (GCP-2), CXCL7 (NAP-2), and CXCL8 (IL-8) in humans (Allen, Crown, & Handel, 2007; F, 2000; Rossi, & Zlotnik, 2000; Sallusto, Mackay, & Lanzavecchia, 2000). These CXC receptors are expressed on the surface of myeloid cells, such as tumor-associated macrophages (TAMS), MDSCs, macrophages, and neutrophils.

Studies by O'Hayer et al. have demonstrated that human embryonic kidney cells with HRASG12V and KRAS$^{G12D}$ mutations can induce the expression of all ELR+ chemokines. In pancreatic cancer patients, higher serum levels of CXCL1 and CXCL7 were observed (O'Hayer, Brady, & Counter, 2009). Similarly, in CRC patients, CRC tissues exhibited elevated levels of CXCL1 and IL-8, which were associated with decreased OS. Mutant KRAS-harboring human CRC cells displayed increased secretion of CXCL1 (le Rolle et al., 2015). In the CC10-Cre/LSL-KRAS$^{G12D}$ model of lung adenocarcinoma (LAC), KRAS$^{G12D}$ was found to induce the expression of CXCL1, CXCL2, and CXCL5, leading to the infiltration of macrophages and neutrophils in the tumor, thus promoting tumor progression (Ji et al., 2006). Moreover, Steele and colleagues established a connection between KRAS-mediated chemokine upregulation and the establishment of an immune-suppressive TME. They observed that elevated levels of CXCL1, CXCL2, and CXCL5 in KRAS mutant cells correlated with heightened CXCR2 expression in the myeloid compartment of KRAS$^{G12D}$/+p53$^{R172H}$/+ (KPC) mice. Blocking CXCR2, depleting MDSCs and neutrophils, or deleting CXCR2 increased T cell infiltration into the tumor, thereby removing the immunosuppressive effect (Steele et al., 2016). While the precise regulatory mechanisms of ELR+ chemokines and their receptors remain to be fully elucidated, the consistent link between elevated serum chemokines and increased expression of chemokine receptors in Ras mutant tumors suggests an intriguing target for therapeutic interventions.

### 7.1.3 Signaling associates of Ras in regulating cytokine signaling- NF-κB, MAPK/Raf, PI3K and activated

The previous sections have extensively covered the intricate network of Ras-controlled cytokines and chemokines. Mutant Ras proteins have the capability to trigger the expression of diverse cytokines through the NF-κB, MAPK, or PI3K pathways. Notably, NF-κB acts as a pivotal signaling mediator in the regulation of CXCL8 by HRAS$^{G12V}$ (Sparmann & Bar-Sagi, 2004). In the context of KRAS$^{G12D}$-driven pancreatic cancer models, KRAS orchestrates an elevation in GM-CSF levels via downstream PI3K and MAPK pathways. Inhibition of either of these pathways pharmacologically leads to the suppression of GM-CSF expression in KRAS$^{G12D}$-positive pancreatic ductal epithelial cells (PDEC). Furthermore, mutant KRAS also exerts control over the expression of IL-8, TGF-β, IL-10, and chemokines like CXCL1 through the MAP/ERK and PI3K pathways, independently of the NF-κB pathway (Sunaga et al., 2021; Suzuki et al., 2007; Tsubaki et al., 2015; Zdanov et al., 2016). Collectively, the KRAS signaling network and its collaborative interaction with NF-κB, MAPK, and PI3K pathways, contribute not only to the stabilization of cancer progression but also to the modulation of the TME.

## 7.2 Modulation of immune cells in the TME by Ras signaling

The solid TME is characterized by the presence of various immune cells, each with distinct roles in tumor clearance. Immune cells like cytotoxic CD8 T cells, NK cells, and professional antigen-presenting cells (APCs) such as DCs play a crucial role in recognizing and eliminating tumor cells. However, alongside these anti-tumor immune responses, there are also several immunosuppressive cell types present, including Tumor-Associated Macrophages (TAMs), MDSCs, Tregs, neutrophils, and mast cells. These cells contribute to the creation of a cytokine milieu that favors immune escape of tumor cells.

The immunosuppressive effects orchestrated by these cell types involve the secretion of various cytokines that upregulate immune checkpoints on T cells and NK cells, eventually making them exhausted and dysfunctional. In this discussion, we will delve into the various ways through which the Ras signaling pathway regulates immune cell functions in the TME, shedding light on the complex interplay between tumor cells and the immune system and exploring potential therapeutic strategies to overcome immunosuppression and enhance anti-tumor immune responses.

Among the three most common mutations in KRAS (KRAS$^{G12D}$, KRAS$^{G12V}$, and KRAS$^{G12C}$), current pharmacological research is predominantly focused on targeting KRAS$^{G12C}$ (Salem et al., 2022). Intriguingly, the KRAS$^{G12C}$ variant is associated with a higher RR to immune checkpoint blockade (ICB) therapy and improved PFS. Several factors contribute to this phenomenon, including higher TMB, increased PD-L1 expression, and even elevated tumor-specific PD-L1 expression, collectively creating an immune-hot environment (Borghaei et al., 2015; Liu et al., 2022; Watterson & Coelho, 2023).

### 7.2.1 Suppressing immune responses from lymphocytes via PD-1/PD-L1 pathway

Several studies have provided evidence of a noteworthy correlation between the mutational status of KRAS and the response to PD-1 blockade therapy. When investigating the use of Nivolumab, a PD-1 blocker, in patients with NSCLC, it was observed that individuals with mutant KRAS experienced improved OS. However, patients with wild-type KRAS did not exhibit the same benefit (Garon et al., 2015). Similarly, in the case of Pembrolizumab treatment, another PD-1 blocker, the RR and PFS were notably better in patients harboring KRAS mutations. This improvement was attributed to an increase in the overall expression of PD-L1 on the tumors (Sumimoto, Takano, Teramoto, & Daigo, 2016). Several studies have highlighted a strong correlation between KRAS expression and the upregulation of PD-1 on tumor cells (D'Incecco et al., 2015), as well as the overall and CD8-specific PD-L1 expression (Dong et al., 2017) in tumors harboring KRAS mutations.

The underlying mechanism of how KRAS regulates PD-L1 was elucidated by Coelho and colleagues. They found that RAS-MEK signaling plays a role in stabilizing the PD-L1 mRNA, leading to higher PD-L1 expression (Coelho et al., 2017). This finding was further supported by other studies that demonstrated an association between KRAS alterations and increased PD-L1 expression mediated by ERK in patients with lung adenocarcinoma (Chen et al., 2017; Schoenfeld et al., 2020) and regulated by PI3K/AKT/mTOR in human LACS and human squamous cell carcinoma (Lastwika et al., 2016). By enhancing the expression of PD-L1, KRAS also influences the fate and activity of immune cells. This includes inducing apoptosis of CD3+ T cells through PD-1/PD-L1 signaling (Chen et al., 2017), eventually contributing to immune escape of the tumor cells. Contrastingly, another study indicates that in the context of tumors bearing mutant KRAS, the inflammatory and highly immunogenic TME ensures better CD8 T cell infiltration compared to tumors harboring wild-type KRAS (Liu et al., 2020b). This indicates that despite KRAS's immunosuppressive effects on immune cells, the inflammatory TME in KRAS mutant tumors might still foster a conducive environment for T cell infiltration. This phenomenon can be effectively influenced through therapeutic interventions, such as the use of

PD-1 targeted antibodies. Consistent with the same principle, a significant study in 2016 revealed that in NSCLC patients with approximately 50% of tumors expressing PD-L1, pembrolizumab, an anti-PD-1 antibody, provided substantial benefits. These benefits were observed in terms of extended PFS and improved OS after treatment. Consequently, this trial led to the approval of pembrolizumab as a first-line treatment option for metastatic NSCLC (Reck et al., 2016). Canon and colleagues, explored the possibility of combining the novel KRAS$^{G12C}$ inhibitor AMG 510 with immune-checkpoint blockade (ICB). The results demonstrated that the combination of AMG 510 with anti-PD-1 blockade led to effective tumor cell killing and enhanced the responsiveness of the Timor Immune microenvironment (TIM) to immunotherapy (Canon et al., 2019). Further supported by several other studies indicating that the combination of ICB therapy targeting PD-1 with KRAS$^{G12C}$ inhibitors can be advantageous primarily in highly immunogenic tumors with a high mutational burden (Mugarza et al., 2022; Stein et al., 2021). However, it was also noted that some of these tumors may eventually develop resistance to either or both of the therapies (Li, van der Merwe, & Sivakumar, 2022) Considering, KRAS$^{G12C}$ being the most responsive to ICB, (among the most common mutations G12, G12V, G12D) many trials are currently underway to combine KRAS$^{G12C}$ inhibitors with PD-1 pathway inhibitors, encompassing all PDAC, NSCLC and CRC (Li et al., 2022).

### 7.2.2 Promoting immunosuppressive TME

The tumor sustains a proinflammatory milieu throughout its development, which ultimately transitions into an immunosuppressive state to evade immune destruction in subsequent stages. This alteration in the TME creates a more favorable condition for tumorigenesis while concurrently rendering it more resistant to immunotherapeutic interventions in an advanced stage pf the disease. Among several mechanisms of immuno-suppression in the TME the most prevalent methods are recruitment of Tregs and MDSCs.

Zdanov and colleagues investigated the impact of KRAS mutations on tumor cells in promoting Treg formation. Their findings revealed a mechanism driven by mutant KRAS$^{G12V}$, leading to the secretion of IL-10 and TGFβ from tumor cells through the MEK-ERK-AP-1 pathway (Zdanov et al., 2016). IL-10 and TGFβ are major contributors to the formation and activation of Foxp3+Tregs. Additionally, the researchers demonstrated that mutant KRAS tumor-derived exosomes (TDE) containing mutant KRAS cDNA could convert naive CD4 T cells into Tregs independently of tumor-secreted IL-10 or TGFβ in non-small cell lung cancer (NSCLC) cells (Kalvala et al., 2019). Moreover, in lung cancer, it was observed that the genetic ablation of Tregs in mutant KRAS transgenic mice resulted in fewer lung tumors, indicating that Tregs play a crucial role in lung tumorigenesis (Granville et al., 2009). Moo and colleagues presented further intriguing evidence of KRAS and TGFβ-dependent Treg formation as discussed earlier under section of cytokine regulation by KRAS (Moo-Young et al., 2009).

MDSCs constitute a subset of immature myeloid cells that significantly contribute to immune suppression within the TME. Several proinflammatory cytokines previously mentioned play a role in the development of MDSCs, including GM-CSF, IL-6, IL-10, IL-1β, and M-CSF (Groth et al., 2019; Nakamura & Smyth, 2020; Yin, Xia, Rui, Wang, & Wang, 2020). The migration of MDSCs to the TME is facilitated by chemokines secreted by tumor cells, which interact with corresponding receptors on MDSCs. Notable examples include CCR2-CCL2, CCL26-CX3CR1, CXCL12, CXCL8, and CXCL3-CXCR2 (Groth et al., 2019). MDSCs can exert an immunosuppressive effect by promoting the formation of Foxp3+Treg cells through IL-10 and TGFβ signaling. They also hinder lymphocyte homing by downregulating the cell adhesion molecule L-selectin on CD4 and CD8 T cells. Additionally, MDSCs contribute to immune cell dysfunction within the TME by depleting essential metabolites required for the maintenance and proper functioning of immune cells (Groth et al., 2019; Nakamura & Smyth, 2020).

In the context of CRC with KRAS$^{G12D}$ mutation, KRAS was found to contribute to an immunosuppressive TME by recruiting MDSCs (myeloid-derived suppressor cells). KRAS facilitates MDSC migration and infiltration into the tumor by suppressing IRF2, a molecule that otherwise suppresses the CXCR2-CXCL3 chemokine axis, to prevent MDSC migration (Liao et al., 2019). Significantly, Pyleyeva-Gupta has illustrated the phenomenon wherein KRAS$^{G12D}$ orchestrates an augmentation of GM-CSF in a model of pancreatic cancer. This event can subsequently attract immunosuppressive Gr1+CD11b+ myeloid cells within the pancreatic ductal environment, consequently dampening the effectiveness of CD8+ T cell-mediated immunity and thereby fostering the progression of the tumor (Pylayeva-Gupta, Lee, Hajdu, Miller, & Bar-Sagi, 2012). Furthermore, an elevated level of GM-CSF expression is observed in human PanIN lesions. In a cumulative context, KRAS mutations across distinct cancer types collectively contribute to the establishment of an immunosuppressive milieu.

### 7.2.3 Evasion of immune targeting by the cytotoxic CD8 T cells by downregulation of MHCI

Tumors employ several common mechanisms to evade immune cell-mediated killing. As previously discussed, one such mechanism involves the recruitment of immunosuppressive cells, including Tregs and MDSCs, ultimately leading to a reduced reactivity of immune cells against the tumor. Another evasion tactic is the reduction of tumor cell visibility to the immune system. The MHC complex on tumor cells plays a crucial role in presenting tumor-associated antigens and neoantigens to helper CD4 and cytotoxic CD8 T cell subsets (Pylayeva-Gupta, Grabocka, & Bar-Sagi, 2011). However, the oncogenic Ras pathway interferes with this process. Oncogenic Ras downregulates the expression of MHC I, leading to diminished antigen presentation on the MHC I complex. Consequently, the immunogenicity of the Ras-transformed cells is decreased (Ehrlich et al., 1993; Lohmann, Wollscheid, Huber, & Seliger, 1996; Seliger et al., 1996; Sers et al., 2009). The downregulation of MHCI by oncogenic Ras occurs through multiple regulatory layers and may be independent of the promoter activity at the MHC loci. Instead, it appears to be influenced by the impact on other components of the MHC machinery. Understanding these evasion mechanisms is crucial for developing effective cancer immunotherapy strategies.

### 7.2.4 Th17- friend or foe in Ras driven tumor

Another less well-understood component of the tumor's immune environment is a recently discovered lineage known as Th17 cells. These cells are a subset of T helper cells and are characterized by their production of the cytokine IL17A. Th17 cells play a significant role in autoimmune diseases and are differentiated in response to the STAT3 activating cytokines IL-6, IL-21, IL-23 and also IL-1β, TGFβ (Korn et al., 2007). Functionally, Th17 cells can exhibit both regulatory and inflammatory characteristics, as they secrete cytokines like IL-17A and IL-22, which possess both pro- and anti-tumor properties (Littman & Rudensky, 2010). However, their precise function within the tumor context remains not fully elucidated.

Despite the lack of complete understanding, numerous malignancies have exhibited a higher proportion of Th17 cells within the tumor compared to the peripheral blood, indicating their recruitment to the TME (Armstrong, Chang, Lazarus, Corry, & Kheradmand, 2019; Bailey et al., 2014; Zou & Restifo, 2010). The dichotomous role of Th17 and associated cytokines has been highlighted in several studies in KRAS driven lung and pancreatic cancer models. Several research groups have observed a significant enrichment of Th17 cells in KRAS mutant lung and pancreatic tumorigenesis models (Chang et al., 2014; Marshall et al., 2016; McAllister et al., 2014). In KRAS-driven tumor models, IL-17A was found to be essential for the growth of KRAS$^{G12D}$-driven tumors, but it did not show the same effect in non-KRAS-driven tumors (Akbay et al., 2017). Additionally, IL-17A was associated with resistance to anti-PD-1 therapy in lung tumors. Conversely, Th17-derived IL-17A was found to be necessary for preventing early oncogenesis in non-KRAS-driven tumor models, suggesting a role in recruiting DCs critical for cytotoxic CD8 T cell activation (You et al., 2018). Furthermore, in a mouse model of KRAS$^{G12D}$-driven pancreatic tumors, a similar infiltration of IL-17A-producing cells was observed (Marshall et al., 2016). Th17 cells have also been shown to promote metastasis in pre-clinical models of KRAS-driven NSCLC (Li et al., 2012; Yoon et al., 2010).

Although the link between KRAS-driven tumors and Th17 cells is well established, the specific mechanisms driving Th17 formation and recruitment have not been fully elucidated. Given that IL-6 and several STAT3-related cytokines are crucial for Th17 formation and are also known targets of KRAS, it is possible that the IL-6-STAT3 circuit regulated by KRAS may play a key role in Th17 development in KRAS-driven tumors. In KRAS-driven tumors, Th17 cells have largely been considered as mediators of immunosuppression. Studies have shown that genetic deletion of the CD4 T cell population, including Th17 cells, resulted in the de-repression of stromal cytotoxic CD8 T cells, which contributed to anti-tumor immunity (Zhang et al., 2014). Similarly, IL-6 blockade in KRAS-driven lung cancer models reduced protumor Th17 responses, further supporting the involvement of KRAS-regulated IL-6-STAT3 circuit in Th17 cell formation and function (Brooks et al., 2016).

Therefore, it is essential to gain a deeper understanding of the specific subpopulations of Th17 cells that play pro- or anti-tumor roles and to investigate how KRAS mediates the regulation of these populations.

## 7.3 Alternate targets for Ras mutant cancers in the immune-community

### 7.3.1 Targeting KRAS with adoptive T cell transfer therapy

Although we have previously discussed the detrimental impact of KRAS in promoting immune escape, it is well-established that KRAS mutant cells harbor a significant number of neoantigens, which present a promising target for cytotoxic T cells, thereby facilitating effective tumor control. Numerous studies have demonstrated the generation of anti-tumoral T cell responses through adoptive T cell transfer, following prior exposure to KRAS mutants.

In the first study, patients with metastatic disease and KRAS proto-oncogene mutation at codon 12 were vaccinated with peptides spanning the three major mutations of KRAS (G12D, G12V, and G12C). Among the eight enrolled patients, three showed a notable MHCI or MHCII bound response from CD8 or CD4 T cells post-vaccination, as indicated by the production of T cell lines resulting from the vaccination (Abrams et al., 1997; Khleif et al., 1999).

In another study conducted by Tran and colleagues, regression of metastatic CRC was observed following adoptive transfer of T cells targeting KRAS$^{G12D}$ NeoEpitopes (Tran et al., 2016). Remarkably, this study identified HLA-C*8:02-restricted T cell receptors (TCRs) targeting KRAS$^{G12D}$, presenting a valuable opportunity for T cell receptor therapy with engineered T cells (Tran et al., 2016). Later, Leidner and colleagues constructed genetically engineered T cells expressing two allogenic HLA-C*08:02-restricted TCRs targeting KRAS$^{G12D}$. A patient with progressive metastatic pancreatic cancer underwent adoptive transfer with these cells, resulting in regression of visceral metastasis. Furthermore, the transferred engineered T cells remained detectable in the patient's peripheral blood even six months after the transfer (Leidner et al., 2022).

### 7.3.2 Immune checkpoints and cytokine signaling targets

Given the active modulation of the immune system by mutant Ras, a predominant alternative strategy for therapeutically addressing Ras mutant tumors involves alleviating immune suppression and rendering the TME more immunogenic to elicit potent anti-tumor responses. Among these interventions, a noteworthy approach entails enhancing the responsiveness of cytotoxic T cells by blocking inhibitory receptors such as PD-1, CTLA-4, TIGIT, and members of the immunosuppressive adenosine pathway utilizing antibodies to revitalize cytotoxic CD8 T cells from a state of near exhaustion or targeting cytokines that have been closely associated with unfavorable prognostic outcomes and OS, such as IL-1β and IL-6 (Dutta, Ganguly, Chatterjee, Spada, & Mukherjee, 2023; Mamdani, Matosevic, Khalid, Durm, & Jalal, 2022).

In conjunction with the extensively targeted immune checkpoints PD-1/PD-L1 and CTLA-4, emerging targets like TIGIT, LAG3, the adenosine pathway-associated ectoenzyme CD73, expressed on Tregs, CD8 T cells, and NK cells, have demonstrated efficacy in augmenting anti-tumor responses through therapeutic targeting. Preliminary findings from a phase I trial of the anti-TIGIT antibody Vibostolimab, in combination with the anti-PD1 antibody Pembrolizumab, have exhibited clinical effectiveness in patients with NSCLC refractory to PD-1/PD-L1 blockade, with an ongoing phase III trial (NCT04738487) currently underway (Niu et al., 2022). Co-inhibition of PD-L1 with Atelizumab and TIGIT with Tiragolumab, in NSCLC patients with high PD-L1 expression, has yielded substantial overall response and survival rates. This success has led to Tiragolumab being designated as a therapeutic option for NSCLC, supported by an ongoing phase III trial (NCT04294810).

In the realm of ongoing investigations, there is a focus on inhibitory receptor NKG2A, which is present on CD8 and NK cells (Haanen & Cerundolo, 2018). Recently reported early results from a phase II clinical trial evaluating the monoclonal antibody Oleclumab against CD73, in combination with durvalumab following chemoradiation in locally advanced unresectable stage III NSCLC, have indicated improved PFS compared to durvalumab alone, while maintaining a manageable safety profile (Herbst et al., 2022).

The pivotal role played by mutant KRAS in modulating cytokine signaling has warranted exploration of targeting components of this pathway for KRAS mutant tumors. Although a phase III trial assessing Canakinumab (anti-IL-1β) for NSCLC treatment did not meet anticipated results in terms of disease-free and OS, avenues remain open for investigating biomarkers linked to therapy resistance and for exploring alternative approaches (Garon et al., 2015). JAK kinase inhibitors, known to be key participants in cytokine signaling pathways, are also under scrutiny. An example, Momelotinib, targeting TBK1 in the JAK/TBK1/IKKe pathway, has been evaluated in KRAS mutant NSCLC (NCT02258607).

The targeting of cytokine signaling or inhibitory receptors/enzymes not only present a significant opportunity to suppress a crucial aspect of RAS mediated transformation but also to disrupt certain elements of the tumor-supportive immune microenvironment (Fig. 2). This approach offers fresh perspectives for combination therapies with other RAS effector pathways and paves the way to investigate and overcome resistance mechanisms to the Ras targeted therapies. Particularly noteworthy is the potential for

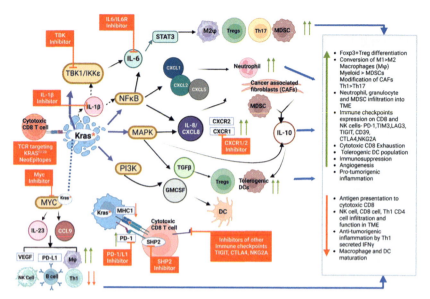

**Fig. 2** Modulation of tumor immune microenvironment by KRAS and possible therapeutic interventions: Oncogenic KRAS signaling regulates cytokine and chemokine signaling pathways as well as immune cell differentiation and function through coordinated activation of mitogen activated protein kinases (MAPK), nuclear factor kappa-light-chain-enhancer of activated B cells (NF-κB), and phosphoinositide 3-kinase (PI3K) cascades and in cooperation with activated Myc signaling. The effects include polarization of CD4+ T cells into regulatory T cells, differentiation and activation of myeloid-derived suppressor cells (MDSCs) and conversion of pro-inflammatory M1 macrophages to anti-inflammatory M2 macrophages, increase in tolerogenic dendritic cells- resulting in an immunosuppressive tumor microenvironment. Additional effects are downregulation of major histocompatibility complex class I-mediated antigen presentation to cytotoxic CD8+ T cells and T cell exhaustion, reduced natural killer cell recruitment and cytotoxicity impacting tumor cell killing. These KRAS-mediated alterations present several opportunities for therapeutic intervention as represented by the red boxes in the figure.

synergistic action and reduced overlapping toxicity through the integration of cytokine network targeting, inhibitory receptor targeting with current strategies focused on inhibiting mitogenic pathways.

## 8. SHP-2 tyrosine phosphatase

### 8.1 Structure and biological function of SHP2

Src homology phosphatase 2 (SHP2), a protein tyrosine phosphatase (PTP), is encoded by PTPN11 and plays a key role in the activation of intracellular signaling pathways such as KRAS pathway. SHP2 has two SH2 domains

(N-SH2 and C-SH2), a protein-tyrosine phosphatase (PTP) domain, and a C-terminal domain with two tyrosine residues (Fedele et al., 2020). The tyrosine residues are the phosphorylation site, such as tyrosine 542 and 580, and acts as a binding site GRB2-SOS1 complex and recruits it to the receptor (Chen et al., 2016; Eulenfeld & Schaper, 2009). Shp2 actively recruits Grb2 to the receptor only once it is phosphorylated on this tyrosine. The autophosphorylation happens when the RTKs such as EGFR are activated. A binding site is created for the phosphor tyrosine-binding domains (SH2) for the adaptor protein GRB2. This binding of GRB2 to the phosphorylated RTKs recruits SOS, a GEF of Ras, to the plasma membrane, rendering inactive RAS to GDP bound active RAS (Dance, Montagner, Salles, Yart, & Raynal, 2008). SHP2 also plays a catalytic role in promoting KRAS activation through dephosphorylation of the substrate (Bunda et al., 2015). When in inactive state, SHP2 stops the PTP domain from binding to the substrate while N-SH2 domain is deformed by PTP, disabling it to bind to the phosphor tyrosyl (pY) peptides. Whereas pY binding to the N-SH2 domain causes it to render into an open state thus engaging RTKs, cytokine receptors and immune checkpoint receptors to the SH2 to domain. This switching of an inactive to active state triggers SHP2 activation in response to signals at appropriate sites (Fedele et al., 2020). This molecular switch is used by many researchers to develop allosteric SHP2 inhibitors (SHP2i). These inhibitors bind to the inactive SHP2 and impedes the N-SH2/loop/C-SH2 movements needed for activation (LaRochelle et al., 2018). Small molecule inhibitors are developed that inhibits SHP2 and thus causing antitumor microenvironment in the tumor and also have antiangiogenic properties (Tang et al., 2022a). The SHP2i in combination with various direct and indirect target inhibitors such as KRASG12Ci and MEKi is under study to improve the antitumor efficacy of monotherapy and also overcomes adaptive resistance to the drugs. This happens as in combination therapy both oncogenic signaling and feedback-induced RTK-mediated RAS activation is co-targeted.

## 8.2 SHP-2 at the interface of KRAS and immune signaling

Understanding the relationship between the KRAS signaling pathway and the immune signaling units was important, given that several targets of KRAS are involved in immunology. A common signaling protein connecting KRAS and the immune signaling pathways needed to be identified to explore potential therapeutic avenues. Therefore, a comprehensive understanding of the overlap between the RAS/ERK MAPK signaling cascade and immune signaling pathways was crucial for the development of novel therapies.

The role of SHP2 in facilitating full activation of Ras has been previously discussed. In T cells, SHP-2 demonstrates an immunomodulatory effect when it is recruited to the cytoplasmic tail of inhibitory immune receptors, PD-1 and BTLA. This recruitment leads to the suppression of T cell activity (Hui et al., 2017; Marasco et al., 2020; Rudd, Taylor, & Schneider, 2009; Sheppard et al., 2004; Yokosuka et al., 2012). Studies utilizing a colon cancer xenograft model have demonstrated that inhibiting SHP2 with SHP099 resulted in tumor regression and enhanced tumor immunity, evident by an increase in the cytotoxic CD8+ T cell population (Zhao et al., 2019). Moreover, Fedele et al. showed that SHP099 effectively overcame resistance to $KRAS^{G12C}$ inhibitors in both NSCLC and PDAC models. Furthermore, a combination of G12C and SHP2is improved CD8+ T cell infiltration and function, while reducing recruitment of MDSCs (Fedele et al.). Quintana et al. investigated the impact of SHP2 blockade in an tumor model and found that it reduced the viability of M2 macrophages by attenuating CSF1R receptor signaling, thereby removing immunosuppression on tumor infiltrating CD8+ T cells. However, in vitro studies confirmed that despite SHP2 acting downstream of PD-1 signaling, SHP2 blockade does not yield similar effects to direct PD-1 signaling blockade (Quintana et al., 2020).

Wang et al. conducted a comprehensive investigation utilizing in-vitro tumor spheroid cultures and human peripheral blood mononuclear cells (PBMCs), as well as in vivo syngeneic mouse models, to explore the impact of Shp-2 blockade on the tumor immune microenvironment (Wang et al., 2021). By employing the SHP099 inhibitor to target SHP2, they observed a noteworthy enhancement in cancer cell IFNγ signaling, subsequently leading to heightened expression of MHCI molecules and consequent augmentation of immunogenicity. This enhancement in immunogenicity was accompanied by an increase in the cytotoxic activity of CD8+ T cells. Additionally, the researchers demonstrated, through a separate experimental approach, that SHP2 blockade directly inhibits the function and differentiation of MDSCs (Wang et al., 2021b). Interestingly, Zhang et al. have provided insights into the paradoxical role of SHP-2 in immunomodulation across distinct stages of melanoma progression (Zhang et al., 2013). Specifically, targeted SHP2 blockade aimed at CD4+ cells during the early stages of melanoma led to effective inhibition of tumor growth. However, in the context of advanced melanoma, the same intervention resulted in an augmented release of IL-6 and accumulation of MDSCs, ultimately fostering tumor progression and metastasis (Zhang et al., 2013). This intriguing dichotomy was attributed to the intricate regulatory interplay between SHP2, STAT3, and PLR3 downstream of IL-6. Notably, this mechanism disrupted the enzymatic activity of SHP2, consequently releasing its inhibitory influence on STAT3 phosphorylation within the context of myeloma (Chong et al., 2019). In a recent study, Tang et al. demonstrated the impact of SHP2 inhibition using a genetically engineered mouse models (GEMMs) of KRAS-mutant or EGFR mutant NSCLC. Their findings revealed an augmentation in the infiltration of T and B cells, a reduction in alveolar or M2-like macrophage accumulation, and a concurrent elevation in the accumulation of potentially immunosuppressive granular MDSCs upon SHP2 inhibition (Tang et al., 2022b).

In light of these contrasting findings pertaining to SHP2's diverse roles in immune regulation and its differential implications across various cancer types, a comprehensive understanding remains elusive. However, given its strategic position downstream of inhibitory receptors in T cells—critical components of the suppressing anti-tumor response—SHP2 presents itself as a promising therapeutic target. This potential extends both independently and in synergy with other agents involved in immune checkpoint blockade therapies. Notably, the combination of SHP2 inhibition with G12C treatment has demonstrated promise in NSCLC, exemplifying the potential of such combined approaches (Wang et al., 2021). Further investigation is warranted to decipher the intricacies of SHP2's multi-faceted contributions to immune modulation and cancer pathogenesis.

## 8.3 SHP2 inhibitors emerge as potential therapy for KRAS-mutant cancers

The SHP2 molecule, an oncogenic tyrosine phosphatase has generated new hope in anti-cancer therapies especially for solid tumours which include that of breast, larynx, stomach, liver, lung and oral cancers as well as certain types of leukaemia (Zhang, Zhang, & Niu, 2015). SHP2 encoded by the proto-oncogene PTPN11 is extensively involved in posttranslational modification and regulates multiple cellular signalling cascades. As a result, SHP2 is multi-functional in a variety of signaling cross-talks especially involving cell survival and immune regulation, and this fact is further substantiated with the association of SHP2 dysregulation in hematologic and solid tumor malignancies. SHP2 not only results in the activation of RAS-RAF-MEK-ERK signaling pathway but also has dual role of both stimulating or antagonizing the PI3K-AKT and JAK-STAT pathways. Moreover, SHP2 is also found to inhibit T cells via the PD-1/PD-L1 pathway (Liu, Gao, Elhassan, Hou, & Fang, 2021; Song, Zhao, Zhang, & Yu, 2022). After the discovery of the RAS oncogene in 1964 and the fact that mutations in a RAS homologue, mainly KRAS, is found in twenty percent of patients, mainly of non-small cell lung cancers, pancreas and colon, lot of focus was being laid on studying the mechanisms of KRAS inhibitors in modulation of cancer progression (Weiss, 2020). There is activation of multiple complex array of pathways via mutant KRAS in which multi-subunit signaling complexes and feedback loops occur frequently (Simanshu, Nissley, & McCormick, 2017; Stephen, Esposito, Bagni, & McCormick, 2014). Although selective and potent inhibitors of the ERK MAPK cascade (RAF/MEK/ERK) to target RAS-mutant malignancies have been used several times, they often failed to exert any effect as their action were negatively modulated by upstream signaling of feedback loops resulting in ERK reactivation (Samatar & Poulikakos, 2014) and it was also noticed that when parallel signaling pathways (PI3K/mTOR + MEK) were targeted, their use was limited by toxicity (Wainberg et al., 2017).

SHP2, known for its diverse biological activities is shown to depend on upstream growth factor signaling involved with differentiation and proliferation of tissues as well as angiogenesis (Bowen, Ayturk, Kurek, Yang, & Warman, 2014; Hao et al., 2019; Ke et al., 2007; Kontaridis et al., 2008; Mannell et al., 2008). The selective allosteric SHP2i named SHP099, has been demonstrated to have promising therapeutic role in cancers dependent on receptor tyrosine kinases (RTKs) (Chen et al., 2016; Garcia Fortanet et al., 2016; Nichols et al., 2018). In cancer cell lines and in GEMMs Inhibition of SHP2 can not only modulate mutant KRAS signalling (Dutta et al., 2023; Fedele et al., 2018; Fedele et al., 2020; Mainardi et al., 2018a; Nichols et al., 2018; Ruess et al., 2018b) but also has an immunomodulatory function as highlighted by confounding evidence. Additionally, SHP2 has been shown to modulate T cell function as it is implicated in suppressing T-cell activation by dephosphorylation of PD1-recruited CD28 (Hui et al., 2017). Interestingly, in $CD4^+$ T-cell specific knockout mouse models, it has also been shown to be quite dispensable for exhaustion of T-cells and for anti-tumor activity by PD1 blockade (Rota et al., 2018). Nevertheless, SHP2is are expected to have tremendous potential and its true potential needs to be realised either as a monotherapy or as part of a combination therapy. The cellular functions of SHP2 are quite unique compared to most cancer therapy targets and studies have pointed out that in addition to having inhibitory effects on cancer cell proliferation, it can be of significant use in immune surveillance for cancer cells [23].

## 8.4 SHP2 inhibitors used as monotherapy

The challenge faced in the treatment of hematologic malignancies is the acquired resistance to TKI therapy. Isoforms of constitutively activated tyrosine kinase BCR-ABL1, derived from the BCR-ABL1 oncogene cause Philadelphia chromosome-positive (Ph+) B-cell lymphoblastic leukemia (B-ALL) and chronic myeloid leukemia (CML). Even though, the entire treatment procedure of these hematologic malignancies have received a significant boost with the use of TKIs targeted to BCR- ABL1, risk of relapse in CML patients have also been reported owing to CML stem cell insensitivity to the TKIs. Much to the worry of clinical oncologists, worse response of TKIs is seen in patients with Ph+ B-ALL, owing to multiple mechanisms of TKI resistance. Recent studies have demonstrated that SHP2 is necessary for BCR-ABL1–initiated myeloid and lymphoid neoplasia (Gu et al., 2018). Studies have also pointed out the involvement of SHP2 is in multiple pathways related to development of hematologic malignancies. Many SHP2is are being studied in clinical trials, but their usefulness as therapy for hematologic malignancies require additional modifications to counteract the resistance of currently used drugs (Kanumuri, Kumar Pasupuleti, Burns, Ramdas, & Kapur, 2022). In case of solid tumours a pattern of SHP2 expression similar to that of hematologic malignancies has been reported. This is exemplified in the activation of SHP2 pathways in breast cancer cells which promote oncogenesis. Moreover, in patients suffering from PDAC, SHP2 expression is linked to outcome: higher expression is linked with poorer outcomes and vice-versa (Zheng et al., 2016). Additionally, it has been shown that SHP2 inhibition induced tumor cell senescence and impairment of tumor growth in a xenograft model of KRAS-mutant NSCLC (Ruess et al., 2018b). Likewise, SHP2 inhibition was found to inhibit tumorigenesis and induce a near-normal cellular phenotype in breast cancer cells (Song et al., 2021). Several clinical trials are currently being undertaken which include phase 1 trials in patients with advanced solid tumors of SHP2i (TNO155) alone or in combination with EGFR TKI nazartinib (NCT03114319) and SHP2i (ET0038)as monotherapy in patients with advanced solid tumors (NCT05354843).

## 9. Combination of SHP2 inhibitors with other drugs in modulating KRAS driven solid tumours

SHP2i are also being tested with various combinations of drugs and inhibitors are being tested in several tumors which show hyperactivity of the RAS/ERK pathway. The immune-modulatory effects of SHP2is can be characterized systematically in genetically defined, immune- competent indigenous or orthotopic tumor models that resemble closely to human cancers might provide important insights into judicious combination of these agents. Many such combination therapies under clinical trials and active research are discussed below:

### 9.1 SHP2 in combination with MEK (mitogen activated ERK kinase) inhibitors

Researchers have earlier demonstrated that treatment of KRAS-mutant cancers with MEK-inhibitor (MEKi) led to a phenomenon of "adaptive resistance", which resulted from upregulation of genes encoding receptor tyrosine kinases (RTKs) and/or their ligands (Hymowitz & Malek, 2018; Manchado et al., 2016; Ryan & Corcoran, 2018). The challenge is, however, that even within a single histotype, there is induction of distinct RTK ligands in different tumors. This makes it quite difficult to get improved outcomes by combination therapies using both RTK and MEK inhibitors. In this regard, SHP2, which is ubiquitous in distribution, lies downstream of majority of RTKs and is vital for RTK-evoked RAS activation (Mohi & Neel, 2007). These properties of SHP2i rendered them as an attractive option as combination therapy particularly for KRAS-driven cancers to augment MEKi action. To work with this idea, several studies (Ahmed et al., 2019; Nichols et al., 2018; Ruess et al., 2018a; Wong et al., 2018), gave pre-clinical data supporting the process of dealing with "adaptive resistance" to MEKi by using SHP2i. Several SHP2is, with multiple agents, have been developed by pharma companies namely RMC-4630, ERAS-601, TNO155, JAB-3068, RLY-1971 and BBP-398, which are currently undergoing phase 1/2 trials. Among them, the compound RMC-4630, has been used in combination with MEKi, responses that have been recorded show that out of the seven patients treated, one had stable disease, one showed partial response, but the main drawback the combined toxicity of the two inhibitors (Bendell et al., 2020). This drawback can be avoided by pairing of a mutant-selective agent with SHP2i as it would decrease the toxicity of "vertical" pathway inhibition observed with the use of MEKi/SHP2i combinations. SHP2is act at the level of SOS1/2 in order to prevent MEK/ERK pathway reactivation when KRAS-mutant cancer cells are concomitantly treated with MEKis (Fedele et al., 2018; Nichols et al., 2018). Studies using combination therapies of MEKi/SHP2i showed that not only SHP2i efficacy was directly proportional to the level of residual GTPase activity (Nichols et al., 2018), but also it showed significant single agent activity especially against KRAS$^{G12C}$-mutant cancer cells (Fedele et al., 2018; Fedele et al., 2020).

The therapeutic potential for the combined action of MEK and SHP2is was explored by Fedele and colleagues by use of KRAS-mutant pancreatic cancer and NSCLC cell culture as well as mouse models (Fedele et al., 2018; Torres-Ayuso & Brognard, 2018). The researchers showed that by depleting cells of SHP2 in presence of MEK inhibitor acts together to inhibit cell growth promote senescence in cancer cells and augment apoptotic pathways. This drug combination has also been shown significantly effective in wild-type and KRAS-amplified gastric cancers by Wong and colleagues (2018). Another interesting observation having important translational implication was that the GTPase activity of the different RAS mutants have a direct correlation with the sensitivity to the allosteric SHP2i SHP099 (Mainardi et al., 2018a). Cancer patients reported to be homozygous for codon 61 RAS mutations (e.g. Q61R) are shown to be very likely to be non-responders to MEK/SHP2 dual-inhibitory therapy, as this mutant demonstrates the lowest intrinsic GTPase activity (Fedele et al., 2018; Mainardi et al., 2018a; Mainardi et al., 2018b). In addition, SHP2i GS-493, which has been shown to target the catalytic site of SHP2, acts synergistically with MEK inhibitors (Ruess et al., 2018b). However, it can be said that efficacy of MEK/SHP2 dual-inhibitory therapy in vivo can be attributed partially to their modification of the surrounding stroma. Reduced tumor vasculature has been observed in animals treated with MEK and SHP2is by Fedele and colleagues (Fedele et al., 2018), along with increased T cells tumor infiltration seen in mice treated with combination therapy (Mainardi et al., 2018a).

## 9.2 SHP2 in combination with KRAS inhibitors

The development of the mutant KRAS-specific inhibitors, especially targeting KRAS$^{G12C}$ which were long considered "undruggable" became possible thanks to the the pioneering work of Kevan Shokat (Ostrem, Peters, Sos, Wells, & Shokat, 2013) which led to incredible developments resulting in the use of several clinical grade KRAS$^{G12C}$ inhibitors. Among these inhibitors, the most advanced have been AMG510 and MRTX849, each of which have demonstrated significant single agent efficacy, especially in NSCLC patients (Canon et al., 2019; Govindan et al., 2019; Hallin et al., 2020; Sheridan, 2020). Drug resistance similar to inhibitors in the RAS/ERK pathway (Ahronian & Corcoran, 2017; Konieczkowski, Johannessen, & Garraway, 2018; Ryan & Corcoran, 2018), will also limit the effects of KRAS$^{G12C}$ inhibitors (abbreviated as G12Cis). Understanding G12Ci resistance mechanisms is important for developing ways to counter their emergence.

Several recent experimental evidences (Amodio et al., 2020; Fedele et al., 2020; Hallin et al., 2020; Misale et al., 2019; Ryan et al., 2020; Xue et al., 2020), reveal the development of adaptive resistance to G12Ci in KRAS-mutant cancer cells. G12Ci efficacy is improved by SHP2is using several mechanisms. One of these mechanisms is the targeting of GDP- bound state of KRAS (KRAS-GDP) byG12Ci which eventually leads to the opposition of G12Ci action by RAS-GEF activity. SHP2is increase the amount of KRAS-GDP by blocking SOS1/2 action, and thus the "target" of G12Cis (Fedele et al., 2020; Nichols et al., 2018). The reactivation of other, wild type RAS isoforms (encoded by the other, normal, KRAS allele and/or H/NRAS) is also known to be inhibited by SHP2is (Ryan et al., 2020).

It is very important to decipher the effects of combination therapies on tumor cells in connection with their TME if we want to progress from responses to cures. This fact is exemplified when we study the effects of inhibitors of targets like SHP2, and understand their action on immune receptor, cytokine signaling along with RTK pathways in multiple TME cell types (Niogret, Birchmeier, & Guarda, 2019).

Characterization of the cell-autonomous and non-autonomous effects of SHP2is alone as well as in combination with G12Cis in syngeneic, GEMMs of NSCLC and PDAC has been done (Fedele et al., 2020) and in agreement to the expected results, each agent/combination had some similar effects in both the models. The G12Ci/SHP2i combination demonstrated appreciable favourable effects on the immune TME of both the models as evidenced by increasing intratumor B cells, reducing myeloid suppressor cells and causing increase of the total and CD8+ T cells.

Data accumulated from experimental work point out that the intrinsic and extrinsic features of the TME as well as the tumor genotype (including mutational burden) and cell-of-origin/histotype play crucial roles in the response to targeted therapies. Multiple ongoing clinical trials have used combinations of AMG510 or MTRX849 with targeted or immune therapies such as SHP2is

(NCT04330664, NCT04699188, NCT04185883), MEKis (NCT04185883), or anti- PD1(NCT04613596, NCT03785249, NCT04185883) along with trials, using MRTX849 in combination with the SOS1/pan-KRAS inhibitor BI 1701963 (announced by Boehringer Ingelheim, December 2020) (Kwan, Piazza, Keeton, & Leite, 2022). However, only time will say how these combinations will modulate the TME as well as the tumor cells themselves. Inhibitors of other mutant KRAS alleles such as MRTX1133-KRAS$^{G12D}$ selective inhibitor are underway for testing and some are in active development (e.g. KRAS-G12D(ON) Inhibitors, Revolution Medicines Inc., September 2020), and thus, it will be important to determine whether the specific mutations also influence the TME response in a similar way and determine what could be the combination(s) most likely to be optimally beneficial for patients.

## 9.3 SHP2 in combination with anti-CXCR1/2

In GEMM of Kras- and Egfr-mutant NSCLC, studies have been conducted by researchers to characterize the autonomous and nonautonomous effects of SHP2 inhibition on tumor cell. These studies have been used to evaluate a novel yet rational combination of SHP2 and CXCR1/2 inhibitors on these tumour cells. SHP2i treatment was shown to promote B and T lymphocyte infiltration in *Kras*- and *Egfr*-mutant NSCLC cells as well as reduce the number of alveolar and M2-like macrophages which lead to enhanced tumor-intrinsic CCL5/CXCL10 secretion. In addition, these inhibitors also positively induced intratumor granulocytic myeloid-derived suppressor cells (gMDSC) through tumor-intrinsic, NFκB-dependent production of CXCR2 ligands. In patients on KRAS$^{G12C}$ inhibitor trials CXCR2 ligands were induced in tumors. Combined SHP2 (SHP099)/CXCR1/2 (SX682) inhibition resulted in the selective decrease of a specific cluster of *S100a8/9*hi gMDSCs, that led to generation of *Klrg1*+ CD8+ effector T cells possessing a powerful cytotoxic phenotype which although expressed the checkpoint receptor NKG2A, demonstrated enhanced survival in *Kras*- and *Egfr*-mutant models (Tang et al., 2022a).

NFκB-dependent recruitment of S100A8hi gMDSCs and upregulation of CXCR2 ligands has been shown with the inhibition of the SHP2/RAS/ERK pathway which leads to triggers for suppression of T cells. The combination of SHP2/CXCR2 inhibitors causes blockade of gMDSC immigration leading to induction of CD8+KLRG1+ effector T cells with raised cytotoxicity, with a concomitant Th1 polarization and improved survival in multiple NSCLC models. Treatment with inhibitors, transcriptional profiling, results from reporter assays and studies on neutralizing antibody have all pointed out the fact that gMDSC immigration results consequently from NFκB-dependent CXCL1/5 production by KP tumor cells. Moreover, treatment of human KRAS-mutant NSCLC lines with SHP2i, G12Ci, or MEKi causes analogous induction of CXCR2 ligands. Along with CXCR2 ligands, it has been shown that inhibition of EGFR/SHP2/KRAS/MEK pathways resulted in NFκB-dependent upregulation of CCL5 and CXCL10 as well as enhance T-cell infiltration in several studies (Canon et al., 2019; Fedele et al., 2021; Hallin et al., 2020; Quintana et al., 2020; Verma et al., 2021; Wang et al., 2021), but the underlying mechanisms remain to be explored.

## 9.4 SHP2 in combination with other drugs

*(a) With ERK signal suppressors.*

In the tumours which are dependent on the ERK pathway, the disruption of negative feedback of RAF or MEK pathways results in RTK upregulation, which in turn leads to RAS activation and rejuvenation of ERK activity, ultimately leading to inhibitor resistance. In RKO xenografts the combination of ERK, MEK, and SHP2is has been found to suppress ERK and SHP2 signals (Liu et al., 2021) This is exemplified with the use of ERK inhibitors dabrafenib (Tafinlar) and trametinib (Mekinist) along with SHP2i SHP099 in an RKO xenograft model in which this combination suppressed tumor growth and similar signal suppression is also shown to modify outcome positively in *BRAF* V600E–mutated colorectal tumors (Ahmed et al., 2019; Liu et al., 2021).

*(b) With ALK inhibitors.*

Tumors which grow resistant to MEK inhibitors may consequently develop ALK(anaplastic lymphoma kinase) inhibitor resistance (e.g. *ALK* rearrangements in NSCLC)and short hairpin RNA gene screening demonstrated SHP2 as a key resistance node. Hence, when SHP2i SHP099 was given in combination with ALK TKI ceritinib, it inhibited the growth of resistant patient- derived tumor cells (Dardaei et al., 2018; Kanumuri et al., 2022).

*(c) With PD-1/PD-L1 targeted agents.*

SHP2 is shown to be a mediator of PD-L1 inhibition of T-cell function by leading to the inactivation of coreceptor CD28. So combination of these drugs presents a more orthogonal approach to revert immunosuppression and modulate TME. Not only does SHP2 acts downstream of RTK, it is also downstream to the PD-1 pathway, involved with anergy and T-cell suppression (Kanumuri et al., 2022). In xenograft models of colon cancer, SHP099 combined with a PD-1 inhibitor, controlled tumor growth much better than when these inhibitors were administered as a monotherapy (Zhao et al., 2019).

## 10. Future directions

Safety and efficacy pose the biggest challenge in the use of SHP2i as a monotherapy or as part of a combination regimen more so because SHP2is are not well tolerated on their own. As such, evaluation of SHP2is have been done in a number of schedules to improve its tolerability such as: one big dose once per week; two days on followed by several days off;or three days on, 4 days off; or even two weeks on, one week off. The current scenario still depicts that the combination strategy results in more number of adverse effects and hence the trade-off has to be worth for its use mainly in terms of prolonging the amount of time that patients can be on both therapies in combination with a better prognosis. Though till date, we have just initiated to scratch the surface of the clinical utility for SHP2is, more studies are needed to establish the efficacy of combination therapies for improved outcome in curing malignancies.

## References

Abrams, S. I., Khleif, S. N., Bergmann-Leitner, E. S., Kantor, J. A., Chung, Y., Hamilton, J. M., & Schlom, J. (1997). Generation of stable CD4+ and CD8+ T cell lines from patients immunized with ras oncogene-derived peptides reflecting codon 12 mutations. *Cellular Immunology, 182*, 137–151.

Ahearn, I. M., Haigis, K., Bar-Sagi, D., & Philips, M. R. (2011). Regulating the regulator: Post-translational modification of RAS. *Nature Reviews. Molecular Cell Biology, 13*, 39–51.

Ahmed, T. A., Adamopoulos, C., Karoulia, Z., Wu, X., Sachidanandam, R., Aaronson, S. A., & Poulikakos, P. I. (2019). SHP2 drives adaptive resistance to ERK signaling inhibition in molecularly defined subsets of ERK-dependent tumors. *Cell Reports, 26*, 65–78.e65.

Ahrendt, S. A., Decker, P. A., Alawi, E. A., Zhu, Yr. Y. R., Sanchez-Cespedes, M., Yang, S. C., ... Sidransky, D. (2001). Cigarette smoking is strongly associated with mutation of the K-ras gene in patients with primary adenocarcinoma of the lung. *Cancer, 92*, 1525–1530.

Ahronian, L. G., & Corcoran, R. B. (2017). Strategies for monitoring and combating resistance to combination kinase inhibitors for cancer therapy. *Genome Medicine, 9*, 37.

Akbay, E. A., Koyama, S., Liu, Y., Dries, R., Bufe, L. E., Silkes, M., ... Wong, K. K. (2017). Interleukin-17A promotes lung tumor progression through neutrophil attraction to tumor sites and mediating resistance to PD-1 blockade. *Journal of Thoracic Oncology: Official Publication of the International Association for the Study of Lung Cancer, 12*, 1268–1279.

Akhave, N. S., Biter, A. B., & Hong, D. S. (2021). Mechanisms of resistance to KRAS (G12C)-targeted therapy. *Cancer Discovery, 11*, 1345–1352.

Allen, S. J., Crown, S. E., & Handel, T. M. (2007). Chemokine: Receptor structure, interactions, and antagonism. *Annual Review of Immunology, 25*, 787–820.

Amodio, V., Yaeger, R., Arcella, P., Cancelliere, C., Lamba, S., Lorenzato, A., ... Misale, S. (2020). EGFR blockade reverts resistance to KRAS(G12C) inhibition in colorectal cancer. *Cancer Discovery, 10*, 1129–1139.

Ancrile, B., Lim, K.-H., & Counter, C. M. (2007). Oncogenic Ras-induced secretion of IL6 is required for tumorigenesis. *Genes & Development, 21*, 1714–1719.

Apte, R. N., Dotan, S., Elkabets, M., White, M. R., Reich, E., Carmi, Y., ... Voronov, E. (2006). The involvement of IL-1 in tumorigenesis, tumor invasiveness, metastasis and tumor-host interactions. *Cancer and Metastasis Reviews, 25*, 387–408.

Armstrong, D., Chang, C. Y., Lazarus, D. R., Corry, D., & Kheradmand, F. (2019). Lung cancer heterogeneity in modulation of Th17/IL17A responses. *Frontiers in Oncology, 9*, 1384.

Awad, M. M., Liu, S., Rybkin, I. I., Arbour, K. C., Dilly, J., Zhu, V. W., ... Aguirre, A. J. (2021). Acquired resistance to KRAS(G12C) inhibition in cancer. *The New England Journal of Medicine, 384*, 2382–2393.

Awaji, M., Saxena, S., Wu, L., Prajapati, D. R., Purohit, A., Varney, M. L., ... Singh, R. K. (2020). CXCR2 signaling promotes secretory cancer-associated fibroblasts in pancreatic ductal adenocarcinoma. *The FASEB Journal, 34*, 9405–9418.

Bailey, S. R., Nelson, M. H., Himes, R. A., Li, Z., Mehrotra, S., & Paulos, C. M. (2014). Th17 cells in cancer: The ultimate identity crisis. *Frontiers in Immunology, 5*, 276.

Barrera, L., Montes-Servín, E., Barrera, A., Ramírez-Tirado, L. A., Salinas-Parra, F., Bañales-Méndez, J. L., ... Arrieta, Ó. (2015). Cytokine profile determined by data-mining analysis set into clusters of non-small-cell lung cancer patients according to prognosis. *Annals of Oncology, 26*, 428–435.

Beaupre, D. M., Talpaz, M., Marini, F. C., 3rd, Cristiano, R. J., Roth, J. A., Estrov, Z., ... Kurzrock, R. (1999). Autocrine interleukin-1beta production in leukemia: Evidence for the involvement of mutated RAS. *Cancer Research, 59*, 2971–2980.

Bendell, J., Ulahannan, S., Koczywas, M., Brahmer, J., Capasso, A., Eckhardt, S. G., ... Ou, S. H. (2020). Intermittent dosing of RMC-4630, a potent, selective inhibitor of SHP2, combined with the MEK inhibitor cobimetinib, in a phase 1b/2 clinical trial for advanced solid tumors with activating mutations of RAS signaling. *European Journal of Cancer, 138*, S8–S9.

Borghaei, H., Paz-Ares, L., Horn, L., Spigel, D. R., Steins, M., Ready, N. E., ... Brahmer, J. R. (2015). Nivolumab versus docetaxel in advanced nonsquamous non-small-cell lung cancer. *The New England Journal of Medicine, 373*, 1627–1639.

Bourne, H. R., Sanders, D. A., & McCormick, F. (1991). The GTPase superfamily: Conserved structure and molecular mechanism. *Nature, 349*, 117–127.

Bowen, M. E., Ayturk, U. M., Kurek, K. C., Yang, W., & Warman, M. L. (2014). SHP2 regulates chondrocyte terminal differentiation, growth plate architecture and skeletal cell fates. *PLoS Genetics, 10*, e1004364.

Brooks, G. D., McLeod, L., Alhayyani, S., Miller, A., Russell, P. A., Ferlin, W., ... Jenkins, B. J. (2016). IL6 trans-signaling promotes KRAS-driven lung carcinogenesis. *Cancer Research, 76*, 866–876.

Buday, L., & Downward, J. (2008). Many faces of Ras activation. *Biochimica et Biophysica Acta (BBA)—Reviews on Cancer, 1786*, 178–187.

Bunda, S., Burrell, K., Heir, P., Zeng, L., Alamsahebpour, A., Kano, Y., ... Ohh, M. (2015). Inhibition of SHP2-mediated dephosphorylation of Ras suppresses oncogenesis. *Nature Communications, 6*, 8859.

Buscail, L., Bournet, B., & Cordelier, P. (2020). Role of oncogenic KRAS in the diagnosis, prognosis and treatment of pancreatic cancer. *Nature Reviews Gastroenterology & Hepatology, 17*, 153–168.

Caetano, M. S., Zhang, H., Cumpian, A. M., Gong, L., Unver, N., Ostrin, E. J., ... Moghaddam, S. J. (2016). IL6 blockade reprograms the lung tumor microenvironment to limit the development and progression of K-ras-mutant lung cancer. *Cancer Research, 76,* 3189–3199.

Campbell, J. D., Alexandrov, A., Kim, J., Wala, J., Berger, A. H., Pedamallu, C. S., ... Meyerson, M. (2016). Distinct patterns of somatic genome alterations in lung adenocarcinomas and squamous cell carcinomas. *Nature Genetics, 48,* 607–616.

Canon, J., Rex, K., Saiki, A. Y., Mohr, C., Cooke, K., Bagal, D., ... Lipford, J. R. (2019). The clinical KRAS(G12C) inhibitor AMG 510 drives anti-tumour immunity. *Nature, 575,* 217–223.

Cantley, L. C. (2002). The phosphoinositide 3-kinase pathway. *Science (New York, N. Y.), 296,* 1655–1657.

Cantrell, D. A. (2003). GTPases and T cell activation. *Immunological Reviews, 192,* 122–130.

Castelli, C., Sensi, M., Lupetti, R., Mortarini, R., Panceri, P., Anichini, A., & Parmiani, G. (1994). Expression of interleukin 1 alpha, interleukin 6, and tumor necrosis factor alpha genes in human melanoma clones is associated with that of mutated N-RAS oncogene. *Cancer Research, 54,* 4785–4790.

Ceddia, S., Landi, L., & Cappuzzo, F. (2022). KRAS-mutant non-small-cell lung cancer: From past efforts to future challenges. *International Journal of Molecular Sciences, 23,* 9391.

Cefalì, M., Epistolio, S., Ramelli, G., Mangan, D., Molinari, F., Martin, V., ... Wannesson, L. (2022). Correlation of KRAS G12C mutation and high PD-L1 expression with clinical outcome in NSCLC patients treated with anti-PD1 immunotherapy. *Journal of Clinical Medicine, 11,* 1627.

Chang, S. H., Mirabolfathinejad, S. G., Katta, H., Cumpian, A. M., Gong, L., Caetano, M. S., ... Dong, C. (2014). T helper 17 cells play a critical pathogenic role in lung cancer. *Proceedings of the National Academy of Sciences, 111,* 5664–5669.

Chaudhry, A., Robert, Treuting, P., Liang, Y., Marina, Heinrich, J.-M., Robert, F., Jens, Müller, W., & Alexander (2011). Interleukin-10 signaling in regulatory T cells is required for suppression of Th17 cell-mediated inflammation. *Immunity, 34,* 566–578.

Chen, H., Huang, D., Lin, G., Yang, X., Zhuo, M., Chi, Y., ... Zhao, J. (2022a). The prevalence and real-world therapeutic analysis of Chinese patients with KRAS-mutant non-small cell lung cancer. *Cancer Medicine, 11,* 3581–3592.

Chen, J., Wei, Y., Yang, W., Huang, Q., Chen, Y., Zeng, K., & Chen, J. (2022b). IL-6: The link between inflammation, immunity and breast cancer. *Frontiers in Oncology, 12,* 903800.

Chen, J. J., Yao, P. L., Yuan, A., Hong, T. M., Shun, C. T., Kuo, M. L., ... Yang, P. C. (2003). Up-regulation of tumor interleukin-8 expression by infiltrating macrophages: Its correlation with tumor angiogenesis and patient survival in non-small cell lung cancer. *Clinical Cancer Research: An Official Journal of the American Association for Cancer Research, 9,* 729–737.

Chen, K., Zhang, Y., Qian, L., & Wang, P. (2021). Emerging strategies to target RAS signaling in human cancer therapy. *Journal of Hematology & Oncology, 14,* 116.

Chen, N., Fang, W., Lin, Z., Peng, P., Wang, J., Zhan, J., ... Zhang, L. (2017). KRAS mutation-induced upregulation of PD-L1 mediates immune escape in human lung adenocarcinoma. *Cancer Immunology, Immunotherapy: CII, 66,* 1175–1187.

Chen, Y. N., LaMarche, M. J., Chan, H. M., Fekkes, P., Garcia-Fortanet, J., Acker, M. G., ... Fortin, P. D. (2016). Allosteric inhibition of SHP2 phosphatase inhibits cancers driven by receptor tyrosine kinases. *Nature, 535,* 148–152.

Cheng, H., Fan, K., Luo, G., Fan, Z., Yang, C., Huang, Q., ... Liu, C. (2019). Kras(G12D) mutation contributes to regulatory T cell conversion through activation of the MEK/ERK pathway in pancreatic cancer. *Cancer Letters, 446,* 103–111.

Chong, P. S. Y., Zhou, J., Lim, J. S. L., Hee, Y. T., Chooi, J. Y., Chung, T. H., ... Chng, W. J. (2019). IL6 promotes a STAT3-PRL3 feedforward loop via SHP2 repression in multiple myeloma. *Cancer Research, 79*, 4679–4688.

Coelho, M. A., de Carne Trecesson, S., Rana, S., Zecchin, D., Moore, C., Molina-Arcas, M., ... Downward, J. (2017). Oncogenic RAS signaling promotes tumor immunoresistance by stabilizing PD-L1 mRNA. *Immunity, 47*, 1083–1099.e1086.

Corcoran, R. B., Contino, G., Deshpande, V., Tzatsos, A., Conrad, C., Benes, C. H., ... Bardeesy, N. (2011). STAT3 plays a critical role in KRAS-induced pancreatic tumorigenesis. *Cancer Research, 71*, 5020–5029.

Cox, A. D., Fesik, S. W., Kimmelman, A. C., Luo, J., & Der, C. J. (2014). Drugging the undruggable RAS: Mission possible? *Nature Reviews: Drug Discovery, 13*, 828–851.

Cullis, J., Das, S., & Bar-Sagi, D. (2018). Kras and tumor immunity: Friend or foe? *Cold Spring Harbor Perspectives in Medicine, 8*.

Dance, M., Montagner, A., Salles, J.-P., Yart, A., & Raynal, P. (2008). The molecular functions of Shp2 in the Ras/Mitogen-activated protein kinase (ERK1/2) pathway. *Cellular Signalling, 20*, 453–459.

Dardaei, L., Wang, H. Q., Singh, M., Fordjour, P., Shaw, K. X., Yoda, S., ... Engelman, J. A. (2018). SHP2 inhibition restores sensitivity in ALK-rearranged non-small-cell lung cancer resistant to ALK inhibitors. *Nature Medicine, 24*, 512–517.

Das, S., Shapiro, B., Vucic, E. A., Vogt, S., & Bar-Sagi, D. (2020). Tumor cell-derived il1beta promotes desmoplasia and immune suppression in pancreatic cancer. *Cancer Research, 80*, 1088–1101.

Dawson, J. C., Munro, A., Macleod, K., Muir, M., Timpson, P., Williams, R. J., ... Carragher, N. O. (2022). Pathway profiling of a novel SRC inhibitor, AZD0424, in combination with MEK inhibitors for cancer treatment. *Molecular Oncology, 16*, 1072–1090.

Delpu, Y., Hanoun, N., Lulka, H., Sicard, F., Selves, J., Buscail, L., ... Cordelier, P. (2011). Genetic and epigenetic alterations in pancreatic carcinogenesis. *Current Genomics, 12*, 15–24.

Deng, S., Clowers, M. J., Velasco, W. V., Ramos-Castaneda, M., & Moghaddam, S. J. (2019). Understanding the complexity of the tumor microenvironment in K-ras mutant lung cancer: Finding an alternative path to prevention and treatment. *Frontiers in Oncology, 9*, 1556.

Dias Carvalho, P., Machado, A. L., Martins, F., Seruca, R., & Velho, S. (2019). Targeting the tumor microenvironment: An unexplored strategy for mutant KRAS tumors. *Cancers, 11*, 2010.

D'Incecco, A., Andreozzi, M., Ludovini, V., Rossi, E., Capodanno, A., Landi, L., ... Cappuzzo, F. (2015). PD-1 and PD-L1 expression in molecularly selected non-small-cell lung cancer patients. *British Journal of Cancer, 112*, 95–102.

Dong, Z. Y., Zhong, W. Z., Zhang, X. C., Su, J., Xie, Z., Liu, S. Y., ... Wu, Y. L. (2017). Potential predictive value of TP53 and KRAS mutation status for response to PD-1 blockade immunotherapy in lung adenocarcinoma. *Clinical Cancer Research: An Official Journal of the American Association for Cancer Research, 23*, 3012–3024.

Duffy, S. A., Taylor, J. M. G., Terrell, J. E., Islam, M., Li, Y., Fowler, K. E., ... Teknos, T. N. (2008). Interleukin-6 predicts recurrence and survival among head and neck cancer patients. *Cancer, 113*, 750–757.

Dutta, S., Ganguly, A., Chatterjee, K., Spada, S., & Mukherjee, S. (2023). Targets of immune escape mechanisms in cancer: Basis for development and evolution of cancer immune checkpoint inhibitors. *Biology (Basel), 12*.

Ehrlich, T., Wishniak, O., Isakov, N., Cohen, O., Segal, S., Rager-Zisman, B., & Gopas, J. (1993). The effect of H-ras expression on tumorigenicity and immunogenicity of Balb/c 3T3 fibroblasts. *Immunology Letters, 39*, 3–8.

Eulenfeld, R., & Schaper, F. (2009). A new mechanism for the regulation of Gab1 recruitment to the plasma membrane. *Journal of Cell Science, 122*, 55–64.
Fakih, M., Desai, J., Kuboki, Y., Strickler, J. H., Price, T. J., Durm, G. A., ... Hong, D. S. (2020). CodeBreak 100: Activity of AMG 510, a novel small molecule inhibitor of KRASG12C, in patients with advanced colorectal cancer. *Journal of Clinical Oncology, 38*, 4018.
Fedele, C., Ran, H., Diskin, B., Wei, W., Jen, J., Geer, M. J., ... Tang, K. H. (2018). SHP2 inhibition prevents adaptive resistance to MEK inhibitors in multiple cancer models. *Cancer Discovery, 8*, 1237–1249.
Fedele, C., Li, S., Teng, K. W., Foster, C. J. R., Peng, D., Ran, H., ... Neel, B. G. (2020). SHP2 inhibition diminishes KRASG12C cycling and promotes tumor microenvironment remodeling. *Journal of Experimental Medicine, 218*.
Fedele, C., Li, S., Teng, K. W., Foster, C. J. R., Peng, D., Ran, H., ... Neel, B. G. (2021). SHP2 inhibition diminishes KRASG12C cycling and promotes tumor microenvironment remodeling. *The Journal of Experimental Medicine, 218*.
Feng, L., Qi, Q., Wang, P., Chen, H., Chen, Z., Meng, Z., & Liu, L. (2018). Serum levels of IL-6, IL-8, and IL-10 are indicators of prognosis in pancreatic cancer. *The Journal of International Medical Research, 46*, 5228–5236.
Fisher, D. T., Appenheimer, M. M., & Evans, S. S. (2014). The two faces of IL-6 in the tumor microenvironment. *Seminars in Immunology, 26*, 38–47.
Fountzilas, G., Bobos, M., Kalogera-Fountzila, A., Xiros, N., Murray, S., Linardou, H., ... Kosmidis, P. (2008). Gemcitabine combined with Gefitinib in patients with inoperable or metastatic pancreatic cancer: A phase II study of the hellenic cooperative oncology group with biomarker evaluation. *Cancer Investigation, 26*, 784–793.
Froesch, P., Mark, M., Rothschild, S. I., Li, Q., Godar, G., Rusterholz, C., ... Früh, M. (2021). Binimetinib, pemetrexed and cisplatin, followed by maintenance of binimetinib and pemetrexed in patients with advanced non-small cell lung cancer (NSCLC) and KRAS mutations. The phase 1B SAKK 19/16 trial. *Lung Cancer (Amsterdam, Netherlands), 156*, 91–99.
Fukuda, A., Sam John, A., Liou, A., Grace Akira, S., Kenneth Matthew, S., & Hebrok, M. (2011). Stat3 and MMP7 contribute to pancreatic ductal adenocarcinoma initiation and progression. *Cancer Cell, 19*, 441–455.
Fung, A. S., Graham, D. M., Chen, E. X., Stockley, T. L., Zhang, T., Le, L. W., ... Leighl, N. B. (2021). A phase I study of binimetinib (MEK 162), a MEK inhibitor, plus carboplatin and pemetrexed chemotherapy in non-squamous non-small cell lung cancer. *Lung Cancer (Amsterdam, Netherlands), 157*, 21–29.
Garcia Fortanet, J., Chen, C. H., Chen, Y. N., Chen, Z., Deng, Z., Firestone, B., ... LaMarche, M. J. (2016). Allosteric inhibition of SHP2: Identification of a potent, selective, and orally efficacious phosphatase inhibitor. *Journal of Medicinal Chemistry, 59*, 7773–7782.
Garon, E. B., Rizvi, N. A., Hui, R., Leighl, N., Balmanoukian, A. S., Eder, J. P., ... Investigators, K.- (2015). Pembrolizumab for the treatment of non-small-cell lung cancer. *The New England Journal of Medicine, 372*, 2018–2028.
Govindan, R., Fakih, M. G., Price, T. J., Falchook, G. S., Desai, J., Kuo, J. C., ... Hong, D. S. (2019). 446PD—Phase I study of AMG 510, a novel molecule targeting KRAS G12C mutant solid tumours. *Annals of Oncology, 30*, v163–v164.
Granville, C. A., Memmott, R. M., Balogh, A., Mariotti, J., Kawabata, S., Han, W., ... Dennis, P. A. (2009). A central role for Foxp3+ regulatory T cells in K-Ras-driven lung tumorigenesis. *PLoS One, 4*, e5061.
Groth, C., Hu, X., Weber, R., Fleming, V., Altevogt, P., Utikal, J., & Umansky, V. (2019). Immunosuppression mediated by myeloid-derived suppressor cells (MDSCs) during tumour progression. *British Journal of Cancer, 120*, 16–25.

Grusch, M., Petz, M., Metzner, T., Ozturk, D., Schneller, D., & Mikulits, W. (2010). The crosstalk of RAS with the TGF-β family during carcinoma progression and its implications for targeted cancer therapy. *Current Cancer Drug Targets, 10,* 849–857.

Gu, S., Sayad, A., Chan, G., Yang, W., Lu, Z., Virtanen, C., ... Neel, B. G. (2018). SHP2 is required for BCR-ABL1-induced hematologic neoplasia. *Leukemia: Official Journal of the Leukemia Society of America, Leukemia Research Fund, U. K, 32,* 203–213.

Haanen, J. B., & Cerundolo, V. (2018). NKG2A, a new kid on the immune checkpoint block. *Cell, 175,* 1720–1722.

Hallin, J., Engstrom, L. D., Hargis, L., Calinisan, A., Aranda, R., Briere, D. M., ... Christensen, J. G. (2020). The KRASG12C inhibitor MRTX849 provides insight toward therapeutic susceptibility of KRAS-mutant cancers in mouse models and patients. *Cancer Discovery, 10,* 54–71.

Hao, H. X., Wang, H., Liu, C., Kovats, S., Velazquez, R., Lu, H., ... Mohseni, M. (2019). Tumor intrinsic efficacy by SHP2 and RTK inhibitors in KRAS-mutant cancers. *Molecular Cancer Therapeutics, 18,* 2368–2380.

Herbst, R. S., Majem, M., Barlesi, F., Carcereny, E., Chu, Q., Monnet, I., ... Martinez-Marti, A. (2022). COAST: An open-label, phase II, multidrug platform study of durvalumab alone or in combination with oleclumab or monalizumab in patients with unresectable, stage III non-small-cell lung cancer. *Journal of Clinical Oncology: Official Journal of the American Society of Clinical Oncology, 40,* 3383–3393.

Ho, C. S. L., Tüns, A. I., Schildhaus, H. U., Wiesweg, M., Grüner, B. M., Hegedus, B., ... Oeck, S. (2021). HER2 mediates clinical resistance to the KRAS(G12C) inhibitor sotorasib, which is overcome by co-targeting SHP2. *European Journal of Cancer, 159,* 16–23.

Holmer, R., Goumas, F. A., Waetzig, G. H., Rose-John, S., & Kalthoff, H. (2014). Interleukin-6: A villain in the drama of pancreatic cancer development and progression. *Hepatobiliary & Pancreatic Diseases International, 13,* 371–380.

Holohan, C., Van Schaeybroeck, S., Longley, D. B., & Johnston, P. G. (2013). Cancer drug resistance: An evolving paradigm. *Nature Reviews: Cancer, 13,* 714–726.

Hou, P., Ma, X., Yang, Z., Zhang, Q., Wu, C. J., Li, J., ... DePinho, R. A. (2021). USP21 deubiquitinase elevates macropinocytosis to enable oncogenic KRAS bypass in pancreatic cancer. *Genes & Development, 35,* 1327–1332.

Huang, L., Guo, Z., Wang, F., & Fu, L. (2021). KRAS mutation: From undruggable to druggable in cancer. *Signal Transduction and Targeted Therapy, 6,* 386.

Hui, E., Cheung, J., Zhu, J., Su, X., Taylor, M. J., Wallweber, H. A., ... Vale, R. D. (2017). T cell costimulatory receptor CD28 is a primary target for PD-1-mediated inhibition. *Science (New York, N. Y.), 355,* 1428–1433.

Hymowitz, S. G., & Malek, S. (2018). Targeting the MAPK pathway in RAS mutant cancers. *Cold Spring Harbor Perspectives in Medicine, 8.*

Indini, A., Rijavec, E., Ghidini, M., Cortellini, A., & Grossi, F. (2021). Targeting KRAS in solid tumors: Current challenges and future opportunities of novel KRAS inhibitors. *Pharmaceutics, 13.*

Jančík, S., Drábek, J., Radzioch, D., & Hajdúch, M. (2010). Clinical relevance of KRAS in human cancers. *Journal of Biomedicine and Biotechnology, 2010,* 150960.

Jänne, P. A., Shaw, A. T., Pereira, J. R., Jeannin, G., Vansteenkiste, J., Barrios, C., ... Crinò, L. (2013). Selumetinib plus docetaxel for KRAS-mutant advanced non-small-cell lung cancer: A randomised, multicentre, placebo-controlled, phase 2 study. *The Lancet Oncology, 14,* 38–47.

Jänne, P. A., Riely, G. J., Gadgeel, S. M., Heist, R. S., Ou, S. I., Pacheco, J. M., ... Spira, A. I. (2022). Adagrasib in non-small-cell lung cancer harboring a KRAS(G12C) mutation. *The New England Journal of Medicine, 387,* 120–131.

Ji, H., Houghton, A. M., Mariani, T. J., Perera, S., Kim, C. B., Padera, R., ... Wong, K. K. (2006). K-ras activation generates an inflammatory response in lung tumors. *Oncogene, 25,* 2105–2112.

Ji, J., Wang, C., & Fakih, M. (2022). Targeting KRAS (G12C)-mutated advanced colorectal cancer: Research and clinical developments. *OncoTargets and Therapy, 15*, 747–756.

Johnson, D. E., O'Keefe, R. A., & Grandis, J. R. (2018). Targeting the IL-6/JAK/STAT3 signalling axis in cancer. *Nature Reviews Clinical Oncology, 15*, 234–248.

Jonckheere, N., Vasseur, R., & Van Seuningen, I. (2017). The cornerstone K-RAS mutation in pancreatic adenocarcinoma: From cell signaling network, target genes, biological processes to therapeutic targeting. *Critical Reviews in Oncology/Hematology, 111*, 7–19.

Jung, H. R., Oh, Y., Na, D., Min, S., Kang, J., Jang, D., ... Cho, S. Y. (2021). CRISPR screens identify a novel combination treatment targeting BCL-X(L) and WNT signaling for KRAS/BRAF-mutated colorectal cancers. *Oncogene, 40*, 3287–3302.

Kalvala, A., Wallet, P., Yang, L., Wang, C., Li, H., Nam, A., ... Salgia, R. (2019). Phenotypic switching of naive T cells to immune-suppressive treg-like cells by mutant KRAS. *Journal of Clinical Medicine, 8*.

Kanumuri, R., Kumar Pasupuleti, S., Burns, S. S., Ramdas, B., & Kapur, R. (2022). Targeting SHP2 phosphatase in hematological malignancies. *Expert Opinion on Therapeutic Targets, 26*, 319–332.

Kapoor, A., Yao, W., Ying, H., Hua, S., Liewen, A., Wang, Q., ... DePinho, R. A. (2014). Yap1 activation enables bypass of oncogenic kras addiction in pancreatic cancer. *Cell, 158*, 185–197.

Ke, Y., Zhang, E. E., Hagihara, K., Wu, D., Pang, Y., Klein, R., ... Feng, G. S. (2007). Deletion of Shp2 in the brain leads to defective proliferation and differentiation in neural stem cells and early postnatal lethality. *Molecular and Cellular Biology, 27*, 6706–6717.

Keegan, A., Ricciuti, B., Garden, P., Cohen, L., Nishihara, R., Adeni, A., ... Walt, D. R. (2020). Plasma IL-6 changes correlate to PD-1 inhibitor responses in NSCLC. *Journal for ImmunoTherapy of Cancer, 8*.

Khleif, S. N., Abrams, S. I., Hamilton, J. M., Bergmann-Leitner, E., Chen, A., Bastian, A., ... Schlom, J. (1999). A phase I vaccine trial with peptides reflecting ras oncogene mutations of solid tumors. *Journal of Immunotherapy (Hagerstown, Md.: 1997), 22*, 155–165.

Kim, H. J., Lee, H. N., Jeong, M. S., & Jang, S. B. (2021). Oncogenic KRAS: Signaling and drug resistance. *Cancers (Basel), 13*.

Konieczkowski, D. J., Johannessen, C. M., & Garraway, L. A. (2018). A convergence-based framework for cancer drug resistance. *Cancer Cell, 33*, 801–815.

Kontaridis, M. I., Yang, W., Bence, K. K., Cullen, D., Wang, B., Bodyak, N., ... Neel, B. G. (2008). Deletion of Ptpn11 (Shp2) in cardiomyocytes causes dilated cardiomyopathy via effects on the extracellular signal-regulated kinase/mitogen-activated protein kinase and RhoA signaling pathways. *Circulation, 117*, 1423–1435.

Korn, T., Bettelli, E., Gao, W., Awasthi, A., Jager, A., Strom, T. B., ... Kuchroo, V. K. (2007). IL-21 initiates an alternative pathway to induce proinflammatory T(H)17 cells. *Nature, 448*, 484–487.

Kortlever, R. M., Sodir, N. M., Wilson, C. H., Burkhart, D. L., Pellegrinet, L., Brown Swigart, L., ... Evan, G. I. (2017). Myc cooperates with Ras by programming inflammation and immune suppression. *Cell, 171*, 1301–1315.e1314.

Krygowska, A. A., & Castellano, E. (2018). PI3K: A crucial piece in the RAS signaling puzzle. *Cold Spring Harbor Perspectives in Medicine, 8*.

Kwan, A. K., Piazza, G. A., Keeton, A. B., & Leite, C. A. (2022). The path to the clinic: A comprehensive review on direct KRASG12C inhibitors. *Journal of Experimental & Clinical Cancer Research, 41*, 27.

Laino, A. S., Woods, D., Vassallo, M., Qian, X., Tang, H., Wind-Rotolo, M., & Weber, J. (2020). Serum interleukin-6 and C-reactive protein are associated with survival in melanoma patients receiving immune checkpoint inhibition. *Journal for ImmunoTherapy of Cancer, 8*.

Lan, F., Zhang, L., Wu, J., Zhang, J., Zhang, S., Li, K., ... Lin, P. (2011). IL-23/IL-23R: Potential mediator of intestinal tumor progression from adenomatous polyps to colorectal carcinoma. *International Journal of Colorectal Disease, 26*, 1511–1518.

Landskron, G., De La Fuente, M., Thuwajit, P., Thuwajit, C., & Hermoso, M. A. (2014). Chronic inflammation and cytokines in the tumor microenvironment. *Journal of Immunology Research, 2014*, 1–19.

LaRochelle, J. R., Fodor, M., Vemulapalli, V., Mohseni, M., Wang, P., Stams, T., ... Blacklow, S. C. (2018). Structural reorganization of SHP2 by oncogenic mutations and implications for oncoprotein resistance to allosteric inhibition. *Nature Communications, 9*, 4508.

Laskovs, M., Partridge, L., & Slack, C. (2022). Molecular inhibition of RAS signalling to target ageing and age-related health. *Disease Models & Mechanisms, 15*.

Lastwika, K. J., Wilson, W. 3rd, Li, Q. K., Norris, J., Xu, H., Ghazarian, S. R., ... Dennis, P. A. (2016). Control of PD-L1 expression by oncogenic activation of the AKT-mTOR pathway in non-small cell lung cancer. *Cancer Research, 76*, 227–238.

Le Rolle, A. F., Chiu, T. K., Fara, M., Shia, J., Zeng, Z., Weiser, M. R., ... Chiu, V. K. (2015). The prognostic significance of CXCL1 hypersecretion by human colorectal cancer epithelia and myofibroblasts. *Journal of Translational Medicine, 13*, 199.

Leidner, R., Sanjuan Silva, N., Huang, H., Sprott, D., Zheng, C., Shih, Y. P., ... Tran, E. (2022). Neoantigen T-cell receptor gene therapy in pancreatic cancer. *The New England Journal of Medicine, 386*, 2112–2119.

Lemmon, M. A., & Schlessinger, J. (2010). Cell signaling by receptor tyrosine kinases. *Cell, 141*, 1117–1134.

Lesina, M., Magdalena, Ludes, K., Rose-John, S., Treiber, M., Klöppel, G., ... Algül, H. (2011). Stat3/Socs3 activation by IL-6 transsignaling promotes progression of pancreatic intraepithelial neoplasia and development of pancreatic cancer. *Cancer Cell, 19*, 456–469.

Li, H., van der Merwe, P. A., & Sivakumar, S. (2022). Biomarkers of response to PD-1 pathway blockade. *British Journal of Cancer, 126*, 1663–1675.

Li, J., Lau, G., Chen, L., Yuan, Y.-F., Huang, J., Luk, J. M., ... Guan, X.-Y. (2012). Interleukin 23 promotes hepatocellular carcinoma metastasis via NF-kappa B induced matrix metalloproteinase 9 expression. *PLoS One, 7*, e46264.

Li, M. O., Wan, Y. Y., Sanjabi, S., Robertson, A.-K. L., & Flavell, R. A. (2006). Transforming growth factor-β regulation of immune responses. *Annual Review of Immunology, 24*, 99–146.

Liao, W., Overman, M. J., Boutin, A. T., Shang, X., Zhao, D., Dey, P., ... DePinho, R. A. (2019). KRAS-IRF2 axis drives immune suppression and immune therapy resistance in colorectal cancer. *Cancer Cell, 35*, 559–572.e557.

Littman, D. R., & Rudensky, A. Y. (2010). Th17 and regulatory T cells in mediating and restraining inflammation. *Cell, 140*, 845–858.

Liu, C., Zheng, S., Jin, R., Wang, X., Wang, F., Zang, R., ... He, J. (2020a). The superior efficacy of anti-PD-1/PD-L1 immunotherapy in KRAS-mutant non-small cell lung cancer that correlates with an inflammatory phenotype and increased immunogenicity. *Cancer Letters, 470*, 95–105.

Liu, C., Zheng, S., Jin, R., Wang, X., Wang, F., Zang, R., ... He, J. (2020b). The superior efficacy of anti-PD-1/PD-L1 immunotherapy in KRAS-mutant non-small cell lung cancer that correlates with an inflammatory phenotype and increased immunogenicity. *Cancer Letters, 470*, 95–105.

Liu, C., Zheng, S., Wang, Z., Wang, S., Wang, X., Yang, L., ... He, J. (2022). KRAS-G12D mutation drives immune suppression and the primary resistance of anti-PD-1/PD-L1 immunotherapy in non-small cell lung cancer. *Cancer Communications, 42*, 828–847.

Liu, M., Gao, S., Elhassan, R. M., Hou, X., & Fang, H. (2021). Strategies to overcome drug resistance using SHP2 inhibitors. *Acta Pharmaceutica Sinica B, 11*, 3908–3924.

Liu, Q., Li, A., Tian, Y., Wu, J. D., Liu, Y., Li, T., ... Wu, K. (2016). The CXCL8-CXCR1/2 pathways in cancer. *Cytokine & Growth Factor Reviews, 31*, 61–71.

Liu, S., Iaria, J., Simpson, R. J., & Zhu, H.-J. (2018). Ras enhances TGF-β signaling by decreasing cellular protein levels of its type II receptor negative regulator SPSB1. *Cell Communication and Signaling, 16*.

Liu, W. N., Yan, M., & Chan, A. M. (2017). A thirty-year quest for a role of R-Ras in cancer: From an oncogene to a multitasking GTPase. *Cancer Letters, 403*, 59–65.

Lohmann, S., Wollscheid, U., Huber, C., & Seliger, B. (1996). Multiple levels of MHC class I down-regulation by ras oncogenes. *Scandinavian Journal of Immunology, 43*, 537–544.

Mainardi, S., Mulero-Sánchez, A., Prahallad, A., Germano, G., Bosma, A., Krimpenfort, P., ... Bernards, R. (2018a). SHP2 is required for growth of KRAS-mutant non-small-cell lung cancer in vivo. *Nature Medicine, 24*, 961–967.

Mainardi, S., Mulero-Sánchez, A., Prahallad, A., Germano, G., Bosma, A., Krimpenfort, P., ... Bernards, R. (2018b). SHP2 is required for growth of KRAS-mutant non-small-cell lung cancer in vivo. *Nature Medicine, 24*, 961–967.

Mamdani, H., Matosevic, S., Khalid, A. B., Durm, G., & Jalal, S. I. (2022). Immunotherapy in lung cancer: Current landscape and future directions. *Frontiers in Immunology, 13*, 823618.

Manchado, E., Weissmueller, S., Morris, J. P. T., Chen, C. C., Wullenkord, R., Lujambio, A., ... Lowe, S. W. (2016). A combinatorial strategy for treating KRAS-mutant lung cancer. *Nature, 534*, 647–651.

Mannell, H., Hellwig, N., Gloe, T., Plank, C., Sohn, H. Y., Groesser, L., ... Krotz, F. (2008). Inhibition of the tyrosine phosphatase SHP-2 suppresses angiogenesis in vitro and in vivo. *Journal of Vascular Research, 45*, 153–163.

Marais, R., Light, Y., Paterson, H. F., & Marshall, C. J. (1995). Ras recruits Raf-1 to the plasma membrane for activation by tyrosine phosphorylation. *The EMBO Journal, 14*, 3136–3145.

Marasco, M., Berteotti, A., Weyershaeuser, J., Thorausch, N., Sikorska, J., Krausze, J., ... Carlomagno, T. (2020). Molecular mechanism of SHP2 activation by PD-1 stimulation. *Science Advances, 6* eaay4458.

Marshall, E. A., Ng, K. W., Kung, S. H. Y., Conway, E. M., Martinez, V. D., Halvorsen, E. C., ... Lam, W. L. (2016). Emerging roles of T helper 17 and regulatory T cells in lung cancer progression and metastasis. *Molecular Cancer, 15*.

Masuya, D., Huang, C., Liu, D., Kameyama, K., Hayashi, E., Yamauchi, A., ... Yokomise, H. (2001). The intratumoral expression of vascular endothelial growth factor and interleukin-8 associated with angiogenesis in nonsmall cell lung carcinoma patients. *Cancer, 92*, 2628–2638.

McAllister, F., Bailey, J. M., Alsina, J., Nirschl, C. J., Sharma, R., Fan, H., ... Leach, S. D. (2014). Oncogenic Kras activates a hematopoietic-to-epithelial IL-17 signaling axis in preinvasive pancreatic neoplasia. *Cancer Cell, 25*, 621–637.

McBride, O. W., Swan, D. C., Tronick, S. R., Gol, R., Klimanis, D., Moore, D. E., & Aaronson, S. A. (1983). Regional chromosomal localization of N-ras, K-ras-1, K-ras-2 and myb oncogenes in human cells. *Nucleic Acids Research, 11*, 8221–8236.

McLoed, A. G., Sherrill, T. P., Cheng, D.-S., Han, W., Saxon, J. A., Gleaves, L. A., ... Blackwell, T. S. (2016). Neutrophil-Derived IL-1β Impairs the Efficacy of NF-κB Inhibitors against Lung Cancer. *Cell Reports, 16*, 120–132.

Misale, S., Fatherree, J. P., Cortez, E., Li, C., Bilton, S., Timonina, D., ... Benes, C. H. (2019). KRAS G12C NSCLC models are sensitive to direct targeting of KRAS in combination with PI3K inhibition. *Clinical Cancer Research, 25*, 796–807.

Mizukami, Y., Jo, W.-S., Duerr, E.-M., Gala, M., Li, J., Zhang, X., ... Chung, D. C. (2005). Induction of interleukin-8 preserves the angiogenic response in HIF-1α-deficient colon cancer cells. *Nature Medicine, 11*, 992–997.

Mohi, M. G., & Neel, B. G. (2007). The role of Shp2 (PTPN11) in cancer. *Current Opinion in Genetics & Development, 17*, 23–30.

Molina, J. R., & Adjei, A. A. (2006). The Ras/Raf/MAPK Pathway. *Journal of Thoracic Oncology, 1*, 7–9.

Moodie, S. A., Willumsen, B. M., Weber, M. J., & Wolfman, A. (1993). Complexes of Ras·GTP with Raf-1 and mitogen-activated protein kinase kinase. *Science (New York, N. Y.), 260*, 1658–1661.

Moo-Young, T. A., Larson, J. W., Belt, B. A., Tan, M. C., Hawkins, W. G., Eberlein, T. J., ... Linehan, D. C. (2009). Tumor-derived TGF-beta mediates conversion of CD4+Foxp3+ regulatory T cells in a murine model of pancreas cancer. *Journal of Immunotherapy (Hagerstown, Md.: 1997), 32*, 12–21.

Mugarza, E., van Maldegem, F., Boumelha, J., Moore, C., Rana, S., Llorian Sopena, M., ... Downward, J. (2022). Therapeutic KRAS(G12C) inhibition drives effective interferon-mediated antitumor immunity in immunogenic lung cancers. *Science Advances, 8* eabm8780.

Muzumdar, M. D., Chen, P. Y., Dorans, K. J., Chung, K. M., Bhutkar, A., Hong, E., ... Jacks, T. (2017). Survival of pancreatic cancer cells lacking KRAS function. *Nature Communications, 8*, 1090.

Nakamura, K., & Smyth, M. J. (2020). Myeloid immunosuppression and immune checkpoints in the tumor microenvironment. *Cellular & Molecular Immunology, 17*, 1–12.

Nakayama, A., Nagashima, T., Nishizono, Y., Kuramoto, K., Mori, K., Homboh, K., ... Shimazaki, M. (2022). Characterisation of a novel KRAS G12C inhibitor ASP2453 that shows potent anti-tumour activity in KRAS G12C-mutated preclinical models. *British Journal of Cancer, 126*, 744–753.

Nichols, R. J., Haderk, F., Stahlhut, C., Schulze, C. J., Hemmati, G., Wildes, D., ... Bivona, T. G. (2018). RAS nucleotide cycling underlies the SHP2 phosphatase dependence of mutant BRAF-, NF1- and RAS-driven cancers. *Nature Cell Biology, 20*, 1064–1073.

Niogret, C., Birchmeier, W., & Guarda, G. (2019). SHP-2 in lymphocytes' cytokine and inhibitory receptor signaling. *Frontiers in Immunology, 10*, 2468.

Niu, J., Maurice-Dror, C., Lee, D. H., Kim, D. W., Nagrial, A., Voskoboynik, M., ... Ahn, M. J. (2022). First-in-human phase 1 study of the anti-TIGIT antibody vibostolimab as monotherapy or with pembrolizumab for advanced solid tumors, including non-small-cell lung cancer. *Annals of Oncology: Official Journal of the European Society for Medical Oncology/ESMO, 33*, 169–180.

O'Garra, A., Vieira, P. L., Vieira, P., & Goldfeld, A. E. (2004). IL-10–producing and naturally occurring CD4+ Tregs: Limiting collateral damage. *Journal of Clinical Investigation, 114*, 1372–1378.

O'Hayer, K. M., Brady, D. C., & Counter, C. M. (2009). ELR+ CXC chemokines and oncogenic Ras-mediated tumorigenesis. *Carcinogenesis, 30*, 1841–1847.

Ostrem, J. M., Peters, U., Sos, M. L., Wells, J. A., & Shokat, K. M. (2013). K-Ras(G12C) inhibitors allosterically control GTP affinity and effector interactions. *Nature, 503*, 548–551.

Ozga, A. J., Chow, M. T., & Luster, A. D. (2021). Chemokines and the immune response to cancer. *Immunity, 54*, 859–874.

Padua, D., & Massagué, J. (2009). Roles of TGFβ in metastasis. *Cell Research, 19*, 89–102.

Pawson, T. (2004). Specificity in signal transduction: From phosphotyrosine-SH2 domain interactions to complex cellular systems. *Cell, 116*, 191–203.

Petanidis, S., Anestakis, D., Argyraki, M., Hadzopoulou-Cladaras, M., & Salifoglou, A. (2013). Differential expression of IL-17, 22 and 23 in the progression of colorectal cancer in patients with K-RAS mutation: Ras signal inhibition and crosstalk with GM-CSF and IFN-γ. *PLoS One, 8*, e73616.

Pirlog, R., Piton, N., Lamy, A., Guisier, F., Berindan-Neagoe, I., Sabourin, J.-C., & Marguet, F. (2022). Morphological and molecular characterization of KRAS G12C-mutated lung adenocarcinomas. *Cancers, 14*, 1030.

Porru, M., Artuso, S., Salvati, E., Bianco, A., Franceschin, M., Diodoro, M. G., ... Leonetti, C. (2015). Targeting G-quadruplex DNA structures by EMICORON has a strong antitumor efficacy against advanced models of human colon cancer. *Molecular Cancer Therapeutics, 14*, 2541–2551.

Purohit, A., Varney, M., Rachagani, S., Ouellette, M. M., Batra, S. K., & Singh, R. K. (2016). CXCR2 signaling regulates KRAS(G12D)-induced autocrine growth of pancreatic cancer. *Oncotarget, 7*, 7280–7296.

Pylayeva-Gupta, Y., Grabocka, E., & Bar-Sagi, D. (2011). RAS oncogenes: Weaving a tumorigenic web. *Nature Reviews. Cancer, 11*, 761–774.

Pylayeva-Gupta, Y., Lee, K. E., Hajdu, C. H., Miller, G., & Bar-Sagi, D. (2012). Oncogenic Kras-induced GM-CSF production promotes the development of pancreatic neoplasia. *Cancer Cell, 21*, 836–847.

Qu, L., Pan, C., He, S.-M., Lang, B., Gao, G.-D., Wang, X.-L., & Wang, Y. (2019). The Ras superfamily of small GTPases in non-neoplastic cerebral diseases. *Frontiers in Molecular Neuroscience, 12*.

Qu, Z., Sun, F., Zhou, J., Li, L., Shapiro, S. D., & Xiao, G. (2015). Interleukin-6 prevents the initiation but enhances the progression of lung cancer. *Cancer Research, 75*, 3209–3215.

Quintana, E., Schulze, C. J., Myers, D. R., Choy, T. J., Mordec, K., Wildes, D., ... Smith, J. A. M. (2020). Allosteric inhibition of SHP2 stimulates antitumor immunity by transforming the immunosuppressive environment. *Cancer Research, 80*, 2889–2902.

Reck, M., Rodriguez-Abreu, D., Robinson, A. G., Hui, R., Csoszi, T., Fulop, A., ... Investigators, K.- (2016). Pembrolizumab versus chemotherapy for PD-L1-positive non-small-cell lung cancer. *The New England Journal of Medicine, 375*, 1823–1833.

Reita, D., Pabst, L., Pencreach, E., Guérin, E., Dano, L., Rimelen, V., ... Beau-Faller, M. (2022). Direct targeting KRAS mutation in non-small cell lung cancer: Focus on resistance. *Cancers (Basel), 14*.

Riely, G. J., Kris, M. G., Rosenbaum, D., Marks, J., Li, A., Chitale, D. A., ... Ladanyi, M. (2008). Frequency and distinctive spectrum of KRAS mutations in never smokers with lung adenocarcinoma. *Clinical Cancer Research, 14*, 5731–5734.

Rossi, D., & Zlotnik, A. (2000). The biology of chemokines and their receptors. *Annual Review of Immunology, 18*, 217–242.

Rota, G., Niogret, C., Dang, A. T., Barros, C. R., Fonta, N. P., Alfei, F., ... Guarda, G. (2018). Shp-2 is dispensable for establishing T cell exhaustion and for PD-1 signaling in vivo. *Cell Reports, 23*, 39–49.

Rudd, C. E., Taylor, A., & Schneider, H. (2009). CD28 and CTLA-4 coreceptor expression and signal transduction. *Immunological Reviews, 229*, 12–26.

Ruess, D. A., Heynen, G. J., Ciecielski, K. J., Ai, J., Berninger, A., Kabacaoglu, D., ... Algül, H. (2018a). Mutant KRAS-driven cancers depend on PTPN11/SHP2 phosphatase. *Nature Medicine, 24*, 954–960.

Ruess, D. A., Heynen, G. J., Ciecielski, K. J., Ai, J., Berninger, A., Kabacaoglu, D., ... Algül, H. (2018b). Mutant KRAS-driven cancers depend on PTPN11/SHP2 phosphatase. *Nature Medicine, 24*, 954–960.

Ryan, M. B., & Corcoran, R. B. (2018). Therapeutic strategies to target RAS-mutant cancers. *Nature Reviews Clinical Oncology, 15*, 709–720.

Ryan, M. B., Fece de la Cruz, F., Phat, S., Myers, D. T., Wong, E., Shahzade, H. A., ... Corcoran, R. B. (2020). Vertical pathway inhibition overcomes adaptive feedback resistance to KRAS(G12C) inhibition. *Clinical Cancer Research: An Official Journal of the American Association for Cancer Research, 26*, 1633–1643.

Salem, M. E., El-Refai, S. M., Sha, W., Puccini, A., Grothey, A., George, T. J., ... Tie, J. (2022). Landscape of KRAS(G12C), associated genomic alterations, and interrelation with immuno-oncology biomarkers in KRAS-mutated cancers. *JCO Precision Oncology, 6*, e2100245.

Sallusto, F. (2000). The role of chemokine receptors in primary, effector, and memory immune responses. *Annual Review of Immunology, 18*, 593–620.

Sallusto, F., Mackay, C. R., & Lanzavecchia, A. (2000). The role of chemokine receptors in primary, effector, and memory immune responses. *Annual Review of Immunology, 18*, 593–620.

Samatar, A. A., & Poulikakos, P. I. (2014). Targeting RAS-ERK signalling in cancer: Promises and challenges. *Nature Reviews. Drug Discovery, 13*, 928–942.

Sanjabi, S., Oh, S. A., & Li, M. O. (2017). Regulation of the immune response by TGF-β: From conception to autoimmunity and infection. *Cold Spring Harbor Perspectives in Biology, 9*, a022236.

Scheffler, M., Ihle, M. A., Hein, R., Merkelbach-Bruse, S., Scheel, A. H., Siemanowski, J., ... Wolf, J. (2019). K-ras mutation subtypes in NSCLC and associated co-occuring mutations in other oncogenic pathways. *Journal of Thoracic Oncology, 14*, 606–616.

Schlessinger, J. (2000). Cell signaling by receptor tyrosine kinases. *Cell, 103*, 211–225.

Schlessinger, J. (2002). Ligand-induced, receptor-mediated dimerization and activation of EGF receptor. *Cell, 110*, 669–672.

Schoenfeld, A. J., Rizvi, H., Bandlamudi, C., Sauter, J. L., Travis, W. D., Rekhtman, N., ... Hellmann, M. D. (2020). Clinical and molecular correlates of PD-L1 expression in patients with lung adenocarcinomas. *Annals of Oncology: Official Journal of the European Society for Medical Oncology/ESMO, 31*, 599–608.

Seliger, B., Harders, C., Wollscheid, U., Staege, M. S., Reske-Kunz, A. B., & Huber, C. (1996). Suppression of MHC class I antigens in oncogenic transformants: Association with decreased recognition by cytotoxic T lymphocytes. *Experimental Hematology, 24*, 1275–1279.

Sers, C., Kuner, R., Falk, C. S., Lund, P., Sueltmann, H., Braun, M., ... Schafer, R. (2009). Down-regulation of HLA Class I and NKG2D ligands through a concerted action of MAPK and DNA methyltransferases in colorectal cancer cells. *International Journal of Cancer, 125*, 1626–1639.

Sheffels, E., & Kortum, R. L. (2021). The role of wild-type RAS in oncogenic RAS transformation. *Genes (Basel), 12*.

Sheppard, K. A., Fitz, L. J., Lee, J. M., Benander, C., George, J. A., Wooters, J., ... Chaudhary, D. (2004). PD-1 inhibits T-cell receptor induced phosphorylation of the ZAP70/CD3zeta signalosome and downstream signaling to PKCtheta. *FEBS Letters, 574*, 37–41.

Sheridan, C. (2020). Grail of RAS cancer drugs within reach. *Nature Biotechnology, 38*, 6–8.

Silva, E. M., Mariano, V. S., Pastrez, P. R. A., Pinto, M. C., Castro, A. G., Syrjanen, K. J., & Longatto-Filho, A. (2017). High systemic IL-6 is associated with worse prognosis in patients with non-small cell lung cancer. *PLoS One, 12*, e0181125.

Simanshu, D. K., Nissley, D. V., & McCormick, F. (2017). RAS proteins and their regulators in human disease. *Cell, 170*, 17–33.

Singh, A., Greninger, P., Rhodes, D., Koopman, L., Violette, S., Bardeesy, N., & Settleman, J. (2009). A gene expression signature associated with "K-Ras addiction" reveals regulators of EMT and tumor cell survival. *Cancer Cell, 15*, 489–500.

Skoulidis, F., Li, B. T., Dy, G. K., Price, T. J., Falchook, G. S., Wolf, J., ... Govindan, R. (2021). Sotorasib for lung cancers with KRAS p.G12C mutation. *The New England Journal of Medicine, 384*, 2371–2381.

Song, Y., Zhao, M., Zhang, H., & Yu, B. (2022). Double-edged roles of protein tyrosine phosphatase SHP2 in cancer and its inhibitors in clinical trials. *Pharmacology & Therapeutics, 230*, 107966.

Song, Z., Wang, M., Ge, Y., Chen, X. P., Xu, Z., Sun, Y., & Xiong, X. F. (2021). Tyrosine phosphatase SHP2 inhibitors in tumor-targeted therapies. *Acta Pharmaceutica Sinica B, 11*, 13–29.

Sparmann, A., & Bar-Sagi, D. (2004). Ras-induced interleukin-8 expression plays a critical role in tumor growth and angiogenesis. *Cancer Cell, 6*, 447–458.

Steele, C. W., Karim, S. A., Leach, J. D. G., Bailey, P., Upstill-Goddard, R., Rishi, L., ... Morton, J. P. (2016). CXCR2 inhibition profoundly suppresses metastases and augments immunotherapy in pancreatic ductal adenocarcinoma. *Cancer Cell, 29*, 832–845.

Stein, A., Simnica, D., Schultheiss, C., Scholz, R., Tintelnot, J., Gokkurt, E., ... Binder, M. (2021). PD-L1 targeting and subclonal immune escape mediated by PD-L1 mutations in metastatic colorectal cancer. *Journal for ImmunoTherapy of Cancer, 9*.

Stephen, A. G., Esposito, D., Bagni, R. K., & McCormick, F. (2014). Dragging ras back in the ring. *Cancer Cell, 25*, 272–281.

Stewart, C. A., Metheny, H., Iida, N., Smith, L., Hanson, M., Steinhagen, F., ... Trinchieri, G. (2013). Interferon-dependent IL-10 production by Tregs limits tumor Th17 inflammation. *Journal of Clinical Investigation, 123*, 4859–4874.

Sumimoto, H., Takano, A., Teramoto, K., & Daigo, Y. (2016). RAS-mitogen-activated protein kinase signal is required for enhanced PD-L1 expression in human lung cancers. *PLoS One, 11*, e0166626.

Sunaga, N., Miura, Y., Kasahara, N., & Sakurai, R. (2021). Targeting oncogenic KRAS in non-small-cell lung cancer. *Cancers, 13*, 5956.

Suzuki, K., Wilkes, M. C., Garamszegi, N., Edens, M., & Leof, E. B. (2007). Transforming growth factor beta signaling via Ras in mesenchymal cells requires p21-activated kinase 2 for extracellular signal-regulated kinase-dependent transcriptional responses. *Cancer Research, 67*, 3673–3682.

Suzuki, S., Yonesaka, K., Teramura, T., Takehara, T., Kato, R., Sakai, H., ... Nakagawa, K. (2021). KRAS inhibitor resistance in MET-amplified KRAS (G12C) non-small cell lung cancer induced by RAS- and Non-RAS-mediated cell signaling mechanisms. *Clinical Cancer Research: An Official Journal of the American Association for Cancer Research, 27*, 5697–5707.

Tan, X., Carretero, J., Chen, Z., Zhang, J., Wang, Y., Chen, J., ... Wong, K.-K. (2013). Loss of p53 attenuates the contribution of IL-6 deletion on suppressed tumor progression and extended survival in Kras-driven murine lung cancer. *PLoS One, 8*, e80885.

Tanaka, N., Lin, J. J., Li, C., Ryan, M. B., Zhang, J., Kiedrowski, L. A., ... Corcoran, R. B. (2021). Clinical acquired resistance to KRAS(G12C) inhibition through a novel KRAS switch-II pocket mutation and polyclonal alterations converging on RAS-MAPK reactivation. *Cancer Discovery, 11*, 1913–1922.

Tanaka, T., Narazaki, M., & Kishimoto, T. (2014). IL-6 in inflammation, immunity, and disease. *Cold Spring Harbor Perspectives in Biology, 6*, a016295.

Tang, K. H., Li, S., Khodadadi-Jamayran, A., Jen, J., Han, H., Guidry, K., ... Neel, B. G. (2022a). Combined inhibition of SHP2 and CXCR1/2 promotes antitumor T-cell response in NSCLC. *Cancer Discovery, 12*, 47–61.

Tang, K. H., Li, S., Khodadadi-Jamayran, A., Jen, J., Han, H., Guidry, K., ... Neel, B. G. (2022b). Combined inhibition of SHP2 and CXCR1/2 promotes antitumor T-cell response in NSCLC. *Cancer Discovery, 12*, 47–61.

Tartour, E., Dorval, T., Mosseri, V., Deneux, L., Mathiot, C., Brailly, H., ... Fridman, W. H. (1994). Serum interleukin 6 and C-reactive protein levels correlate with resistance to IL-2 therapy and poor survival in melanoma patients. *British Journal of Cancer, 69*, 911–913.

Tas, F., Duranyildiz, D., Oguz, H., Camlica, H., Yasasever, V., & Topuz, E. (2006). Serum vascular endothelial growth factor (VEGF) and interleukin-8 (IL-8) levels in small cell lung cancer. *Cancer Investigation, 24*, 492–496.

Tobin, R. P., Jordan, K. R., Kapoor, P., Spongberg, E., Davis, D., Vorwald, V. M., ... McCarter, M. D. (2019). IL-6 and IL-8 are linked with myeloid-derived suppressor cell accumulation and correlate with poor clinical outcomes in melanoma patients. *Frontiers in Oncology, 9*, 1223.

Torres-Ayuso, P., & Brognard, J. (2018). Shipping out MEK inhibitor resistance with SHP2 inhibitors. *Cancer Discovery, 8*, 1210–1212.

Tran, E., Robbins, P. F., Lu, Y. C., Prickett, T. D., Gartner, J. J., Jia, L., ... Rosenberg, S. A. (2016). T-cell transfer therapy targeting mutant KRAS in cancer. *The New England Journal of Medicine, 375*, 2255–2262.

Tsai, F. D., Lopes, M. S., Zhou, M., Court, H., Ponce, O., Fiordalisi, J. J., ... Philips, M. R. (2015). K-Ras4A splice variant is widely expressed in cancer and uses a hybrid membrane-targeting motif. *Proceedings of the National Academy of Sciences, 112*, 779–784.

Tsubaki, M., Takeda, T., Sakamoto, K., Shimaoka, H., Fujita, A., Itoh, T., ... Nishida, S. (2015). Bisphosphonates and statins inhibit expression and secretion of MIP-1α via suppression of Ras/MEK/ERK/AML-1A and Ras/PI3K/Akt/AML-1A pathways. *American Journal of Cancer Research, 5*, 168–179.

Uehara, Y., Watanabe, K., Hakozaki, T., Yomota, M., & Hosomi, Y. (2022). Efficacy of first-line immune checkpoint inhibitors in patients with advanced NSCLC with KRAS, MET, FGFR, RET, BRAF, and HER2 alterations. *Thoracic Cancer, 13*, 1703–1711.

Verma, V., Jafarzadeh, N., Boi, S., Kundu, S., Jiang, Z., Fan, Y., ... Khleif, S. N. (2021). MEK inhibition reprograms CD8+ T lymphocytes into memory stem cells with potent antitumor effects. *Nature Immunology, 22*, 53–66.

Vetter, I. R., & Wittinghofer, A. (2001). The guanine nucleotide-binding switch in three dimensions. *Science (New York, N. Y.), 294*, 1299–1304.

Vigil, D., Cherfils, J., Rossman, K. L., & Der, C. J. (2010). Ras superfamily GEFs and GAPs: Validated and tractable targets for cancer therapy? *Nature Reviews: Cancer, 10*, 842–857.

Wainberg, Z. A., Alsina, M., Soares, H. P., Braña, I., Britten, C. D., Del Conte, G., ... Tabernero, J. (2017). A multi-arm phase I study of the PI3K/mTOR inhibitors PF-04691502 and fedratolisib (PF-05212384) plus irinotecan or the MEK inhibitor PD-0325901 in advanced cancer. *Targeted Oncology, 12*, 775–785.

Wang, Y., Mohseni, M., Grauel, A., Diez, J. E., Guan, W., Liang, S., ... Goldoni, S. (2021a). SHP2 blockade enhances anti-tumor immunity via tumor cell intrinsic and extrinsic mechanisms. *Scientific Reports, 11*, 1399.

Watterson, A., & Coelho, M. A. (2023). Cancer immune evasion through KRAS and PD-L1 and potential therapeutic interventions. *Cell Communication and Signaling: CCS, 21*, 45.

Waugh, D. J., & Wilson, C. (2008). The interleukin-8 pathway in cancer. *Clinical Cancer Research: An Official Journal of the American Association for Cancer Research, 14*, 6735–6741.

Weiss, R. A. (2020). A perspective on the early days of RAS research. *Cancer Metastasis Reviews, 39*, 1023–1028.

West, H. J., McCleland, M., Cappuzzo, F., Reck, M., Mok, T. S., Jotte, R. M., ... Socinski, M. A. (2022). Clinical efficacy of atezolizumab plus bevacizumab and chemotherapy in KRAS-mutated non-small cell lung cancer with STK11, KEAP1, or TP53 comutations: Subgroup results from the phase III IMpower150 trial. *Journal for ImmunoTherapy of Cancer, 10*, e003027.

Wong, G. S., Zhou, J., Liu, J. B., Wu, Z., Xu, X., Li, T., ... Bass, A. J. (2018). Targeting wild-type KRAS-amplified gastroesophageal cancer through combined MEK and SHP2 inhibition. *Nature Medicine, 24*, 968–977.

Wu, C., Xu, B., Zhou, Y., Ji, M., Zhang, D., Jiang, J., & Wu, C. (2016). Correlation between serum IL-1β and miR-144-3p as well as their prognostic values in LUAD and LUSC patients. *Oncotarget, 7*, 85876–85887.

Xu, L., Pathak, P. S., & Fukumura, D. (2004). Hypoxia-induced activation of p38 mitogen-activated protein kinase and phosphatidylinositol 3'-kinase signaling pathways contributes to expression of interleukin 8 in human ovarian carcinoma cells. *Clinical Cancer Research: An Official Journal of the American Association for Cancer Research, 10*, 701–707.

Xue, J. Y., Zhao, Y., Aronowitz, J., Mai, T. T., Vides, A., Qeriqi, B., ... Lito, P. (2020). Rapid non-uniform adaptation to conformation-specific KRAS(G12C) inhibition. *Nature, 577*, 421–425.

Yang, Y., Zhang, H., Huang, S., & Chu, Q. (2023). KRAS mutations in solid tumors: Characteristics, current therapeutic strategy, and potential treatment exploration. *Journal of Clinical Medicine, 12*.

Yin, K., Xia, X., Rui, K., Wang, T., & Wang, S. (2020). Myeloid-derived suppressor cells: A new and pivotal player in colorectal cancer progression. *Frontiers in Oncology, 10*, 610104.

Yokosuka, T., Takamatsu, M., Kobayashi-Imanishi, W., Hashimoto-Tane, A., Azuma, M., & Saito, T. (2012). Programmed cell death 1 forms negative costimulatory microclusters that directly inhibit T cell receptor signaling by recruiting phosphatase SHP2. *The Journal of Experimental Medicine, 209*, 1201–1217.

Yoon, Y. K., Kim, H. P., Han, S. W., Oh, D. Y., Im, S. A., Bang, Y. J., & Kim, T. Y. (2010). KRAS mutant lung cancer cells are differentially responsive to MEK inhibitor due to AKT or STAT3 activation: Implication for combinatorial approach. *Molecular Carcinogenesis, 49*, 353–362.

You, R., DeMayo, F. J., Liu, J., Cho, S. N., Burt, B. M., Creighton, C. J., ... Kheradmand, F. (2018). IL17A regulates tumor latency and metastasis in lung adeno and squamous SQ.2b and AD.1 cancer. *Cancer Immunology Research, 6*, 645–657.

Yuan, A., Yang, P. C., Yu, C. J., Chen, W. J., Lin, F. Y., Kuo, S. H., & Luh, K. T. (2000). Interleukin-8 messenger ribonucleic acid expression correlates with tumor progression, tumor angiogenesis, patient survival, and timing of relapse in non-small-cell lung cancer. *American Journal of Respiratory and Critical Care Medicine, 162*, 1957–1963.

Yuan, B., Clowers, M. J., Velasco, W. V., Peng, S., Peng, Q., Shi, Y., ... Moghaddam, S. J. (2022). Targeting IL-1β as an immunopreventive and therapeutic modality for K-ras–mutant lung cancer. *JCI Insight, 7*.

Yuan, T. L., Fellmann, C., Lee, C.-S., Ritchie, C. D., Thapar, V., Lee, L. C., ... Lowe, S. W. (2014). Development of siRNA payloads to target KRAS-mutant cancer. *Cancer Discovery, 4*, 1182–1197.

Zdanov, S., Mandapathil, M., Abu Eid, R., Adamson-Fadeyi, S., Wilson, W., Qian, J., ... Khleif, S. N. (2016). Mutant KRAS conversion of conventional T cells into regulatory T cells. *Cancer Immunology Research, 4*, 354–365.

Zhang, B., Zhang, Y., Zhang, J., Liu, P., Jiao, B., Wang, Z., & Ren, R. (2021). Focal adhesion kinase (FAK) inhibition synergizes with KRAS G12C inhibitors in treating cancer through the regulation of the FAK-YAP signaling. *Advanced Science, 8* e2100250.

Zhang, J., Zhang, F., & Niu, R. (2015). Functions of Shp2 in cancer. *Journal of Cellular and Molecular Medicine, 19*, 2075–2083.

Zhang, Y., Yan, W., Collins, M. A., Bednar, F., Rakshit, S., Zetter, B. R., ... di Magliano, M. P. (2013). Interleukin-6 is required for pancreatic cancer progression by promoting MAPK signaling activation and oxidative stress resistance. *Cancer Research, 73*, 6359–6374.

Zhang, Y., Yan, W., Mathew, E., Bednar, F., Wan, S., Collins, M. A., ... Di Magliano, M. P. (2014). CD4+ T lymphocyte ablation prevents pancreatic carcinogenesis in mice. *Cancer Immunology Research, 2*, 423–435.

Zhao, M., Guo, W., Wu, Y., Yang, C., Zhong, L., Deng, G., ... Sun, Y. (2019). SHP2 inhibition triggers anti-tumor immunity and synergizes with PD-1 blockade. *Acta Pharmaceutica Sinica B, 9*, 304–315.

Zheng, J., Huang, S., Huang, Y., Song, L., Yin, Y., Kong, W., ... Ouyang, X. (2016). Expression and prognosis value of SHP2 in patients with pancreatic ductal adenocarcinoma. *Tumour Biology: The Journal of the International Society for Oncodevelopmental Biology and Medicine, 37*, 7853–7859.

Zheng, X., Luo, J., Liu, W., Ashby, C. R. Jr, Chen, Z. S., & Lin, L. (2022). Sotorasib: A treatment for non-small cell lung cancer with the KRAS G12C mutation. *Drugs Today (Barc), 58*, 175–185.

Zhu, Y. M., Webster, S. J., Flower, D., & Woll, P. J. (2004). Interleukin-8/CXCL8 is a growth factor for human lung cancer cells. *British Journal of Cancer, 91*, 1970–1976.

Zhu, Z., Aref, A. R., Cohoon, T. J., Barbie, T. U., Imamura, Y., Yang, S., ... Barbie, D. A. (2014). Inhibition of KRAS-driven tumorigenicity by interruption of an autocrine cytokine circuit. *Cancer Discovery, 4*, 452–465.

Zocche, D. M., Ramirez, C., Fontao, F. M., Costa, L. D., & Redal, M. A. (2015). Global impact of KRAS mutation patterns in FOLFOX treated metastatic colorectal cancer. *Frontiers in Genetics, 6*.

Zou, W., & Restifo, N. P. (2010). T(H)17 cells in tumour immunity and immunotherapy. *Nature Reviews. Immunology, 10*, 248–256.

CHAPTER SIX

# Mitochondria driven innate immune signaling and inflammation in cancer growth, immune evasion, and therapeutic resistance

Sanjay Pandey[a,*], Vandana Anang[b], and Michelle M. Schumacher[a,c]
[a]Department of Radiation Oncology, Montefiore Medical Center, Bronx, NY, United States
[b]International Center for Genetic Engineering and Biotechnology (ICGEB), New Delhi, India
[c]Department of Pathology, Albert Einstein College of Medicine, Bronx, NY, United States
*Corresponding author. e-mail address: sanjay.pandey@einsteinmed.edu

## Contents

| | |
|---|---|
| 1. Introduction | 224 |
| 2. Mitochondrial outer membrane permeabilization (MOMP) driven inflammation and therapeutic resistance | 226 |
| 3. Mitochondrial ROS and the mitochondrial DAMPS in cancer | 227 |
| 4. mtDNA as immune activator and cytosolic DAMP | 229 |
| 5. Immune sensing of mtDNA by cGAS sting pathway | 230 |
| 6. ATP and mtDNA activate inflammaomes | 231 |
| 7. mtDNA drives TLR9 signaling | 232 |
| 8. mtRNAsensing by RIG-1 signaling | 233 |
| 9. Cardiolipin (CL) in innate inflammation | 233 |
| 10. Formyl peptides driven inflammation | 234 |
| 11. Innate immune signaling and cancer growth and acquired resistance | 235 |
| 12. Modulation of mitochondria-associated inflammation and future directions | 237 |
| 13. Conclusion | 241 |
| References | 242 |

## Abstract

Mitochondria play an important and multifaceted role in cellular function, catering to the cell's energy and biosynthetic requirements. They modulate apoptosis while responding to diverse extracellular and intracellular stresses including reactive oxygen species (ROS), nutrient and oxygen scarcity, endoplasmic reticulum stress, and signaling via surface death receptors. Integral components of mitochondria, such as mitochondrial DNA (mtDNA), mitochondrial RNA (mtRNA), Adenosine triphosphate (ATP), cardiolipin, and formyl peptides serve as major damage-associated molecular patterns (DAMPs). These molecules activate multiple innate immune pathways both in the cytosol [such as Retionoic Acid-Inducible Gene-1 (RIG-1) and Cyclic GMP-AMP Synthase (cGAS)] and on the cell surface [including Toll-like receptors (TLRs)]. This activation cascade leads to the release of various cytokines, chemokines, interferons, and other inflammatory molecules and oxidative species. The innate immune pathways further induce chronic inflammation in the tumor microenvironment which either promotes survival and proliferation or promotes epithelial to mesenchymal transition (EMT), metastasis and therapeutic resistance in the cancer cell's.

Chronic activation of innate inflammatory pathways in tumors also drives immunosuppressive checkpoint expression in the cancer cells and boosts the influx of immune-suppressive populations like Myeloid-Derived Suppressor Cells (MDSCs) and Regulatory T cells (Tregs) in cancer. Thus, sensing of cellular stress by the mitochondria may lead to enhanced tumor growth. In addition to that, the tumor microenvironment also becomes a source of immunosuppressive cytokines. These cytokines exert a debilitating effect on the functioning of immune effector cells, and thus foster immune tolerance and facilitate immune evasion. Here we describe how alteration of the mitochondrial homeostasis and cellular stress drives innate inflammatory pathways in the tumor microenvironment.

## 1. Introduction

In addition to being a major source of energy, mitochondria play crucial roles in various cellular functions. These include providing precursors for generating structural and functional components and coordinating with the endoplasmic reticulum (ER) to maintain calcium ion ($Ca^{+2}$) homeostasis within cells (Duchen, 2000; Spinelli & Haigis, 2018; Tanwar, Singh, & Motiani, 2021). Mitochondria regulate the intrinsic apoptotic pathway, responding to signals of damage and stress. Furthermore, mitochondria are central in both the innate sensing of cellular stress and in producing damage-associated molecular patterns (DAMPs). When cells encounter stressors like disrupted $Ca^{+2}$ balance, reactive oxygen species (ROS), pathogenic threats, or the compromised integrity of mitochondrial membranes, the structural components of the membrane are either oxidized or released into the cytosol where they become detectable by cytosolic or cell surface receptors associated with innate and inflammatory responses (Brookes, Yoon, Robotham, Anders, & Sheu, 2004; Murakami et al., 2012; Zhao et al., 2020).

During the early stages of cellular transformation, the host immune system actively removes precancerous or transformed cells. This initial elimination phase is succeeded by the establishment of quiescent tumors in an equilibrium state, where tumor growth is held in check by potent antitumor forces. However, prolonged inflammation within the tumor microenvironment and the manipulation of the immune response by the tumor cells can lead to diminishing antitumor immune functions (Maiorino, Daßler-Plenker, Sun, & Egeblad, 2022; Mittal, Gubin, Schreiber, & Smyth, 2014). Critical in shaping this subdued tumor microenvironment are mitochondrial DAMPs including Adenosine triphosphate (ATP), mitochondrial DNA (mtDNA), and mitochondrial RNA (mtRNA). These molecules trigger multiple innate inflammatory signaling pathways such as NFκB and IRF-3/7, leading to the release of pro-inflammatory cytokines and chemokines. When these inflammatory signals persist in the tumor microenvironment they can drive the infiltration of immunosuppressive immune cells like myeloid-derived suppressor cells (MDSCs) into the tumor milieu (Riley & Tait, 2020; Zhang, Zhang, et al., 2022).

The production of mitochondrial reactive oxygen species (mtROS) can inflict DNA damage, leading to the accumulation of genetic instabilities and mutations that promote immune tolerance within tumor cells. Mutations occurring within antigen processing pathways have been identified as drivers of immune tolerance and evasion. Furthermore, the instability of nuclear and mitochondrial genomes due to mtROS-induced damage can result in the release of double-stranded DNA (dsDNA) into the cytosol. Both mtDNA and cytosolic DNA in the cytosol are detected by the cGAS-STING (cyclic GMP-AMP synthase-stimulator of interferon genes) pathway, triggering the release of interferons via interferon regulatory factors (IRFs) (Marchi, Guilbaud, Tait, Yamazaki, & Galluzzi, 2022; Orekhov et al., 2022; Zhang, Zhang, et al., 2022). While interferon signaling initially stimulates antitumor functions within the tumor microenvironment, it can also lead to elevated levels of immune checkpoint molecules like PD-L1. These factors may further contribute to immune anergy and resistance to immunotherapies.

While the link between inflammation and tumorigenesis is poorly understood, there have been multiple studies which state that prolonged inflammation increases the likelihood of mutations and chromatin instability. Additionally, chronic inflammation, via the mitochondrial DAMPs sensing, promotes cancer invasion and metastasis. There is a multitude of clinical evidence linking chronic inflammation to poor prognosis in various cancers including colon cancer, prostate cancer, and liver cancer. This article discusses the innate sensing of the mitochondrial components which may contribute to chronic inflammation, cancer growth, and resistance to therapy.

## 2. Mitochondrial outer membrane permeabilization (MOMP) driven inflammation and therapeutic resistance

Cancer treatment therapies, unfavorable tumor microenvironment conditions, metabolic stress, and immune cytotoxic effector T cell molecules like Fas ligand (Fasl) can activate cellular stress responses leading to ROS, hypoxia inducible factors (HIFs), and $Ca^{+2}$ level imbalances. Response to these unfavorable conditions may activate B cell lymphoma 2 (BCL-2) homology 3 (BH3)-only proteins. Consequently, BCL-2-associated X protein (BAX) and BCL-2 antagonist or killer (BAK) are activated, triggering MOMP. Additionally, elevated cytosolic stress and $Ca^{+2}$ levels induce the formation of the mitochondrial permeability transition pore (MPTP) complex, leading to the release of mtDNA, mtRNA, and other components into the cytosol (Heilig, Lee, & Tait, 2023; Redza-Dutordoir & Averill-Bates, 2016). While MOMP driven by BAX and BAK is pivotal for apoptosis, insufficient MOMP allows tumor cells to survive. An insufficient MOMP can set off CASP3-dependent genomic instability via the DNA fragmentation factor subunit beta (DFFβ) (Liu, Zou, Slaughter, & Wang, 1997).

Though genomic instability can result in the accumulation of mutations that enable tumors to evade immune detection (for eg. mutations in the antigen presentation pathways), it can also lead to the escape of mitotic DNA into the cytosol, which activates cGAS and RIG-1 pathways (Kwon & Bakhoum, 2020). Genomic instability in tumor cells, such as an increase in c-MYC copies, has been associated with immune suppression and an increase in the release of immunosuppressive cytokines and chemokines like CCL9 and IL-23, resulting in the suppression of immune responses in the tumor microenvironment (TME) (Kortlever et al., 2017). This suboptimal MOMP can also drive survival pathways and enhance resistance to therapies through a process independent of CASP activation. Additionally, it can enhance the metastatic potential in cancer cells by engaging the integrated stress response (ISR) through eukaryotic translation initiation factor 2 alpha kinase 1 (EIF2AK1) (Lines, Mcgrath, Dorwart, & Conn, 2023; Ljujic et al., 2016; Timosenko et al., 2016). The ISR amplifies ATF4-driven transcription, promoting the expression of indoleamine 2,3-dioxygenase (IDO) in tumor cells. Overexpressed IDO curbs T cell activation and effector functions by limiting the availability of arginine within the TME. Furthermore,

ATF4-induced upregulation of amino acid transporters, including SLC1A5 in tumor cells, benefits their survival while inducing T cell anergy and diminishing their effectiveness (Yang et al., 2023).

## 3. Mitochondrial ROS and the mitochondrial DAMPS in cancer

In non-phagocytic cells, almost 95% of ROS are generated in the electron transport chain (Turrens, 1997). ROS can damage DNA, proteins, and lipids, leading to gene mutations, apoptosis, and necrosis (Redza-Dutordoir & Averill-Bates, 2016). ROS-induced DNA damage can drive genomic instability by oxidizing nuclear and mtDNA. Genomic instability can result in the accumulation of mutations, promoting tumor growth and immune evasion. The accumulation of loss-of-function mutations in the P53 gene can make cells prone to acquiring additional mutations. Similarly, mutations in the antigen processing pathways, such as human leukocyte antigen (HLA) in tumors or antigen-presenting cells, can prevent appropriate antigen presentation and CD8 activation by tumor cells. For instance, downregulation of the transporter associated with antigen processing 1 (TAP1) is known to elicit immune escape in colorectal cancer. NLRC5, a key transcriptional co-activator of gene expression of MHC class I antigen (MHCI) and other genes in the MHCI presentation pathway, can be affected by epigenetic and genetic alterations in various types of cancers, leading to compromised antigen presentation and immune escape of tumor cells (Ling et al., 2017; Yoshihama et al., 2016). Recent studies have suggested that loss-of-function in Janus-activated kinase (JAK) 1 mutations can prevent cells from responding to IFNs and can drive resistance to immunotherapy. Further, mitochondrial metabolism via the pentose phosphate shunt and mitochondrial ROS production from the Qo site of complex III can drive hypoxia by inducing the expression of HIFs (Chandel et al., 2000; Liemburg-Apers, Willems, Koopman, & Grefte, 2015). Hypoxic conditions in solid tumors stimulate the expression of HIF family transcription factors, which coordinates the molecular response to oxygen unavailability and inflammation. The stabilization of HIF-a drives upregulation of glycolysis and mitochondrial dysfunction, which leads to the release of excessive ROS and mitochondrial DAMPs like mtDNA and ATP into the cytosol and extracellular surface (Bouhamida et al., 2022). The roles of mtDNA and ATP in the upcoming sections will be further discussed (Fig. 1).

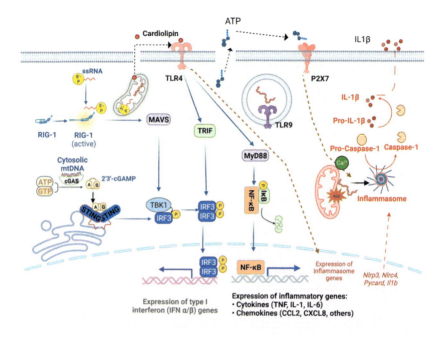

Fig. 1 Innate recognition of mitochondrial DAMPs and inflammation. The release of mitochondrial components into the cytosol (mtDNA, mtRNA) and ROS drives pathogen-sensing innate pathways such as TLR and RIG-1 signaling in tumor cells and APCs. Extracellular release or accumulation of other mitochondria-associated damage-associated molecular patterns (DMAPs), such as cardiolipin, formyl peptides, and ATP, can also activate TLRs and NLRs through various signals. All these inflammatory signaling processes can further amplify chronic inflammation in the tumor microenvironment (TME) by inducing the expression of inflammatory genes via NFKb and IRF3. Additionally, autocrine, and paracrine binding of inflammatory cytokines, such as IFNs and interleukins like IL-1beta and IL-6, to their receptors plays a role in tumor cell survival and chronic inflammation in TME.

Upon release from dysfunctional mitochondria, mtDNA and ROS can drive IL-1β and IL-18 secretion via inflammasome activation in the cytosol (Riley & Tait, 2020; Shimada et al., 2012). Additionally, ATP from dying cells is released into the TME through lysosomal secretion and pannexin 1, exerting immunostimulatory effects on tumor cells and antigen-presenting cells (APCs). The release of oxidized mtDNA during proapoptotic events is also a key event for NLRP3 activation, leading to ATP depletion and mitochondrial dysfunction in tumor cells (Fig. 1). ATP, when engaged to purinergic receptors P2RY2 and P2RX7, also contributes to the inflammasomes activation (Ayna et al., 2012). Increased levels of 8-OHdG, the most common DNA base change arising from oxidative stress, are linked with various cancers and neurodegenerative diseases (Ralph, Rodríguez-Enríquez, Neuzil, Saavedra, & Moreno-Sánchez, 2010; Van Houten, Woshner, & Santos, 2006; Weinberg & Chandel, 2009). The next few sections will describe the role of mitochondrial DAMPs and structural elements of mitochondrial membranes, such as cardiolipin and formyl peptides, in modulating the inflammatory and immune signaling pathways in the TME (Fig. 1).

## 4. mtDNA as immune activator and cytosolic DAMP

The presence of certain bacterial nucleic acid sequences and unique methylation patterns give distinct molecular identity to mtDNA. However, there have been conflicting claims regarding the existence of CpG methylation in mtDNA (Boguszewska, Szewczuk, Kaźmierczak-Barańska, & Karwowski, 2020; Chatterjee, Das, & Chakrabarti, 2022). While the presence of DNA methyltransferases (DNMT1 and DNMT3b) in mitochondria supports the notion that CpG methylation may be present in mtDNA, other reports indicate that CpG methylation is either absent or methylation follows a distinct pattern in mtDNA (Liu et al., 2016; Mechta, Ingerslev, Fabre, Picard, & Barrès, 2017). The presence of distinct methylation patterns, such as 5-methylcytosine (5mC) and 5-hydroxymethylcytosine (5hmC) at CpG islands in mtDNA, contribute to its recognition by the DNA sensors like cGASs and Pattern Recognition Receptors (PRRs) such as TLR9 (Toll-Like Receptors 9) (Liu et al., 2016; Riley & Tait, 2020). MtDNA located in the endolysosome can act as a source of endogenous activation ligands for DNA-sensing innate pathways. Other DAMPs that can activate innate signaling through PRRs include double-stranded RNA, uncapped mRNAs, and RNA-DNA hybrids (Riley & Tait, 2020). Absence of nucleotide excision repair makes mtDNA susceptible to DNA damage accumulation and thus may lead to its innate sensing.

MtDNA is packaged into protein–DNA complexes with mitochondrial transcription factor A (TFAM), which defines the overall nucleoid structure in mitochondria. While TFAM bound to mtDNA can dampen mtDNA recognition by PRRs, TFAM itself can act as a DAMP (Riley & Tait, 2020). There are reports indicating that TFAM can promote mtDNA-driven innate signaling in plasmacytoid dendritic cells (pDCs) through TLR9 and RAGE (Julian et al., 2012; Julian, Shao, Vangundy, Papenfuss, & Crouser, 2013).

Additionally, mtDNA released into the cytosol as a result of optimal or suboptimal apoptotic induction can drive ATP-dependent activation of NLRP3 inflammasomes (Riley & Tait, 2020).

## 5. Immune sensing of mtDNA by cGAS sting pathway

MtDNA serves as a potent activator of cyclic GMP–AMP synthase (cGAS). Upon binding to mtDNA, cGAS dimerizes and drives the activation of cGAS and the synthesis of 2′3′ cyclic GMP–AMP (cGAMP). Subsequently, cGAMP binds to Stimulator of interferon genes (STING) on the ER membrane, promoting STING oligomerization and its release from the ER membrane. STING dimerization and subsequent activation leads to the activation of Tank Binding Kinase (TBK1), which, in turn, phosphorylates and recruits IRF3 (dimer form) to the nucleus. Interferon regulatory factors (IRFs) then activate the expression of antiviral type 1 interferons (Riley & Tait, 2020; Vashi & Bakhoum, 2021). These type 1 interferons, via autocrine or paracrine signaling and STAT1 activation, drive the expression of interferon-stimulated genes (ISGs), immune-activating inflammatory mediators, pro-apoptotic genes, and chemokines (Van Vugt & Parkes, 2022). Interferon receptor signaling enhances antigen presentation and T cell effector functions. While early cGAS-STING responses have an immune-stimulating effect, persistent activation of STING may lead to immune suppression and evasion. Studies have shown that STING drives the infiltration of MDSCs into tumors (Riley & Tait, 2020; Van Vugt & Parkes, 2022). Moreover, chronic interferon signaling upregulate negative regulatory pathways and immune-suppressive genes through feedback mechanisms (Benci et al., 2016). Constitutive expression of cGAS-STING activation in targeted cancer management therapies induces noncanonical NF-κB/RelB pathways (Hou et al., 2018). RelB signaling suppresses type 1 IFN production in dendritic cells following radiation treatment (Hou et al., 2018). Most of the information on cGAS-STING signaling in cancer originates from genomically unstable tumors, such as those carrying BRCA mutations. Another mechanism for immune suppression through constant STING stimulation in BRCA mutated cancers involves the cytosolic activation of transmembrane pyrophosphatase Ectonucleotide pyrophosphatase/phosphodiesterase 1 (ENPP1). ENPP1 negatively regulates cGAS by hydrolyzing cyclic dinucleotides like 2′3′cGAMP to generate AMP (Carozza et al., 2022; Cogan & Bakhoum, 2020). AMP is further broken down into immunosuppressive adenosine by cell surface CD73 and is released into the TME (Regateiro, Cobbold, & Waldmann, 2013; Zhang, 2010). Additionally, ENPP1 upregulation inhibits cGAMP transport to neighboring tumor or immune cells, thereby preventing cGAS-STING activation (Carozza et al., 2022). STING-mediated upregulation of CCL2, CCL7, and CCL12 also results in the infiltration of MDSCs, promoting immunosuppression and therapeutic resistance (Van Vugt & Parkes, 2022).

## 6. ATP and mtDNA activate inflammaomes

Dying cells release mitochondrial DMAPs, ATPs, and oxidized mtDNA which, upon release to TME and cytosol, can coordinate the activation of NLRP3 inflammasomes. NLRP3 inflammasomes are the intracellular macromolecular protein assembly which sense and detect a broad range of microbial PAMPS, endogenous danger signals, and ion imbalances. The activation of the inflammasomes requires the interaction of caspase recruitment domain (CARD), the pyrin domains of ASC (Apoptosis-associated speck-like protein), and the CARD of NLRP3, leading to the proteolytic cleavage of procaspase 1 to caspase 1 and its subsequent activation. Caspase 1 then cleaves pro IL-1β and pro IL-18 to promote the release of mature IL-1β and IL-18 (Riley & Tait, 2020; Shimada et al., 2012; Xian et al., 2022).

NLRP3 inflammasome activation involves two signals. The first signal is provided by proinflammatory cytokines or PRR signaling, resulting in the NF-κB-driven upregulation of NLRP3 assembly genes. The second signal is triggered by the binding of ATP or other NLRP3-activating DAMPs (extracellular ATP, CPPD crystals monosodium urate crystals, asbestos, silica, and β-amyloid). ATP binding to the P2X7 receptors leads to the activation and translocation of phospholipase C (PLC), which in turn catalyzes the production of inositol triphosphate (IP3). IP3 destabilizes the ER membrane, causing the release of $Ca^{2+}$. Various theories explain the activation of NLRP3 by $Ca^{2+}$, but it is widely believed that $Ca^{2+}$-driven mitochondrial dysfunction and the release of oxidized mtDNA into the cytosol drive NLRP3 activation (Gombault, Baron, & Couillin, 2013; He, Hara, & Núñez, 2016; Lee et al., 2012; Yang, Wang, Kouadir, Song, & Shi, 2019). This idea is supported by a study showing that blocking the mitochondrial calcium uniporter (MCU) abolishes NLRP3-dependent IL-1β release (Triantafilou, Hughes, Triantafilou, & Morgan, 2013).

Additionally, the mitochondrial phospholipid cardiolipin provides a platform for inflammasome assembly and activation. Apart from releasing IL-1β and IL-18, Caspase 1 also cleaves Gasdermin D (GSDMD), which then translocates and attaches to the cytoplasmic membrane, forming pores and inducing pyroptosis (He et al., 2016). Furthermore, decreased cellular cyclic AMP (cAMP) and increased intracellular $Ca^{2+}$ and other CASR agonists can activate the NLRP3 inflammasome even in the absence of exogenous ATP (Yang et al., 2019).

## 7. mtDNA drives TLR9 signaling

The TLRs play a crucial role in sensing microbial pathogen-associated molecular patterns (PAMPs) and initiating innate inflammatory pathways. In addition to microbial PAMPs, TLRs can also detect DAMPs released during cellular death. These receptors are found on the cell surface, endosomal membranes, and ER, allowing them to respond to diverse bacterial features and trigger innate immunity (Pandey et al., 2015). TLR9 specifically recognizes DNA with hypomethylated CpG motifs. mtDNA, being hypomethylated at CpG motifs, serves as a natural and potent inducer of TLR9 (Takeshita et al., 2001).

TLR9 is primarily expressed in plasmacytoid dendritic cells (pDCs) and some cancer cells and resides in the ER outer membrane; however, the recognition of DNA occurs in the endo-lysosomes. Interaction between mtDNA and TLR9 activates the MyD88 adapter protein, which subsequently triggers IRAK4 activation. This activation leads to the translocation of NF-κB and IRF3 to the nucleus (Heilig et al., 2023). NF-κB induces the release of proinflammatory mediators in the TME, while IRF3 leads to the release of interferons. The release of mtDNA into the cytosol or the endocytosis/phagocytosis of dying cells by APCs drives proinflammatory signaling in APCs through receptors such as RAGE and TLR9 (Caielli et al., 2016; Heilig et al., 2023; Van Houten et al., 2006).

The activation of TLR9 by mtDNA has been extensively studied in the pathogenesis of chronic liver pathologies and has been suggested as a potential serological biomarker for non-small cell lung cancer (Lai, Liu, Huang, Chau, & Wu, 2019). Additionally, TLR9 signaling associated with mitochondrial DNA has been implicated in hepatic chronic inflammatory diseases and cancer (Riley & Tait, 2020). Mitochondrial Cardiolipin, when extracellular, can bind to the TLR4 receptor and can drive TLR4 induction leading to the NF-κB and IRF3 translocation to the nucleus through the MyD88-TRAF6 and TRIF adapter cascades (Pandey et al., 2015; Pizzuto & Pelegrin, 2020).

## 8. mtRNAsensing by RIG-1 signaling

As described in an earlier section, TME stress signaling drives MOMP and mitochondrial herniation mediated by BAX and BAK (Chipuk, Bouchier-Hayes, & Green, 2006; Guilbaud & Galluzzi, 2022). MOMP leads to the escape of mtRNA and mtDNA into the cytosol. RLR (RIG-I-like receptor) signaling pathways, such as RIG-I and Melanoma differentiation-associated protein 5 (MDA5), can be activated by immunostimulatory RNA and DNA (Jiang et al., 2023). Binding to the nucleic acid induces conformational changes in the RLRs, followed by the multimerization of their CARD domains. RLR CARDs then interact with the Mitochondrial Antiviral Signaling protein (MAVS) CARD domain. The MOMP initiator proteins BAX–BAK1 mediate the release of ERAL1, an RNA chaperone that promotes MAVS stabilization at the mitochondrial surface (Marchi et al., 2022). Besides the outer mitochondrial membrane (OMM), MAVS can also be anchored in mitochondrial-associated membranes (MAMs) and peroxisomes. MAVS activation relays the signal to TBK1 and IKKε, which, in turn, drive the IRF3 and IRF7-driven transcription of type I interferons and other genes (Bender et al., 2015; Brubaker, Bonham, Zanoni, & Kagan, 2015). Mutant RIG-I forming circular RIG-I has been shown to promote colitis and promote colitis-associated colon cancer development by activating DDX3X (Dead-box helicase 3 x-linked) signaling cascade, type I IFN production, and severe inflammatory tissue damage. While there have been very few studies suggesting a tumor-promoting role of the RIG-1 via interferons, RIG-1 has been associated with inflammasome activation and associated IL1-β production which are well documented to promote tumor growth. Besides, CD8 specific RIG1 edificant was seen to promote antitumor-functions suggesting a negative regulatory function of the RIG-1 pathway (Iurescia, Fioretti, & Rinaldi, 2020).

## 9. Cardiolipin (CL) in innate inflammation

CL is a phospholipid that is normally present in the inner mitochondria membrane (IMM) under physiological conditions. However, under the influence of death signals and cellular stress, CL undergoes oxidation, dissociates from the IMM, and translocates to the cytosolic side of the OMM. Interestingly, unsaturated cardiolipins can competitively inhibit TLR4 signaling while saturated CLs can activate TLR4 signaling, promoting the release of proinflammatory cytokines through NF-κB and IRF signaling pathways. The stimulation of TLR4 triggers inflammatory responses via MyD88-TRAF-6-NF-κB and TRIF-IRF3 signaling cascades. Moreover, CL also plays a critical role in the activation of the NLRP3 inflammasome. After LPS/DAMPs-driven TLR4 activation, CL translocates to the OMM and acts as a platform for NLRP3 assembly, facilitating the production of IL-1β through NLRP3 activation (Pizzuto & Pelegrin, 2020). CL metabolism has been implicated in cancer but the role of its inflammatory functions on cancer have not been extensively explored. It has been proposed that TLR4-dependent proinflammatory properties of CL might be associated with chronic inflammation in Barth syndrome (Pizzuto & Pelegrin, 2020; Pizzuto et al., 2019).

## 10. Formyl peptides driven inflammation

In the tumor microenvironment, the accumulation of N-formyl peptides following cell death can lead to neutrophil activation and infiltration. This infiltration of immune cells is associated with the activation of innate receptors such as TLR9, AGER (Advanced Glycosylation End-Product Specific Receptor), and formyl peptide receptor 1 (FPR1) by the formyl peptides. Extracellular CL can also be taken up by and presented on CD1d (an MHC class I-like molecule) through cross-presentation, resulting in the activation of γδ T cells. Formyl-Peptide Receptors (FPRs), members of the GPCR superfamily, are associated with chronic inflammatory pathologies, including cardiovascular diseases. FPRs can also regulate oxidative stress by activating NADPH oxidase and generating ROS. Like bacterial formylated peptides, mitochondrial formyl peptides can also induce a proinflammatory cellular response through FPR1 activation. While expression of FPR1 in non-immune human cells is low, FPR1 is highly expressed in high grade glioblastoma, neuroblastoma, colon, breast, and bladder cancers (Ahmet et al., 2020). In gastric cancers, FPR1 expression has been associated with poor prognosis. FPRs result in the activation of the AKT and the MAP Kinases pathways in the tumor cells, imparting survival benefits and EMT (Weiß & Kretschmer, 2018). FPR signaling promotes chemotaxis in cancer cells and neutrophils. Formyl peptide receptor 2 (FPR2) activation can elicit both pro- and anti-inflammatory responses, depending on the nature of the conformational changes upon ligand binding (Lee, Han, & Jung, 2023; Li et al., 2016). Ligands such as Lipoxin A4 and Annexin A1 are FPR2 ligands that drive anti-inflammatory responses. FPR-1 and FPR-2 induced inflammation have been implicated in various inflammatory diseases and cancers (Tian et al., 2020).

## 11. Innate immune signaling and cancer growth and acquired resistance

Innate signaling pathways driven by mitochondrial components and DAMPs exhibit a dual role in tumor growth modulation. These signaling cascades result in the expression of chemokines and cytokines, which facilitate the infiltration of immune cells into cold tumors, rendering them susceptible to the systemic impacts of targeted therapies. However, prolonged activation of these pathways can promote the accumulation of immunosuppressive cells within the tumor microenvironment. This influx subsequently fosters conditions conducive to tumor escape, proliferation, and immune exhaustion (Fig. 2). One such example is the role of TLR4s in cancer. While TLR-agonists and IFNs are being used extensively in cancer management therapies, many preclinical studies and clinical studies have suggested their chronic stimulation induces resistance in cancer cells (Huang, Zhao, Unkeless, Feng, & Xiong, 2008; Huang et al., 2020; John & Darcy, 2016; Musella, Galassi, Manduca, & Sistigu, 2021). Similarly, persistent stimulation of interferons within the tumor microenvironment has been observed to drive the expression of crucial immune checkpoint PD-L1 and other checkpoint molecules.

Chronic stimulation of PRRs and other signaling by mitochondria-associated innate signaling molecular mediators of these inflammatory signals include NF-kB, STAT3 and IRF3/7, TLRs, cGAS/STING, and MAPKs (Cohen, 2014; Hirano, 2021; Rébé & Ghiringhelli, 2020; Samson & Ablasser, 2022). Cytokines, chemokines, and interferons act as the target genes of these innate signaling by paracrine or autocrine engagement with their receptors and may either confer a protective effect on the tumor or impart resistance to therapies. These soluble mediators further amplify the inflammatory response and enhance the infiltration of M2 macrophages, regulatory T cells (Tregs), and MDSCs (Greten & Grivennikov, 2019).

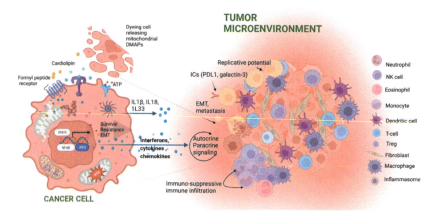

**Fig. 2** Mitochondria-associated inflammation promotes tumor growth, epithelial-mesenchymal transition (EMT), and metastasis. Chronic stimulation of innate inflammatory pathways, induced by cellular stress and the release of damage-associated molecular patterns (DAMPs) due to cell death in the tumor microenvironment (TME), results in enhanced survival, therapeutic resistance, EMT, and metastasis of tumor cells. The release of chemokines, interleukins, and interferons (IFNs) in the TME may contribute to the upregulation of immune checkpoints and the infiltration of immunosuppressive immune cells, thereby further promoting therapeutic resistance.

These immune cells, fueled by hypoxic and inflammatory conditions, subsequently release immunosuppressive cytokines (e.g., IL-6, IL-10 and TGF-β) as part of the feedback mechanisms inherent to inflamed tumors. In the presence of these inflammatory cytokines, immune cells within the hypoxic and nutrient-deprived environment can undergo polarization into a suppressive phenotype, such as tumor-associated macrophages (TAMs), dendritic cell regulatory subsets (DCregs), and tumor-associated neutrophils (TANs) (Veglia & Gabrilovich, 2017; Yan et al., 2022). This intricate orchestration of immunosuppressive signaling and polarization of immune cells leads to immune evasion and contributes to the survival, proliferation, and metastatic phenotype of cancer (Yan et al., 2022). Cancer management therapies like chemotherapy or radiotherapy also instigate a suppressive state which leads to the downregulation of host anti-tumor immune responses and a state of immune exhaustion. Excessive release of inflammatory mediators by these inflammatory cells, like nitric oxide (NO), reactive oxygen species (ROS), cyclooxygenase-2 (COX-2), prostaglandin E2 (PGE2), have been shown to promote angiogenesis and induce EMT in tumors (Fig. 2) (Greten & Grivennikov, 2019).

IL-1β, upon binding to its receptors, triggers proinflammatory and pro-survival signaling via MAPK pathways (p38 and JNK). Target genes in canonical IL-1β signaling (IL-1α, IL-8, IL-6, MCP-1, COX-2, and IκBα) have been implicated in tumor progression and serve as prognostic markers for therapeutic outcomes. Also, IL-1β upregulation induces miR-181a expression in human colon cancer cells via NF-κB, which represses PTEN (phosphatase and tensin homolog) to induce proliferation in cancer cells. IL-1β and IL-18 both have been shown to increase infiltration of immunosuppressive populations in the tumor (Pretre, Papadopoulos, Regard, Pelletier, & Woo, 2022; Rébé & Ghiringhelli, 2020; Zhang & Veeramachaneni, 2022). IL-6 is a crucial component of the NF-κB/IL-6/STAT-3 pathway, intricately linked to tumorigenesis. IL-6 stimulates tumor cell proliferation via multiple signaling pathways including Ras/Raf/MEK/MAPK, PI3K/AKT, and JAK/STAT (Kumari, Dwarakanath, Das, & Bhatt, 2016). TNF-α can promote EMT across diverse cancer models, including breast, lung, and hepatocellular carcinoma (Shrestha, Bridle, Crawford, & Jayachandran, 2020). In melanoma, TNF-α increase in the tumor was also found to be concordant with enhanced immune checkpoint molecules (such as PD-L1 and CD73 expression) as the result of NF-κB and c-Jun/AP-1 transcription factor complex activation, respectively (Bertrand et al., 2017; Ji, He, Regev, & Struhl, 2019). H3K4me1 and STAT3 have been associated with epigenetic changes associated with the accessibility of DNA for high activity of the protein transcriptional traffic in inflammatory processes, a characteristic of inflammatory memory (Platanitis & Decker, 2018; Zhang, Nesvick et al., 2022). Besides, cancer cells with constant stimulation from interferons display high PD-L1 and other immune checkpoints. Notably, the induction of IFN through mechanisms like STING or TLR activation in B16 melanoma cells has been found to drive a STAT1-dependent resistance to immune checkpoint blockade (ICBs) (Benci et al., 2016).

## 12. Modulation of mitochondria-associated inflammation and future directions

In the context of cancer, inflammation has a dual role. On one hand, it fuels cancer growth and stands as a prominent hallmark of the disease. On the other hand, therapeutic approaches leveraging inflammation have been harnessed to enhance treatment efficacy. Agents such as TLR agonists,

RIG activators, and STING agonists have found substantial utility as adjuvants in cancer therapies. However, recent studies have illuminated a multifaceted perspective on the innate pathways driven by mitochondrial DAMPs and the IFN response in chronic contexts. Paradoxically, these pathways contribute to tumor promotion and foster resistance (Huang et al., 2008; John & Darcy, 2016; Marchi et al., 2022; Van Vugt & Parkes, 2022; Bender et al., 2015).

Considering the significant role of the mitochodria in the survival of tumor cells and tumor promoting inflammation, it becomes imperative to target the innate signaling induced by mitochondrial damage. One approach could involve dietary adjustments and the implementation of metabolic modifiers aimed at curbing mitochondrial dysfunction. Similarly, antioxidants and ROS modulators can play a crucial role in averting damage from ROS, thereby preventing oxidative harm to mitochondrial DAMPs and critical structural components (Forrester, Kikuchi, Hernandes, Xu, & Griendling, 2018; Liu et al., 2018).

Strategies to evade suboptimal MOMP involve the use of synthetic lethality as well as the consideration of ablative treatments like focused radiotherapy, such as SBRT (Stereotactic body radiotherapy), to ensure efficient elimination (Montero & Haq, 2022). DNA repair inhibitors, ROS regulators, and Bcl2 family regulators are some other interventions which can prevent MOMP-induced survival in cancer cells (Bao, Liu, Li, & Li, 2020; Kalkavan & Green, 2018; Reuter, Gupta, Chaturvedi, & Aggarwal, 2010; Zhang et al., 2016).

Promoting autophagy and mitophagy emerges as a potential strategy for efficiently recycling compromised mitochondria, contributing to the overall health of the cellular milieu (Song, Zhou, & Zhou, 2020; Zhong, Sanchez-Lopez, & Karin, 2016). Mitochondrial stress induces the disruption of mitochondrial membrane potential, initiating the mitophagy process. Mitophagy, in turn, governs immune responses prompted by mitochondrial stress, including inflammasome activation, type I interferon (IFN) responses, and intrinsic apoptosis by selectively degrading dysfunctional mitochondria (Song et al., 2020).

An alternative approach aimed at preventing the tumor-promoting impact of mitochondrial-driven innate signaling involves the deployment of inhibitors or antibodies targeting distinct cytokines like TNF, IL-6, and IL-1β. Targeting tumor promoting inflammation is one of the major prevention and therapeutic approaches in certain cancers (Hou, Karin, & Sun, 2021). Controlling chronic inflammation holds significant importance

Table 1 Innate inflammatory signaling associated with the mitochondrial DAMPs.

| Mitochondrial DAMP | Innate receptor signaling | Subcellular localization | Major effector molecules/pathways | Tumor promoting functions | References |
|---|---|---|---|---|---|
| mtDNA | NLRP3 inflammasomes | Cytosol | IL-1β, IL-18 | M2 polarization in TAMs, Infiltration of MDSCs and Tregs, Activation of pro-survival pathways in tumor cells, enhance metastasis | Rébé and Ghiringhelli (2020), Shimada et al. (2012) |
| | cGAS-STING signaling | Cytosol | IFNs, Adenosine | Increased ICB expression, CCL2, CCL7 Infiltration of MDSCs and Tregs, Adenosine-mediated immune suppression and evasion | Regateiro et al. (2013), Zhang (2010), Van Vugt and Parkes (2022) |
| | TLR9 | Endosome | IFNs, IL-6, TNFα | Promotion of metastasis, Increased ICB expression, Activation of pro-survival pathways in tumor cells | Van Vugt and Parkes (2022), Huang et al. (2008) |
| mtRNA | RIG-1 | Cytosol | IFNs | Increased ICB expression | Van Vugt and Parkes (2022) |

(continued)

Table 1 Innate inflammatory signaling associated with the mitochondrial DAMPs. (cont'd)

| Mitochondrial DAMP | Innate receptor signaling | Subcellular localization | Major effector molecules/ pathways | Tumor promoting functions | References |
|---|---|---|---|---|---|
| Cardiolipin | TLR4 | Cell surface | IFNs, IL-6, TNFα | Promotion of metastasis, Increased ICB expression, Activation of pro-survival pathways in tumor cells | Van Vugt and Parkes (2022), Huang et al. (2008) |
| | NLRP3 inflammasomes | Cytosol | IL-1β, IL-18 | M2 polarization in TAMs, Infiltration of MDSCs and Tregs, Activation of pro-survival pathways in tumor cells, Promotion of metastasis | Rébé and Ghiringhelli (2020), Shimada et al. (2012) |
| Formyl peptides | FPRs | Cell surface | MAP Kinases and AKT activation | Survival and EMT pathways activation, Metastasis, Chemotaxis. | Weiß and Kretschmer (2018) |
| ATP | NLRP3 inflammasomes | Cytosol | IL-1β, IL-18 | M2 polarization in TAMs, Infiltration of MDSCs and Tregs, Activation of pro-survival pathways in tumor cells, Promotion of metastasis | Rébé and Ghiringhelli (2020), Shimada et al. (2012) |

in curbing the immunosuppressive aftermath linked to innate immune pathways and has garnered considerable attention. However, while the application of TNF inhibitors in cancer therapy has been constrained by their adverse impact on systemic immunity, the utilization of IL6 and IL-1β inhibitors has been promising in various cancers (Berraondo et al., 2019; Briukhovetska et al., 2021; Rossi, Lu, Jourdan, & Klein, 2015). Targeting the final phase of tumor-promoting inflammation, which involves immunosuppression, can be achieved through various strategies. ICBs and agents that activate T cells, such as CD27 and CD28 antibodies, as well as IL-15 agonist antibodies, hold promise in this regard. Additionally, to counteract immunosuppression, myeloid activation agents like anti-CD40 antibodies and TLR agonists can be employed. Alternatively, anti-CSF-1 can also serve to deplete immunosuppressive myeloid cells and other innate signaling in immune evasion (Berraondo et al., 2019; Choi et al., 2020; Korman, Garrett-Thomson, & Lonberg, 2022; Vonderheide, 2020).

## 13. Conclusion

Mitochondria serve as sensors for a wide range of stress and are critical in both extrinsic and intrinsic apoptotic pathways. The release of mitochondrial components from their natural form into the cytosol signals stress to the mitochondria. Reactive oxygen species (ROS), generated because of metabolic disturbances and hypoxia, lead to oxidation of macromolecules. The presence of these oxidized macromolecules, or the non-natural presence of the mitochondrial components in the cytosol or on the cell surface can act as a stress signal for innate signaling (Fig. 1; Table 1). In the tumor microenvironment, chronic stimulation of these innate signaling pathways can drive inflammatory responses. Signaling pathways like TLRs, cGAS-STING, and RIG-1 have previously been exploited for enhancing therapy-induced immune functions. Conversely, many studies have also shown that in certain cancers these pathways are overexpressed as well as associated with poor prognosis. Recent studies have also shown that some innate signaling pathways in tumor cells and immune cells lead to different outcomes in the balance between protumor and anti-tumor functions. Therefore, there is a need for further studies and the development of more experimental models, which can help investigators explain the dichotomy of mitochondrial DNA-driven innate inflammatory processes. Mitochondrial damage and suboptimal apoptosis may drive constant activation of the innate pathways in the TME, serving as both potential targets for therapy and contributors to poor prognosis. The complexity of these pathways necessitates a deeper understanding through additional research and the refinement of experimental models to elucidate factors influencing the outcomes of mitochondrial DNA-driven innate inflammatory processes in cancer.

## References

Ahmet, D. S., Basheer, H. A., Salem, A., Lu, D., Aghamohammadi, A., Weyerhäuser, P., ... Afarinkia, K. (2020). Application of small molecule FPR1 antagonists in the treatment of cancers. *Scientific Reports, 10*, 17249.

Ayna, G., Krysko, D. V., Kaczmarek, A., Petrovski, G., Vandenabeele, P., & Fésüs, L. (2012). ATP release from dying autophagic cells and their phagocytosis are crucial for inflammasome activation in macrophages. *PLoS One, 7*, e40069.

Bao, X., Liu, X., Li, F., & Li, C.-Y. (2020). Limited MOMP, ATM, and their roles in carcinogenesis and cancer treatment. *Cell & Bioscience, 10*, 81.

Benci, J. L., Xu, B., Qiu, Y., Wu, T. J., Dada, H., Twyman-Saint Victor, C., ... Minn, A. J. (2016). Tumor interferon signaling regulates a multigenic resistance program to immune checkpoint blockade. *Cell, 167*, 1540–1554.e12.

Bender, S., Reuter, A., Eberle, F., Einhorn, E., Binder, M., & Bartenschlager, R. (2015). Activation of type I and III interferon response by mitochondrial and peroxisomal MAVS and inhibition by hepatitis C virus. *PLoS Pathogens, 11*, e1005264.

Berraondo, P., Sanmamed, M. F., Ochoa, M. C., Etxeberria, I., Aznar, M. A., Pérez-Gracia, J. L., ... Melero, i (2019). Cytokines in clinical cancer immunotherapy. *British Journal of Cancer, 120*, 6–15.

Bertrand, F., Montfort, A., Marcheteau, E., Imbert, C., Gilhodes, J., Filleron, T., ... Meyer, N. (2017). TNFα blockade overcomes resistance to anti-PD-1 in experimental melanoma. *Nature Communications, 8*, 2256.

Boguszewska, K., Szewczuk, M., Kaźmierczak-Barańska, J., & Karwowski, B. T. (2020). The similarities between human mitochondria and bacteria in the context of structure, genome, and base excision repair system. *Molecules (Basel, Switzerland), 25*.

Bouhamida, E., Morciano, G., Perrone, M., Kahsay, A. E., Della Sala, M., Wieckowski, M. R., ... Patergnani, S. (2022). The interplay of hypoxia signaling on mitochondrial dysfunction and inflammation in cardiovascular diseases and cancer: From molecular mechanisms to therapeutic approaches. *Biology (Basel), 11*.

Briukhovetska, D., Dörr, J., Endres, S., Libby, P., Dinarello, C. A., & Kobold, S. (2021). Interleukins in cancer: From biology to therapy. *Nature Reviews. Cancer, 21*, 481–499.

Brookes, P. S., Yoon, Y., Robotham, J. L., Anders, M. W., & Sheu, S. S. (2004). Calcium, ATP, and ROS: a mitochondrial love-hate triangle. *American Journal of Physiology. Cell Physiology, 287*, C817–C833.

Brubaker, S. W., Bonham, K. S., Zanoni, I., & Kagan, J. C. (2015). Innate immune pattern recognition: A cell biological perspective. *Annual Review of Immunology, 33*, 257–290.

Caielli, S., Athale, S., Domic, B., Murat, E., Chandra, M., Banchereau, R., ... Gong, M. (2016). Oxidized mitochondrial nucleoids released by neutrophils drive type I interferon production in human lupus. *Journal of Experimental Medicine, 213*, 697–713.

Carozza, J. A., Cordova, A. F., Brown, J. A., Alsaif, Y., Böhnert, V., Cao, X., ... Li, L. (2022). ENPP1's regulation of extracellular cGAMP is a ubiquitous mechanism of attenuating STING signaling. *Proceedings of the National Academy of Sciences, 119*, e2119189119.

Chandel, N. S., Mcclintock, D. S., Feliciano, C. E., Wood, T. M., Melendez, J. A., Rodriguez, A. M., & Schumacker, P. T. (2000). Reactive oxygen species generated at mitochondrial complex III stabilize hypoxia-inducible factor-1α during hypoxia: A mechanism of $O_2$ sensing*. *Journal of Biological Chemistry, 275*, 25130–25138.

Chatterjee, D., Das, P., & Chakrabarti, O. (2022). Mitochondrial epigenetics regulating inflammation in cancer and aging. *Frontiers in Cell and Developmental Biology, 10*, 929708.

Chipuk, J. E., Bouchier-Hayes, L., & Green, D. R. (2006). Mitochondrial outer membrane permeabilization during apoptosis: The innocent bystander scenario. *Cell Death & Differentiation, 13*, 1396–1402.

Choi, Y., Shi, Y., Haymaker, C. L., Naing, A., Ciliberto, G., & Hajjar, J. (2020). T-cell agonists in cancer immunotherapy. *Journal for Immunotherapy of Cancer, 8*.

Cogan, D., & Bakhoum, S. F. (2020). Re-awakening innate immune signaling in cancer: The development of highly potent ENPP1 inhibitors. *Cell Chemical Biology, 27*, 1327–1328.

Cohen, P. (2014). The TLR and IL-1 signalling network at a glance. *Journal of Cell Science, 127*, 2383–2390.

Duchen, M. R. (2000). Mitochondria and calcium: From cell signalling to cell death. *The Journal of Physiology, 529*(Pt 1), 57–68.

Forrester, S. J., Kikuchi, D. S., Hernandes, M. S., Xu, Q., & Griendling, K. K. (2018). Reactive oxygen species in metabolic and inflammatory signaling. *Circulation Research, 122*, 877–902.

Gombault, A., Baron, L., & Couillin, I. (2013). ATP release and purinergic signaling in NLRP3 inflammasome activation. *Frontiers in Immunology, 3*, 414.

Greten, F. R., & Grivennikov, S. I. (2019). Inflammation and cancer: Triggers, mechanisms, and consequences. *Immunity, 51*, 27–41.

Guilbaud, E., & Galluzzi, L. (2022). Adaptation to MOMP drives cancer persistence. *Cell Research*.

He, Y., Hara, H., & Núñez, G. (2016). Mechanism and regulation of NLRP3 inflammasome activation. *Trends in Biochemical Sciences, 41*, 1012–1021.

Heilig, R., Lee, J., & Tait, S. W. G. (2023). Mitochondrial DNA in cell death and inflammation. *Biochemical Society Transactions, 51*, 457–472.

Hirano, T. (2021). IL-6 in inflammation, autoimmunity and cancer. *International Immunology, 33*, 127–148.

Hou, J., Karin, M., & Sun, B. (2021). Targeting cancer-promoting inflammation—Have anti-inflammatory therapies come of age? *Nature Reviews Clinical Oncology, 18*, 261–279.

Hou, Y., Liang, H., Rao, E., Zheng, W., Huang, X., Deng, L., ... Fu, Y. X. (2018). Non-canonical NF-κB antagonizes STING sensor-mediated DNA sensing in radiotherapy. *Immunity, 49*, 490–503.e4.

Huang, B., Zhao, J., Unkeless, J., Feng, Z., & Xiong, H. (2008). TLR signaling by tumor and immune cells: A double-edged sword. *Oncogene, 27*, 218–224.

Huang, L. S., Hong, Z., Wu, W., Xiong, S., Zhong, M., Gao, X., ... Malik, A. B. (2020). mtDNA activates cGAS signaling and suppresses the YAP-mediated endothelial cell proliferation program to promote inflammatory injury. *Immunity, 52*, 475–486.e5.

Iurescia, S., Fioretti, D., & Rinaldi, M. (2020). The innate immune signalling pathways: Turning RIG-I sensor activation against cancer. *Cancers, 12*, 3158.

Ji, Z., He, L., Regev, A., & Struhl, K. (2019). Inflammatory regulatory network mediated by the joint action of NF-kB, STAT3, and AP-1 factors is involved in many human cancers. *Proceedings of the National Academy of Sciences, 116*, 9453–9462.

Jiang, Y., Zhang, H., Wang, J., Chen, J., Guo, Z., Liu, Y., & Hua, H. (2023). Exploiting RIG-I-like receptor pathway for cancer immunotherapy. *Journal of Hematology & Oncology, 16*, 8.

John, L. B., & Darcy, P. K. (2016). The double-edged sword of IFN-[gamma]-dependent immune-based therapies. *Immunology and Cell Biology, 94,* 527.

Julian, M. W., Shao, G., Bao, S., Knoell, D. L., Papenfuss, T. L., Vangundy, Z. C., & Crouser, E. D. (2012). Mitochondrial transcription factor A serves as a danger signal by augmenting plasmacytoid dendritic cell responses to DNA. *The Journal of Immunology, 189,* 433–443.

Julian, M. W., Shao, G., Vangundy, Z. C., Papenfuss, T. L., & Crouser, E. D. (2013). Mitochondrial transcription factor A, an endogenous danger signal, promotes TNFα release via RAGE-and TLR9-responsive plasmacytoid dendritic cells. *PLoS One, 8,* e72354.

Kalkavan, H., & Green, D. R. (2018). MOMP, cell suicide as a BCL-2 family business. *Cell Death & Differentiation, 25,* 46–55.

Korman, A. J., Garrett-Thomson, S. C., & Lonberg, N. (2022). The foundations of immune checkpoint blockade and the ipilimumab approval decennial. *Nature Reviews. Drug Discovery, 21,* 509–528.

Kortlever, R. M., Sodir, N. M., Wilson, C. H., Burkhart, D. L., Pellegrinet, L., Swigart, L. B., ... Evan, G. I. (2017). Myc cooperates with Ras by programming inflammation and immune suppression. *Cell, 171,* 1301–1315.e14.

Kumari, N., Dwarakanath, B., Das, A., & Bhatt, A. N. (2016). Role of interleukin-6 in cancer progression and therapeutic resistance. *Tumor Biology, 37,* 11553–11572.

Kwon, J., & Bakhoum, S. F. (2020). The cytosolic DNA-sensing cGAS-STING pathway in cancer. *Cancer Discovery, 10,* 26–39.

Lai, Y. H., Liu, H. Y., Huang, C. Y., Chau, Y. P., & Wu, S. (2019). Mitochondrial-DNA-associated TLR9 signalling is a potential serological biomarker for non-small cell lung cancer. *Oncology Reports, 41,* 999–1006.

Lee, C., Han, J., & Jung, Y. (2023). Formyl peptide receptor 2 is an emerging modulator of inflammation in the liver. *Experimental & Molecular Medicine, 55,* 325–332.

Lee, G.-S., Subramanian, N., Kim, A. I., Aksentijevich, I., Goldbach-Mansky, R., Sacks, D. B., ... Chae, J. J. (2012). The calcium-sensing receptor regulates the NLRP3 inflammasome through $Ca^{2+}$ and cAMP. *Nature, 492,* 123–127.

Li, L., Chen, K., Xiang, Y., Yoshimura, T., Su, S., Zhu, J., ... Wang, J. M. (2016). New development in studies of formyl-peptide receptors: Critical roles in host defense. *Journal of Leucocyte Biology, 99,* 425–435.

Liemburg-Apers, D. C., Willems, P. H., Koopman, W. J., & Grefte, S. (2015). Interactions between mitochondrial reactive oxygen species and cellular glucose metabolism. *Archives of Toxicology, 89,* 1209–1226.

Lines, C. L., Mcgrath, M. J., Dorwart, T., & Conn, C. S. (2023). The integrated stress response in cancer progression: A force for plasticity and resistance. *Frontiers in Oncology, 13*.

Ling, A., Löfgren-Burström, A., Larsson, P., Li, X., Wikberg, M. L., Öberg, Å., ... Palmqvist, R. (2017). TAP1 down-regulation elicits immune escape and poor prognosis in colorectal cancer. *OncoImmunology, 6,* e1356143.

Liu, B., Du, Q., Chen, L., Fu, G., Li, S., Fu, L., ... Bin, C. (2016). CpG methylation patterns of human mitochondrial DNA. *Scientific Reports, 6,* 23421.

Liu, X., Zou, H., Slaughter, C., & Wang, X. (1997). DFF, a heterodimeric protein that functions downstream of caspase-3 to trigger DNA fragmentation during apoptosis. *Cell, 89,* 175–184.

Liu, Z., Ren, Z., Zhang, J., Chuang, C.-C., Kandaswamy, E., Zhou, T., & Zuo, L. (2018). Role of ROS and nutritional antioxidants in human diseases. *Frontiers in Physiology, 9,* 477.

Ljujic, M., Samali, A., Pakos-Zebrucka, K., Koryga, I., Mnich, K., & Gorman, A. M. (2016). The integrated stress response. *EMBO Reports*.

Maiorino, L., Daßler-Plenker, J., Sun, L., & Egeblad, M. (2022). Innate immunity and cancer pathophysiology. *Annual Review of Pathology, 17,* 425–457.

Marchi, S., Guilbaud, E., Tait, S. W. G., Yamazaki, T., & Galluzzi, L. (2022). Mitochondrial control of inflammation. *Nature Reviews. Immunology*, 1–15.

Mechta, M., Ingerslev, L. R., Fabre, O., Picard, M., & Barrès, R. (2017). Evidence suggesting absence of mitochondrial DNA methylation. *Frontiers in Genetics, 8*.

Mittal, D., Gubin, M. M., Schreiber, R. D., & Smyth, M. J. (2014). New insights into cancer immunoediting and its three component phases—Elimination, equilibrium and escape. *Current Opinion in Immunology, 27*, 16–25.

Montero, J., & Haq, R. (2022). Adapted to survive: Targeting cancer cells with BH3 mimetics. *Cancer Discovery, 12*, 1217–1232.

Murakami, T., Ockinger, J., Yu, J., Byles, V., Mccoll, A., Hofer, A. M., & Horng, T. (2012). Critical role for calcium mobilization in activation of the NLRP3 inflammasome. *Proceedings of the National Academy of Sciences of the United States of America, 109*, 11282–11287.

Musella, M., Galassi, C., Manduca, N., & Sistigu, A. (2021). The yin and yang of type I IFNs in cancer promotion and immune activation. *Biology, 10*, 856.

Orekhov, A. N., Nikiforov, N. G., Omelchenko, A. V., Sinyov, V. V., Sobenin, I. A., Vinokurov, A. Y., & Orekhova, V. A. (2022). The role of mitochondrial mutations in chronification of inflammation: Hypothesis and overview of own data. *Life (Basel), 12*.

Pandey, S., Singh, S., Anang, V., Bhatt, A. N., Natarajan, K., & Dwarakanath, B. S. (2015). Pattern recognition receptors in cancer progression and metastasis. *Cancer Growth Metastasis, 8*, 25–34.

Pizzuto, M., Lonez, C., Baroja-Mazo, A., Martínez-Banaclocha, H., Tourlomousis, P., Gangloff, M., ... Bryant, C. E. (2019). Saturation of acyl chains converts cardiolipin from an antagonist to an activator of Toll-like receptor-4. *Cellular and Molecular Life Sciences: CMLS, 76*, 3667–3678.

Pizzuto, M., & Pelegrin, P. (2020). Cardiolipin in immune signaling and cell death. *Trends in Cell Biology, 30*, 892–903.

Platanitis, E., & Decker, T. (2018). Regulatory networks involving STATs, IRFs, and NFκB in inflammation. *Frontiers in Immunology, 9*, 2542.

Pretre, V., Papadopoulos, D., Regard, J., Pelletier, M., & Woo, J. (2022). Interleukin-1 (IL-1) and the inflammasome in cancer. *Cytokine, 153*, 155850.

Ralph, S. J., Rodríguez-Enríquez, S., Neuzil, J., Saavedra, E., & Moreno-Sánchez, R. (2010). The causes of cancer revisited: "Mitochondrial malignancy" and ROS-induced oncogenic transformation – Why mitochondria are targets for cancer therapy. *Molecular Aspects of Medicine, 31*, 145–170.

Rébé, C., & Ghiringhelli, F. (2020). Interleukin-1β and cancer. *Cancers, 12*, 1791.

Redza-Dutordoir, M., & Averill-Bates, D. A. (2016). Activation of apoptosis signalling pathways by reactive oxygen species. *Biochimica et Biophysica Acta (BBA)—Molecular Cell Research, 1863*, 2977–2992.

Regateiro, F. S., Cobbold, S. P., & Waldmann, H. (2013). CD73 and adenosine generation in the creation of regulatory microenvironments. *Clinical and Experimental Immunology, 171*, 1–7.

Reuter, S., Gupta, S. C., Chaturvedi, M. M., & Aggarwal, B. B. (2010). Oxidative stress, inflammation, and cancer: how are they linked? *Free Radical Biology & Medicine, 49*, 1603–1616.

Riley, J. S., & Tait, S. W. (2020). Mitochondrial DNA in inflammation and immunity. *EMBO Reports, 21*, e49799.

Rossi, J.-F., Lu, Z.-Y., Jourdan, M., & Klein, B. (2015). Interleukin-6 as a therapeutic target. *Clinical Cancer Research, 21*, 1248–1257.

Samson, N., & Ablasser, A. (2022). The cGAS–STING pathway and cancer. *Nature Cancer, 3*, 1452–1463.

Shimada, K., Crother, T. R., Karlin, J., Dagvadorj, J., Chiba, N., Chen, S., ... Arditi, M. (2012). Oxidized mitochondrial DNA activates the NLRP3 inflammasome during apoptosis. *Immunity, 36*, 401–414.

Shrestha, R., Bridle, K. R., Crawford, D. H., & Jayachandran, A. (2020). TNF-α-mediated epithelial-to-mesenchymal transition regulates expression of immune checkpoint molecules in hepatocellular carcinoma. *Molecular Medicine Reports, 21*, 1849–1860.

Song, Y., Zhou, Y., & Zhou, X. (2020). The role of mitophagy in innate immune responses triggered by mitochondrial stress. *Cell Communication and Signaling, 18*, 186.

Spinelli, J. B., & Haigis, M. C. (2018). The multifaceted contributions of mitochondria to cellular metabolism. *Nature Cell Biology, 20*, 745–754.

Takeshita, F., Leifer, C. A., Gursel, I., Ishii, K. J., Takeshita, S., Gursel, M., & Klinman, D. M. (2001). Cutting edge: Role of Toll-like receptor 9 in CpG DNA-induced activation of human cells. *The Journal of Immunology, 167*, 3555–3558.

Tanwar, J., Singh, J. B., & Motiani, R. K. (2021). Molecular machinery regulating mitochondrial calcium levels: The nuts and bolts of mitochondrial calcium dynamics. *Mitochondrion, 57*, 9–22.

Tian, C., Chen, K., Gong, W., Yoshimura, T., Huang, J., & Wang, J. M. (2020). The G-protein coupled formyl peptide receptors and their role in the progression of digestive tract cancer. *Technology in Cancer Research & Treatment, 19* 1533033820973280.

Timosenko, E., Ghadbane, H., Silk, J. D., Shepherd, D., Gileadi, U., Howson, L. J., ... Cerundolo, V. (2016). Nutritional stress induced by tryptophan-degrading enzymes results in ATF4-dependent reprogramming of the amino acid transporter profile in tumor cells. *Cancer Research, 76*, 6193–6204.

Triantafilou, K., Hughes, T. R., Triantafilou, M., & Morgan, B. P. (2013). The complement membrane attack complex triggers intracellular $Ca^{2+}$ fluxes leading to NLRP3 inflammasome activation. *Journal of Cell Science, 126*, 2903–2913.

Turrens, J. F. (1997). Superoxide production by the mitochondrial respiratory chain. *Bioscience Reports, 17*, 3–8.

Van Houten, B., Woshner, V., & Santos, J. H. (2006). Role of mitochondrial DNA in toxic responses to oxidative stress. *DNA Repair (Amst), 5*, 145–152.

Van Vugt, M. A. T. M., & Parkes, E. E. (2022). When breaks get hot: Inflammatory signaling in BRCA1/2-mutant cancers. *Trends in Cancer, 8*, 174–189.

Vashi, N., & Bakhoum, S. F. (2021). The evolution of STING signaling and its involvement in cancer. *Trends in Biochemical Sciences, 46*, 446–460.

Veglia, F., & Gabrilovich, D. I. (2017). Dendritic cells in cancer: The role revisited. *Current Opinion in Immunology, 45*, 43–51.

Vonderheide, R. H. (2020). CD40 agonist antibodies in cancer immunotherapy. *Annual Review of Medicine, 71*, 47–58.

Weinberg, F., & Chandel, N. S. (2009). Reactive oxygen species-dependent signaling regulates cancer. *Cellular and Molecular Life Sciences: CMLS, 66*, 3663–3673.

Weiß, E., & Kretschmer, D. (2018). Formyl-peptide receptors in infection, inflammation, and cancer. *Trends in Immunology, 39*, 815–829.

Xian, H., Watari, K., Sanchez-Lopez, E., Offenberger, J., Onyuru, J., Sampath, H., ... Karin, M. (2022). Oxidized DNA fragments exit mitochondria via mPTP-and VDAC-dependent channels to activate NLRP3 inflammasome and interferon signaling. *Immunity, 55*, 1370–1385.e8.

Yan, M., Zheng, M., Niu, R., Yang, X., Tian, S., Fan, L., ... Zhang, S. (2022). Roles of tumor-associated neutrophils in tumor metastasis and its clinical applications. *Frontiers in Cell and Developmental Biology, 10*, 938289.

Yang, L., Chu, Z., Liu, M., Zou, Q., Li, J., Liu, Q., ... Wang, B. (2023). Amino acid metabolism in immune cells: Essential regulators of the effector functions, and promising opportunities to enhance cancer immunotherapy. *Journal of Hematology & Oncology, 16*, 59.

Yang, Y., Wang, H., Kouadir, M., Song, H., & Shi, F. (2019). Recent advances in the mechanisms of NLRP3 inflammasome activation and its inhibitors. *Cell Death & Disease, 10,* 128.

Yoshihama, S., Roszik, J., Downs, I., Meissner, T. B., Vijayan, S., Chapuy, B., ... Kobayashi, K. S. (2016). NLRC5/MHC class I transactivator is a target for immune evasion in cancer. *Proceedings of the National Academy of Sciences of the United States of America, 113,* 5999–6004.

Zhang, B. (2010). CD73: A novel target for cancer immunotherapy. *Cancer Research, 70,* 6407–6411.

Zhang, J., & Veeramachaneni, N. (2022). Targeting interleukin-1β and inflammation in lung cancer. *Biomarker Research, 10,* 5.

Zhang, L., Li, J., Zong, L., Chen, X., Chen, K., Jiang, Z., ... Ma, Z. (2016). Reactive oxygen species and targeted therapy for pancreatic cancer. *Oxidative Medicine and Cellular Longevity, 2016,* 1616781.

Zhang, L., Nesvick, C. L., Day, C. A., Choi, J., Lu, V. M., Peterson, T., ... Decker, P. A. (2022). STAT3 is a biologically relevant therapeutic target in H3K27M-mutant diffuse midline glioma. *Neuro-Oncology, 24,* 1700–1711.

Zhang, X., Zhang, H., Zhang, J., Yang, M., Zhu, M., Yin, Y., ... Yu, F. (2022). The paradoxical role of radiation-induced cGAS-STING signalling network in tumour immunity. *Immunology.*

Zhao, S., Chen, F., Yin, Q., Wang, D., Han, W., & Zhang, Y. (2020). Reactive oxygen species interact with NLRP3 inflammasomes and are involved in the inflammation of sepsis: From mechanism to treatment of progression. *Frontiers in Physiology, 11.*

Zhong, Z., Sanchez-Lopez, E., & Karin, M. (2016). Autophagy, inflammation, and immunity: A troika governing cancer and its treatment. *Cell, 166,* 288–298.

9780443235481